Hydrological Hazard: Analysis and Prevention

Hydrological Hazard: Analysis and Prevention

Special Issue Editor

Tommaso Caloiero

MDPI • Basel • Beijing • Wuhan • Barcelona • Belgrade

MDPI

Special Issue Editor
Tommaso Caloiero
Institute for Agricultural and Forest Systems in the Mediterranean (CNR-ISAFOM)
Italy

Editorial Office
MDPI
St. Alban-Anlage 66
4052 Basel, Switzerland

This is a reprint of articles from the Special Issue published online in the open access journal *Geosciences* (ISSN 2076-3263) in 2018 (available at: https://www.mdpi.com/journal/geosciences/special_issues/Hydrogeological_Hazard_Prevention)

For citation purposes, cite each article independently as indicated on the article page online and as indicated below:

LastName, A.A.; LastName, B.B.; LastName, C.C. Article Title. *Journal Name* **Year**, *Article Number, Page Range.*

ISBN 978-3-03897-374-4 (Pbk)
ISBN 978-3-03897-375-1 (PDF)

Contents

About the Special Issue Editor

Tommaso Caloiero graduated in 2002 in Civil Engineer (specializing in Hydraulic) at the University of Calabria (Italy); in 2005, he received a Second level Master's Degree in Mathematical Modelling of Hydrogeological Disaster from the same University, and in 2009 he obtained a Ph.D. in Hydraulic Engineering at the Politecnico of Milan. Since 2011, he has been a researcher at the National Research Council—Institute for Agricultural and Forest Systems in the Mediterranean (CNR-ISAFOM), in Rende (CS), Italy. His preferred research topics are hydrology, climatology, climate change, natural hazards, hydrologic and water resource modelling and simulation, environmental engineering, ecological engineering, land-use change, and forest ecology. He has developed original works in these areas, and he is the author of about 150 scientific papers published in national and international academic journals and as contributions to national and international conferences proceedings. He has worked for different regional institutions in the Calabria region (Southern Italy) such as the Regional Agency for Environmental Protection (ARPA), Civil Protection and Basin Authority. He has been a consultant for the Institute for the Industrial Promotion (IPI). He has collaborated as a scientific consultant with the Research Institute for Geo-Hydrological Protection (IRPI), the Institute of Atmospheric Sciences and Climate (ISAC), and the Institute for Agricultural and Forest Systems in the Mediterranean (ISAFOM) of the National Research Council (CNR), and with the Department of Soil Defense of the University of Calabria and with the Department of Environmental, Hydraulic, Infrastructures, and Surveying Engineering (DIIAR) of the Politecnico of Milan.

Preface to "Hydrological Hazard: Analysis and Prevention"

This book presents a print version of the Special Issue of the journal Geosciences dedicated to "Hydrological Hazard: Analysis and Prevention". The overall goal of this Special Issue was to consider innovative approaches to the analysis, prediction, prevention, and mitigation of hydrological extremes. In particular, innovative modelling methods for flood hazards, regional flood and drought analysis, and the use of satellite and climate data for drought analysis were the main research and practice targets that the papers published in this Special Issue aimed to address. These original objectives were achieved, and in the thirteen papers collected in this volume readers will find a collection of scientific contributions providing a sample of the state-of-the-art and forefront research in these fields. Among the articles published in the Special Issue, one is a technical note, one is a case report, and eleven are original research articles. Thirty-nine authors from three different continents (North America, Europe, and Oceania) contributed to the Special Issue, showing results of case studies and demonstration sites involving five continents (North America, Europe, Africa, Asia, and Oceania). The geographic distribution of the case studies is wide enough to attract the interest of an international audience of readers. The articles collected here will hopefully provide different, useful insights into advancements in emerging technologies for the monitoring of key hydrological variables, highlighting new ideas, approaches, and innovations in the analysis of various types of droughts (e.g., meteorological, agricultural, and hydrological droughts) and various types of flood (e.g., fluvial, coastal, and pluvial).

Tommaso Caloiero
Special Issue Editor

geosciences

MDPI

Editorial
Hydrological Hazard: Analysis and Prevention

Tommaso Caloiero

National Research Council—Institute for Agricultural and Forest Systems in Mediterranean (CNR-ISAFOM), Via Cavour 4/6, 87036 Rende (CS), Italy; tommaso.caloiero@isafom.cnr.it; Tel.: +39-0984-841-464

Received: 22 October 2018; Accepted: 25 October 2018; Published: 26 October 2018

Abstract: As a result of the considerable impacts of hydrological hazard on water resources, on natural environments and human activities, as well as on human health and safety, climate variability and climate change have become key issues for the research community. In fact, a warmer climate, with its heightened climate variability, will increase the risk of hydrological extreme phenomena, such as droughts and floods. The Special Issue "Hydrological Hazard: Analysis and Prevention" presents a collection of scientific contributions that provides a sample of the state-of-the-art and forefront research in this field. In particular, innovative modelling methods for flood hazards, regional flood and drought analysis, and the use of satellite and climate data for drought analysis were the main topics and practice targets that the papers published in this Special Issue aimed to address.

Keywords: catchment; climate; drought; flood; forecast; hazards; landslide; modelling; precipitation; temperature

1. Introduction

As a result of economic and population growth in the world, the fifth Intergovernmental Panel on Climate Change (IPCC) report [1] evidenced an increase in anthropogenic greenhouse gas emissions (carbon dioxide, methane, and nitrous oxide) whose atmospheric concentrations reached values never touched in at least the past 800,000 years. Consequently, the IPCC report showed an increase of about 0.9 °C in the Earth's surface temperature in the twentieth century and forecasted a further increase for the twenty-first century, with natural and anthropic consequences [1]. In fact, anthropic systems and terrestrial ecosystems are becoming more vulnerable to environmental phenomena and an increase in floods, heat waves, forest fires, and droughts can be expected [2,3]. Within such a purview, scholarly investigation has primarily focused on multiple analyses of meteorological, hydrological, and climatological variables based on different methodologies.

Given the above scenario, the call for papers for publication in the Special Issue "Hydrological Hazard: Analysis and Prevention", which was launched in October 2017, aimed to consider innovative approaches to the analysis, prediction, prevention, and mitigation of hydrological extremes. With this aim, interdisciplinary original research articles highlighting new ideas, approaches, and innovations in the analysis of various types of droughts (e.g., meteorological, agricultural, and hydrological drought) and various types of floods (e.g., fluvial, coastal, and pluvial) were welcomed.

Potential topics of this Special Issue of Geosciences included, but were not limited to, the following:

- Regional flood and drought analysis
- Case studies and comparative studies in different parts of the world
- Analyses of regional/global patterns and trends
- Effects of land-use or land-cover change on hydrological extremes
- Prediction and prevention of hydrological extremes
- Use of satellite and climate data for drought analysis
- Innovative modelling methods for flood hazards

- Strategies for reducing the vulnerability to hydrological extremes
- Climate change and hydrogeological risk

2. Some Data of the Special Issue

From early January 2018 to late September 2018, a total of 18 papers have been submitted for consideration for publication in the Special Issue. After a rigorous editorial check and peer-review processes, which involved external and independent experts in the field, 4 papers were rejected, 1 paper has been withdrawn, and 13 papers have been accepted, with an acceptance rate of about 72%.

Among the 13 articles published in the Special Issue, 1 is a Technical Note (Terranova et al. [4]), 11 are Research Articles [5–15], and one is a Case Report (Schmid-Breton et al. [16]).

Figure 1 compares the geographic distribution of the authors and research teams publishing in the Special Issue (Figure 1a), as well as of the case studies and demonstration sites (Figure 1b). The analysis of this figure allows one to have an idea of the scientific community working on hydrological hazards, although it is just a sample and thus is not an exhaustive representation. Thirty-nine authors from three different continents (North America, Europe, and Oceania) contributed to the Special Issue, showing results of case studies and demonstration sites involving five continents (North America, Europe, Africa, Asia, and Oceania).

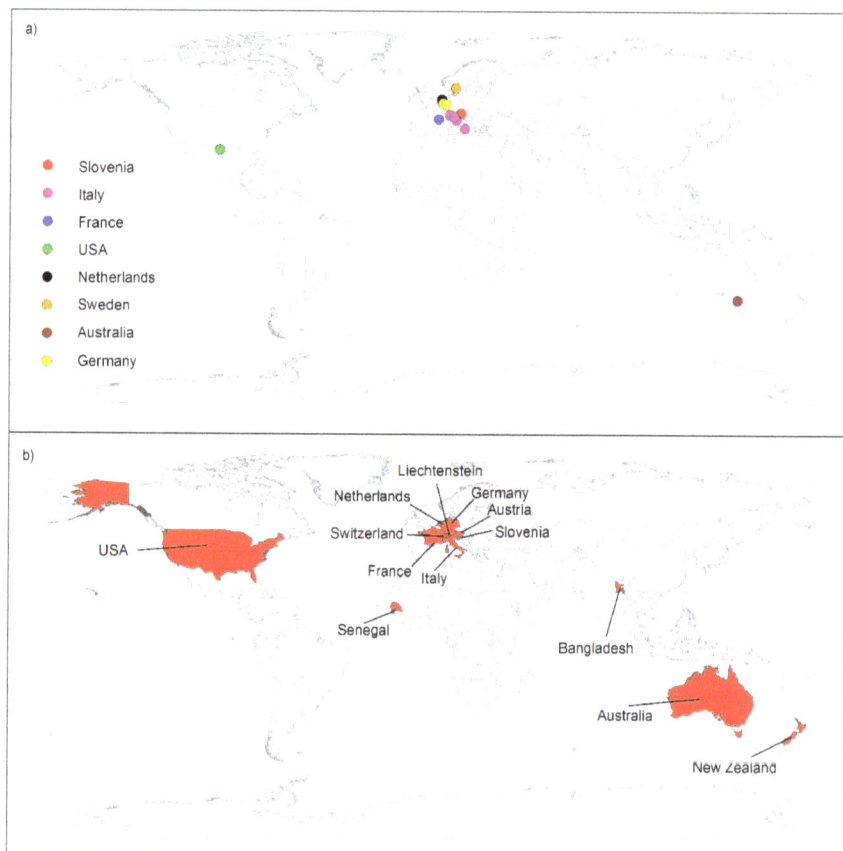

Figure 1. Geographic distribution of (**a**) authors and research teams publishing in the Special Issue; (**b**) case studies and demonstration sites that are discussed in the papers.

Figure 2 shows the word cloud of the keywords published in the papers of the Special Issue. From the analysis of the word cloud, it can be easily seen that "Flood" is the predominant keyword, cited in 8 out of 13 articles, followed by "Modelling" (6 out of 13), and "Catchment" (or basin) and "Precipitation" (or rainfall), which have each been cited in 3 papers.

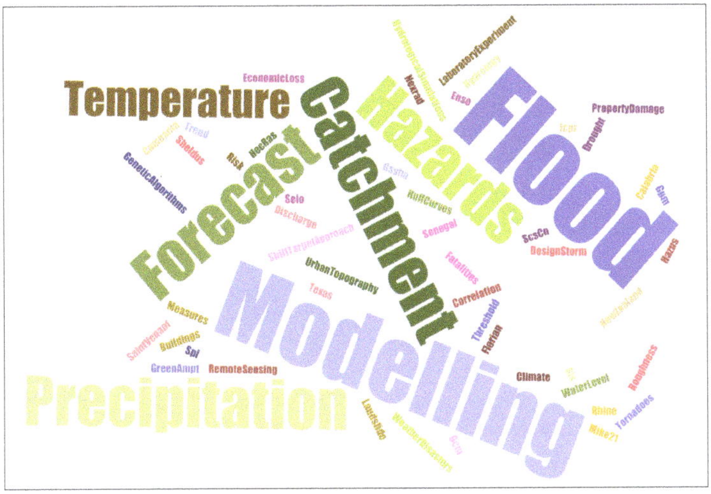

Figure 2. Word cloud of the keywords published in the papers [4–14] of this Special Issue.

3. Overview of the Special Issue Contributions

Terranova et al. [4] applied the GASAKe, an empirical-hydrological model that aims at forecasting the time of occurrence of landslides, in four case studies in two different regions of Italy—three rock slides in Calabria and one soil slip in Campania. As a result, for two of the Calabrian rock slides, the activation dates were correctly predicted by the model, probably thanks to an accurate knowledge of the activation history of the landslide and a proper hydrological characterization of the site. For such cases, GASAKe could be applied to predict the timing of activation of future landslide activations in the same areas. In the other two cases, weaker model performances have been detected, probably because of an inaccurate knowledge of the activation dates and/or rainfall series.

Bezak et al. [5] investigated the impact of the design rainfall on the combined 1D/2D hydraulic modelling results in the Glinščica Stream catchment (Slovenia), which is ungauged in terms of discharges. In particular, Bezak et al. [5] evaluated 10 different design rainfall events (scenarios) that were used as inputs to the hydrological model. Using calibrated and validated hydrological models, the inputs for the hydraulic model were determined. The results indicated that the selection of the design rainfall event should be regarded as an important step, as the hydraulic modelling results for different scenarios differ significantly. As an example, the maximum flooded area extent was twice as large as the minimum one, and the maximum water velocity over flooded areas was more than 10 times larger than the minimum one. This means that the design rainfall definition can significantly influence the hydraulic modelling results, leading to the production of very different flood hazard maps, and consequently the planning of very different flood protection schemes.

Beretta et al. [6] tested three different methods to simulate the influence of buildings on flood inundation by performing a number of laboratory experiments carried out with a simplified urban district physical model, and reconstructing results with a hydraulic mathematical model considering both the solution of the full shallow water equations and the diffusive simplification. Simplified methods were also tested for the simulation of a real flood event, which occurred in 2013 in the city of

Olbia, Italy. The results showed that use of a 2D diffusive model and setting a high friction instead of detailed building geometry are effective methods to assess flood inundation extent.

Lombardi et al. [7] suggested a low computational cost method to produce a probabilistic flood prediction system using a single forecast precipitation scenario perturbed via a spatial shift. The method was applied to three basins located in the northern part of Milan city (northern Italy): Seveso, Olona, and Lambro. To produce hydro-meteorological simulations and forecasts, a flood forecasting system, which comprises the physically-based rainfall-runoff hydrological model FEST-WB and the MOLOCH meteorological model, has been used. In particular, the performance of the shift-target approach was compared with the "unperturbed" MOLOCH forecast over a period of four years. The results showed how the shift-target approach complements the deterministic MOLOCH-based flood forecast for warning purposes.

Bouvier et al. [8] analyzed the skill of two well-known event-based models, the Soil Conservation Service model and the Green-Ampt model, in reproducing the flood processes in a semi-arid agricultural catchment of Senegal (Ndiba). In particular, twenty-eight flood events have been extracted and modelled. As a result, both the models were able to reproduce the flood events after calibration, but they had to account for the fact that the infiltration processes are highly dependent on the tillage of the soils and the growing of the crops during the rainy season, which made the initialization of the event-based models difficult. Specifically, the Soil Conservation Service model performed better than the Green-Ampt model, because the latter was very sensitive to the variability of the hydraulic conductivity at saturation. The variability of the parameters of the models highlights the complexity of this kind of cultivated catchment, with highly non-stationary conditions.

Caloiero [9] studied dry and wet periods in New Zealand using the Standardized Precipitation Index (SPI) and by means of a new graphical technique, the Innovative Trend Analysis (ITA), which allows trend identification of the low, medium, and high values of a series. The results show that, in every area currently subject to drought, an increase of this phenomenon can be expected. Specifically, the results of the paper highlighted that agricultural regions on the eastern side of the South Island, as well as the north-eastern regions of the North Island, are the most consistently vulnerable areas. In fact, in these regions, the trend analysis mainly showed a general reduction in all the values of the SPI; that is, a tendency toward heavier droughts and weaker wet periods.

Paul et al. [10] analyzed the fatality rates caused by hydrometeorological disasters in Texas for the period 1959–2016 in an effort to identify counties and metropolitan areas that have a greater risk for particular hydrometeorological disasters. The study examined temporal trends, spatial variations, and demographic characteristics of the victims from 1959–2016. The results showed that the number of hydrometeorological fatalities in Texas has increased over the 58-year study period, but the per capita fatalities have significantly decreased. Moreover, seasonal and monthly stratification identifies spring and summer as the deadliest seasons, with the month of May registering the highest number of total fatalities dominated by flooding and tornado fatalities. Finally, demographic trends of hydrometeorological disaster fatalities indicated approximately twice the amount of male fatalities than female fatalities from 1959–2016 and that adults are the highest fatality risk group overall.

Hossain et al. [11] assessed the efficiency of a non-linear regression technique in predicting long-term seasonal rainfall. The non-linear models were developed using the lagged (past) values of the climate drivers, which have a significant correlation with rainfall. More specifically, the capabilities of south-eastern Indian Ocean and El Nino Southern Oscillation were assessed in reproducing the rainfall characteristics using the non-linear regression approach. Three rainfall stations located in the Australian Capital Territory were selected as a case study. The analysis suggested that the predictors that have the highest correlation with the predictands do not necessarily produce the least errors in rainfall forecasting. The outcomes of the analysis could help the watershed management authorities to adopt an efficient modelling technique by predicting long-term seasonal rainfall.

Furl et al. [12] investigated the performance of several satellite precipitation products with respect to gauge corrected ground-based radar estimations for nine moderate to high magnitude events across the Guadalupe River system in south Texas. The analysis was conducted across three nested watersheds (with area ranging from 200 to 10,000 km^2) to capture and quantify the effect of the scale on the propagation of the error. In order to understand the propagation of rainfall error into the predicted runoff, hydrologic model simulations were implemented. In particular, the Gridded Surface Subsurface Hydrologic Analysis, a physically-based fully distributed hydrologic model, forced with those ten satellite-based precipitation products, was used to simulate the rainfall-runoff relationship for the basins. The results showed that the satellite-based precipitation products provide very high spatiotemporal resolution precipitation estimates. However, the estimates lack accuracy, especially at a local scale. The products underestimate heavy storm events significantly, and the errors were amplified in the runoff hydrographs generated.

Islam et al. [13] assessed the present and future water level and discharge in the Betna River (Bangladesh) by applying a process-based hydrodynamic model (MIKE 21 FM) to simulate water level and discharge under different future climate conditions. The MIKE 21 FM model for the Betna River was set up, calibrated, and validated using the observed water level and discharge data. The model was then used to project the future (2040s and 2090s) water level and discharge. The modelling results indicated that, compared with the baseline year (2014–2015), both the water level and the monsoon daily maximum discharge are expected to increase by the 2040s and by the 2090s, with the sea level rise mostly responsible for the increase in water level.

Ferrari et al. [14] carried out a joint analysis of temperature and rainfall data by comparing time series recorded in some gauges located in Calabria (Southern Italy) over two distinct 30-year sub-periods (1951–1980 and 1981–2010). In particular, the anomalies of the seasonal values of temperature and precipitation, standardized by means of the mean values and the standard deviations of the period 1961–1990, were analyzed. The series has been selected based on the normality hypothesis. The isocontour lines of the probability density function for the bivariate Gaussian distribution have been considered as ellipses centered on the vector mean of each sub-period. Specifically, the displacements of the ellipses have been quantified and tested for each season, passing from the first sub-period to the following one. The main results concern a decreasing trend of both the temperature and the rainfall anomalies, predominantly in the winter and autumn seasons.

Paul and Sharif [15] tried to verify the assertion that the increase in property damage is a combined contribution of stronger disasters as predicted by climate change models and increases in urban development in risk prone regions such as the Texas Gulf Coast. Within this aim, the study intended to provide a review of historic trends and types of damage and economic losses caused by hydrometeorological disasters impacting the coastal and inland property and infrastructure of Texas from 1960–2016. Spatial analysis of actual and normalized damage, as well as a supplemental assessment of three major disasters causing extensive damage in Texas (Hurricanes Carla 1961, Hurricane Alicia 1983, and Hurricane Ike 2008), highlight the risk as a function of wind or flooding damage and the growth of exposure in hazard prone regions.

Schmid-Breton et al. [16] presented the method and the GIS-tool named "ICPR FloRiAn (Flood Risk Analysis)", developed by the International Commission for the Protection of the Rhine (ICPR) to enable the broad-scale assessment of the effectiveness of flood risk management measures on the Rhine. Moreover, the first calculation results have been also shown. The tool uses flood hazard maps and associated recurrence periods for an overall damage and risk assessment for four receptors: human health, environment, culture heritage, and economic activity. For each receptor, a method is designed to calculate the impact of flooding and the effect of measures. The tool consists of three interacting modules: damage assessment, risk assessment, and measures. Calculations using this tool showed that the flood risk reduction target defined in the Action Plan on Floods of the ICPR in 1998 could be achieved with the measures already taken and those planned until 2030.

Acknowledgments: The Guest Editor thanks all the authors, Geosciences' editors, and reviewers for their great contributions and commitment to this Special Issue. A special thank goes to Daisy Hu, Geoscience's Assistant Editor, for her dedication to this project and her valuable collaboration in the design and setup of the Special Issue.

Conflicts of Interest: The author declares no conflict of interest.

References

1. IPCC. Summary for Policymakers. In *Fifth Assessment Report of the Intergovernmental Panel on Climate Change*; Cambridge University Press: Cambridge, UK, 2013.
2. Estrela, T.; Vargas, E. Drought management plans in the European Union. *Water Resour. Manag.* **2010**, *26*, 1537–1553. [CrossRef]
3. Kreibich, H.; Di Baldassarre, G.; Vorogushyn, S.; Aerts, J.C.J.H.; Apel, H.; Aronica, G.T.; Arnbjerg-Nielsen, K.; Bouwer, L.M.; Bubeck, P.; Caloiero, T.; et al. Adaptation to flood risk: Results of international paired flood event studies. *Earths Future* **2017**, *5*, 953–965. [CrossRef]
4. Terranova, O.; Gariano, S.L.; Iaquinta, P.; Lupiano, V.; Rago, V.; Iovine, G. Examples of Application of GASAKe for Predicting the Occurrence of Rainfall-Induced Landslides in Southern Italy. *Geosciences* **2018**, *8*, 78. [CrossRef]
5. Bezak, N.; Šraj, M.; Rusjan, S.; Mikoš, M. Impact of the Rainfall Duration and Temporal Rainfall Distribution Defined Using the Huff Curves on the Hydraulic Flood Modelling Results. *Geosciences* **2018**, *8*, 69. [CrossRef]
6. Beretta, R.; Ravazzani, G.; Maiorano, C.; Mancini, M. Simulating the Influence of Buildings on Flood Inundation in Urban Areas. *Geosciences* **2018**, *8*, 77. [CrossRef]
7. Lombardi, G.; Ceppi, A.; Ravazzani, G.; Davolio, S.; Mancini, M. From Deterministic to Probabilistic Forecasts: The 'Shift-Target' Approach in the Milan Urban Area (Northern Italy). *Geosciences* **2018**, *8*, 181. [CrossRef]
8. Bouvier, C.; Bouchenaki, L.; Tramblay, Y. Comparison of SCS and Green-Ampt Distributed Models for Flood Modelling in a Small Cultivated Catchment in Senegal. *Geosciences* **2018**, *8*, 122. [CrossRef]
9. Caloiero, T. SPI Trend Analysis of New Zealand Applying the ITA Technique. *Geosciences* **2018**, *8*, 101. [CrossRef]
10. Paul, S.H.; Sharif, H.O.; Crawford, A.M. Fatalities Caused by Hydrometeorological Disasters in Texas. *Geosciences* **2018**, *8*, 186. [CrossRef]
11. Hossain, I.; Esha, R.; Imteaz, M.A. An Attempt to Use Non-Linear Regression Modelling Technique in Long-Term Seasonal Rainfall Forecasting for Australian Capital Territory. *Geosciences* **2018**, *8*, 282. [CrossRef]
12. Furl, C.; Ghebreyesus, D.; Sharif, H.O. Assessment of the Performance of Satellite-Based Precipitation Products for Flood Events across Diverse Spatial Scales Using GSSHA Modeling System. *Geosciences* **2018**, *8*, 191. [CrossRef]
13. Islam, M.M.M.; Hofstra, N.; Sokolova, E. Modelling the Present and Future Water Level and Discharge of the Tidal Betna River. *Geosciences* **2018**, *8*, 271. [CrossRef]
14. Ferrari, E.; Coscarelli, R.; Sirangelo, B. Correlation Analysis of Seasonal Temperature and Precipitation in a Region of Southern Italy. *Geosciences* **2018**, *8*, 160. [CrossRef]
15. Paul, S.H.; Sharif, H.O. Analysis of Damage Caused by Hydrometeorological Disasters in Texas, 1960–2016. *Geosciences* **2018**, *8*, 384. [CrossRef]
16. Schmid-Breton, A.; Kutschera, G.; Botterhuis, T.; ICPR Expert Group 'Flood Risk Analysis'. A Novel Method for Evaluation of Flood Risk Reduction Strategies: Explanation of ICPR FloRiAn GIS-Tool and Its First Application to the Rhine River Basin. *Geosciences* **2018**, *8*, 371. [CrossRef]

geoscienses

MDPI

Article

Impact of the Rainfall Duration and Temporal Rainfall Distribution Defined Using the Huff Curves on the Hydraulic Flood Modelling Results

Nejc Bezak *, Mojca Šraj, Simon Rusjan and Matjaž Mikoš

Faculty of Civil and Geodetic Engineering, University of Ljubljana, 1000 Ljubljana, Slovenia;
mojca.sraj@fgg.uni-lj.si (M.Š.); simon.rusjan@fgg.uni-lj.si (S.R.); matjaz.mikos@fgg.uni-lj.si (M.M.)
* Correspondence: nejc.bezak@fgg.uni-lj.si; Tel.: +386-1476-85-00

Received: 15 January 2018; Accepted: 10 February 2018; Published: 11 February 2018

Abstract: In the case of ungauged catchments, different procedures can be used to derive the design hydrograph and design peak discharge, which are crucial input data for the design of different hydrotechnical engineering structures, or the production of flood hazard maps. One of the possible approaches involves using a hydrological model where one can calculate the design hydrograph through the design of a rainfall event. This study investigates the impact of the design rainfall on the combined one-dimensional/two-dimensional (1D/2D) hydraulic modelling results. The Glinščica Stream catchment located in Slovenia (central Europe) is used as a case study. Ten different design rainfall events were compared for 10 and 100-year return periods, where we used Huff curves for the design rainfall event definition. The results indicate that the selection of the design rainfall event should be regarded as an important step, since the hydraulic modelling results for different scenarios differ significantly. In the presented experimental case study, the maximum flooded area extent was twice as large as the minimum one, and the maximum water velocity over flooded areas was more than 10 times larger than the minimum one. This can lead to the production of very different flood hazard maps, and consequently planning very different flood protection schemes.

Keywords: design storm; hydraulic modelling; flood hazards; Glinščica catchment; hydrological modelling; Huff curves; HEC-RAS

1. Introduction

Floods are one of the natural disasters that cause a large amount of economic damage and endanger human lives all over the world [1]. Moreover, a warming climate may cause more frequent and more extreme river flooding in the future, although a consistent trend over the past 50 years in Europe has not been detected [2]. However, Blöschl et al. [2] showed substantial changes in flood timing of rivers in Europe. Similar conclusions can also be made for Slovenia [3]. Altogether, floods are still one of the natural disasters that cause large amounts of economic damage and have significant direct and indirect consequences for the environment and society; by properly designing different flood protection schemes, one can manage flood risk, and consequently reduce the casualties due to flooding [4].

In order to design either green or grey infrastructure measures to reduce flood risk, the information about the design discharge or design hydrograph is needed. If discharge data is available, one can perform either univariate [5] or multivariate [6] flood frequency analysis in order to define design variables. When no discharge data is available, other approaches can be used to define the design variables. Blöschl et al. [7] made a comprehensive overview of methods that can be used for predictions of different hydrological variables in cases of the so-called ungauged catchments. One of the methods that can be used to estimate design variables in such cases is also the application of a hydrological model to define the design peak discharge or the complete design hydrograph [8,9]. Besides hydrological model parameters that have to be estimated during the calibration of the selected model, a design hyetograph definition has a significant impact on the model results [10–16]. In order to construct a design rainfall event for flood risk assessment, several methods can be applied (e.g., constant intensity method, triangular hyetograph, Natural Resources Conservation Service (NRCS) design storm, frequency-based or alternating block method, and Huff method), most of which are based on intensity–duration–frequency (IDF) relationships, namely on a single point or the entire IDF curve. Using the IDF relationship, we can estimate the frequency or return period of specific rainfall intensity or rainfall amount that can be expected for certain rainfall duration.

However, the same discharge value can be derived from different combinations of storm duration and its return period [13]. In addition to the amount of rainfall with the selected magnitude, the two most important factors related to the design hyetograph selection are the design rainfall duration, and rainfall distribution within the rainfall event (which is also called internal storm structure or temporal rainfall distribution) [15,16]. Šraj et al. [14] have shown that a combination of rainfall duration that is significantly longer than the catchment time of concentration, and constant rainfall intensity within the design rainfall event can yield significantly different (more than 50% smaller) design peak discharges than design hyetographs with a rainfall duration that is approximately equal to the catchment time of concentration and the application of non-uniform (i.e., actual/real) rainfall intensity distribution. The essential differences in the time-to-peak of the resulted hydrographs of the hydrological model and differences in peak discharge can also be the consequence of the maximum rainfall intensity position within the design hyetograph [10,13,14,17].

However, to obtain a typical rainfall distribution within the rainfall event for a region, Huff curves [18] can be used that connect the dimensionless rainfall depth with the dimensionless rainfall duration of an individual rainfall station or region, based upon locally gauged historical data. As such, Huff curves represent typical rainfall characteristics of a region [19,20]. These curves were recently derived for several Slovenian rainfall stations [21]. Dolšak et al. [21] demonstrated that the variability in the Huff curves using different probability levels generally decreases with increasing rainfall duration. The median Huff curve (50%) can be regarded as the most representative, and ought to be used for constructing the design hyetographs [22]. Thus, it appears that a definition of a design hyetograph is one of the most important parts of the hydrograph definition, in cases when hydrological models are used.

In practical engineering applications, design hydrographs are often used as inputs to the hydraulic models in order to determine flooded areas, the impact of the proposed flood protection measures on the flood risk, and similar practical applications. Input hydrographs are one of the most important parameters that can have a significant impact on the hydraulic flood modelling results [23]. Savage et al. [23] have shown that input hydrographs have a significant influence on modelling results, especially during rising limb of the hydrograph. During peak discharge, the channel friction parameter has the largest impact, whereas during the recession part of the hydrograph, the floodplain friction parameter plays an important role. For the predictions of the flood extent, it has been observed that the dominant hydraulic model input factors shift during the flood event. Hall et al. [24], who performed a global sensitivity analysis using flood inundation models, also made similar conclusions. It was found that the Manning roughness coefficient has the dominant impact on uncertainty in the hydraulic model calibration and prediction [24]. The same finding was also reported by Pestotnik et al. [25], who analysed the possibility of using the two-dimensional (2D) model Flo-2D for hydrological modelling for the case of the Glinščica River catchment in Slovenia. Additionally, boundary conditions are also one of the factors that can have a significant impact on hydraulic modelling results [26].

However, the relationship between the design hyetograph selection and hydraulic modelling results remains unclear. Examples of modelling results include the flood extent or flow velocities over floodplains, which can have a significant impact on the stability of a human body or a vehicle in floodwaters [27–30]. Even though some researchers doubt the usefulness of the flood water flow velocities as the appropriate parameter to model flood damages [31], the implementation of the 2007 European Union (EU) Flood Directive governs the determination and zonation of hazards areas using a combination of flood water depths and flow velocities. Different flood hazard zones are then used for the planning of preventive measures, such as the restriction of construction in areas with high flood hazards [32]. Knowing the uncertainty in the assessment of flood hazard and flood risk areas is an important task in flood risk reduction, as the uncertainty in the decision-making process for natural hazards in mountains has been recognised [33,34].

Therefore, the main aim of this study is to explore the relationship between the design hyetograph definition, and hydraulic modelling results. For this purpose, the Glinščica Stream catchment in central Slovenia was selected as the case study. The specific aims are as follows:

(i)   to quantify the effect of rainfall duration on hydraulic modelling results (e.g., flood extent, floodwater velocities);

(ii) to quantify the impact of temporal rainfall distribution within a rainfall event on hydraulic modelling results, and

(iii) to compare the differences between flood modelling results (floodplain extents, velocities, volumes, and water depths) for the events with 10 and 100-year return periods.

2. Data and Methods

2.1. Catchment Description

The Glinščica Stream catchment was selected as the case study in order to investigate the impact of the design rainfall on the hydraulic modelling results. The Glinščica Stream catchment is part of the Gradaščica River catchment that drains into the Ljubljanica River. This river is part of the Sava River catchment; the Glinščica Stream catchment is situated in the central part of Slovenia, and reaches into the eastern part of the urban area of the capital city of Ljubljana (Figure 1). The stream has its source under the southeastern slopes of the hills of Polhograjsko hribovje, and at the village of Podutik, it passes into the flat area of the Ljubljana plain. The topography of the catchment is comprised of hilly areas to the east and west, and a flat plain area in the south. The relief of the Glinščica Stream catchment is diverse, comprising hilly headwater areas, as well as flat plains. The Glinščica Stream catchment is one of the hydrologic experimental catchments in Slovenia [35,36]. Table 1 shows some basic properties of the Glinščica Stream catchment. It has already been studied in some of the previous studies, and a more detailed description of the catchment is provided by Bezak et al., Šraj et al. and Brilly et al. [8,14,37]. The lowland areas of the Gradaščica River, once natural floodplain areas, were partly urbanised in the last couple of decades, which resulted in an elevated flood risk for the area. The last major flood occurred in October 2014, when extensive urbanised areas and more than 1000 houses were flooded.

Table 1. Basic characteristics of the Glinščica Stream catchment.

Catchment Area (km²)	Elevation (m a.s.l.)	Land-Cover	Soil Characteristics (According to Soil Conservation Service (SCS) Classification)	Mean Annual Precipitation (mm)	Time of Concentration (h)
16.85	from 209 to 590	49% forest, 23% agriculture land, 19% urbanised areas	C and D types with generally low infiltration rates	about 1400	about 6

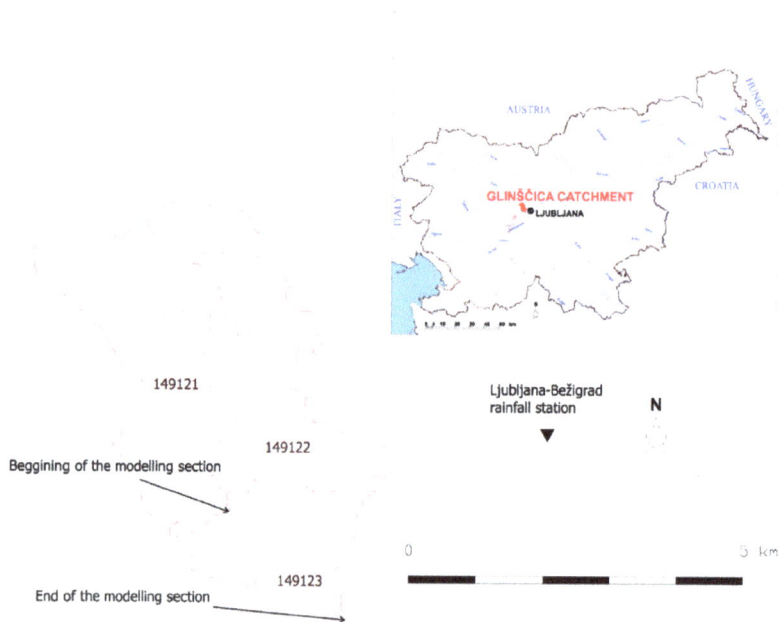

Figure 1. Location of the Glinščica Stream catchment on a map of Slovenia, and the Glinščica River catchment divided into three sub-catchments. The hydraulic modelling was performed in the 149123 sub-catchment from the beginning (confluence of sub-catchments 149121 and 149122) to the end (confluence of the Glinščica Stream and the Gradaščica River) of the river network in this sub-catchment.

The official water level and discharge measurements in Slovenia are performed by the Slovenian Environment Agency (ARSO). However, in the Glinščica Stream catchment, there is currently no discharge gauging station (there is about 15 years of data available before 1970, but catchment has significantly changed during the past 50 years [38]; therefore, this data was omitted in this study). For the purpose of the research projects and investigation of hydrological processes in the experimental catchment, the water station, which was equipped with an ultrasonic Doppler instrument (Starflow Unidata 6526 model), was placed in the channel of the Glinščica Stream. However, it was only placed there for the limited period of time [14,37]. This means that design discharges cannot be determined using the frequency analysis approach [5], the use of a different approach is required in order to derive the design values for this catchment.

2.2. Hydrological Model

The hydrological model HEC-HMS [39], with a combination of the design hyetographs [21], was used in the study in order to compute design hydrographs that were further used as inputs to the hydraulic model. Three different methods were applied in order to construct the design hyetograph: namely, the Huff method, the constant intensity method, and the frequency storm method. Descriptions of the applied methods can be found, for example, in Ball, Alfieri et al., Azli and Rao, Dolšak et al. [11,17,20,21]. Calibration and validation of the hydrological model of the Glinščica Stream catchment was performed by Šraj et al. [14] using measured discharge data obtained as part of the work that has been done in order to investigate the impact of changed land use (urbanisation) on the hydrological and biogeochemical processes in the experimental catchment [14,37]. For modelling purposes, the Glinščica Stream catchment was divided into three sub-catchments that are shown in Figure 1. A detailed description of the calibration and validation of the hydrological model is provided by Šraj et al. [14].

2.3. Hydraulic Model

Results of the hydrological modelling were used as inputs to the hydraulic model. Hydraulic modelling was performed from the begging (confluence of sub-catchments 149121 and 149122) to the end of the sub-catchment 149123 (confluence of the Glinščica Stream and the Gradaščica River) (Figure 1). The Glinščica Stream catchment was modelled with hydraulic model HEC-RAS 5.0.3, which enables one-dimensional (1D) and two-dimensional (2D) unsteady and steady flow simulations [40]. The basic characteristics of the Glinščica Stream catchment hydraulic model are shown in Table 2. Figure 2 shows a graphical representation of the hydraulic model extent. The connection between the river channel (1D) and the 2D flow area was defined as a lateral structure, which is one of the options that can be used to connect 1D flow in a river channel with 2D flow on 2D flow areas [40]. The average cell size on 2D flow areas was 25.2 m^2 and 25.3 m^2 for the right bank and left bank 2D flow areas, respectively (Figure 2). The 2D flow areas were represented by the underlying digital terrain model with a cell size of 1 m × 1 m, which is available for all of Slovenia. The HEC-RAS preprocessor computes several geometric and hydraulic characteristics of each cell face that are important for the hydraulic modelling [40]. The model also includes five bridges that are located in the modelled area [38]. Inflows to the modelled area are indicated with black lines on Figure 2. The upstream boundary condition was the flow hydrograph from catchment 149121 (shown in Figure 1), and the downstream boundary condition was the normal depth and discharge contributions from sub-catchments 149122 and 149123 (shown in Figure 1) were modelled as lateral inflows. Unsteady flow simulations with full momentum equations [40] were used in this study. The computation interval was 20 s, and a 36-h period was considered in simulations. Most of the simulations were computed in less than 10 min. All of the hydraulic parameters in the hydraulic model were kept constant during the simulations of the selected scenarios that are presented in the next sub-section.

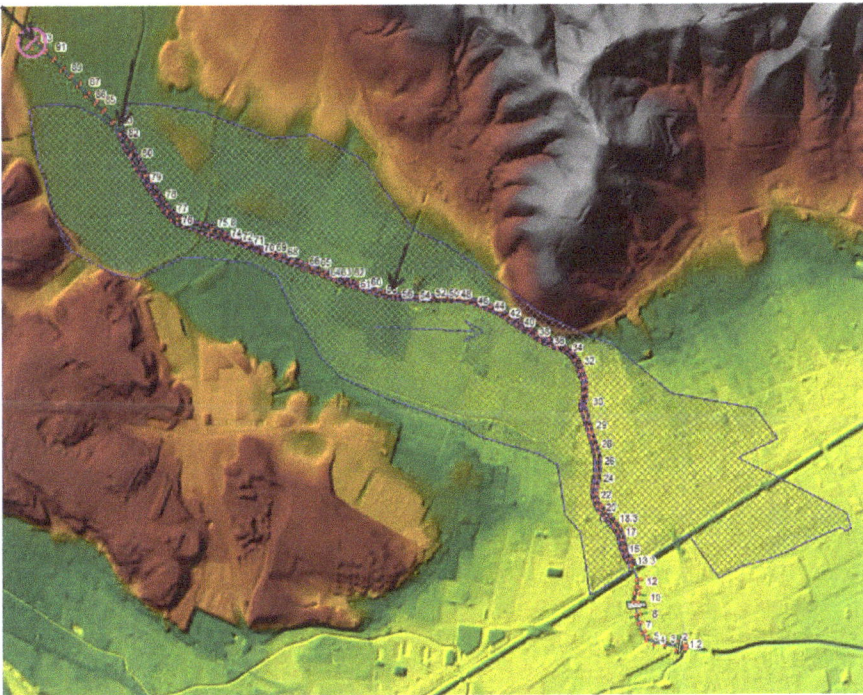

Figure 2. Hydraulic model of the Glinščica Stream catchment with two large 2D flow areas.

Table 2. Basic characteristics of the hydraulic model of the Glinščica Stream catchment. 2D: two-dimensional.

River Length (m)	Number of Cross-Sections	Number of 2D Flow Areas	Size of 2D Flow Areas	Manning's Roughness Coefficients
about 3000	93	2	0.64 km^2 (left) and 0.50 km^2 (right)	Between 0.02 to 0.033 for the river channel, 0.04 for the flood area within the cross-section, and between 0.06 and 0.1 for the 2D flood area

2.4. Scenarios (Design Rainfall Events)

In order to evaluate impact of the design rainfall on the hydraulic modelling results, the following 10 scenarios (design rainfall events) were determined and applied as inputs to the hydrological model that was used to compute the flow hydrographs at outflows from individual sub-catchments:

- Design rainfall was defined based on the 50% (Huff 50%, 6 h), 10% (Huff 10%, 6 h), and 90% (Huff 90%, 6 h) Huff curves with a rainfall duration of 6 h (this duration is approximately equal to the catchment time of concentration);
- Design rainfall was defined based on the 50% Huff curve with a rainfall duration of 2 h (Huff 50%, 2 h), 12 h (Huff 50%, 12 h), and 24 h (Huff 50%, 24 h);
- Design rainfall was defined as constant rainfall intensity and rainfall duration of 6 h (Const., 6 h);
- Design rainfall was defined based on the frequency storm method and peak intensity position at 25% (FreqStorm, peak 25%), 50% (FreqStorm, peak 50%), and 75% (FreqStorm, peak 75%) of rainfall duration.

All 10 scenarios were conducted for rainfall with 10 and 100-year return periods. Thus, in total, 20 different combinations were evaluated and analysed. More information about the methodology used to define the Huff curves [21] and intensity–duration–frequency (IDF) curves that were used in this study is available in Bezak et al. [8]. The Huff and IDF curves from the closest Ljubljana-Bežigrad station (shown in Figure 1) were used. Moreover, also additional information about the frequency storm method that was applied in this study is available in Bezak et al. [8]. This method defines the synthetic design hyetograph using the information from the IDF curves, where for different rainfall durations (e.g., 5 min, 15 min, 1 h, 2 h, 3 h, 6 h) the rainfall amount defined by the IDF curve is used. This approach uses the maximum rainfall amounts of different durations as part of one rainfall event, which is usually not the case in the nature (i.e., there is very low probability that the annual maxima of different rainfall durations occur in the same event). Consequently, application of the frequency storm method often results in higher peak discharge values compared to some other design rainfall definitions (from the engineering perspective, this can be regarded as conservative). The main idea of comparison of different scenarios was to explore the impact of the rainfall duration and temporal rainfall distribution within a rainfall event that was defined using Huff curves that were constructed based on the historical rainfall data for different rainfall durations for several Slovenian stations [21] on the hydraulic modelling results.

For the 10 and 100-year return period events, we compared the maximum floodplain extent area, volume of water flowing on the floodplain areas (selected cross-sections are shown in Figure 3), maximum velocities on floodplains, and outflow hydrographs and maximum water depths for all 10 scenarios.

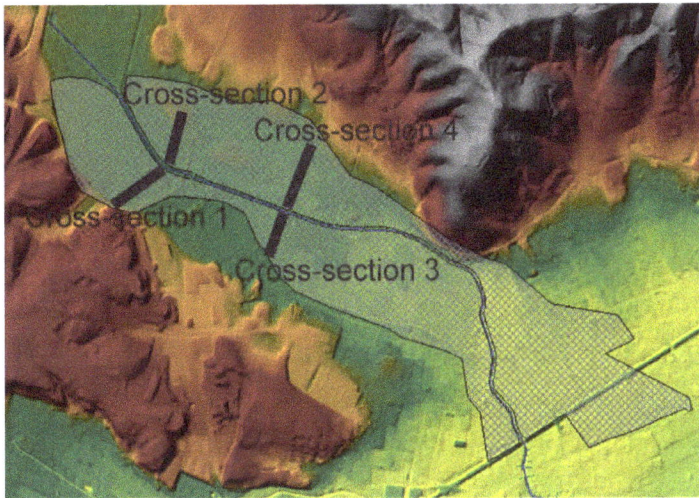

Figure 3. Selected cross-sections (1–4) on the floodplain areas that were used to compare volumes flowing on the floodplains, maximum water velocities on the floodplains, and maximum water depths.

3. Results and Discussion

3.1. 10-Year Return Period Event

In the first step of the study, we obtained hydraulic modelling results for the 10-year return period. Cases for the selected 10 scenarios were computed, and the results were compared. Figure 4 shows a comparison among the outflow hydrographs for the applied scenarios considering the 10-year return period. It can be seen that rainfall duration has a significant influence on the outflow hydrograph. The scenario that represents the 50% Huff curve with a short rainfall duration of 2 h yields smaller peak discharge values than the scenario applying the same Huff curve with a rainfall duration of 6 h. Also, scenarios with longer rainfall durations and the same Huff curve result in smaller peak discharge values compared to the first scenario (Huff 50%, 6 h), where the rainfall duration is approximately equal to the catchment time of the concentration. This finding is consistent with the results from the previous studies, as Šraj et al. [14] documented, which showed that extending the rainfall duration caused increases in the difference in peak discharge and time-to-peak. Furthermore, also, temporal rainfall distribution within a rainfall event has an important impact on the outflow hydrograph, when comparing rainfall events with the same rainfall duration (6 h) (50%, 10%, and 90% Huff curves, constant rainfall intensity and frequency storm method (25%, 50%, and 75% peak position)). It can be seen that the application of the frequency storm method yields larger peak discharge values than the scenario with 6 h of rainfall duration and the 50% Huff curve and the use of constant rainfall intensity within a rainfall event results in smaller peak discharges than the rainfall duration scenario with the Huff 50% curve over 6 h, which has also been reported by other authors [11,13,14,17]. Alfieri et al. [17] argued that the adoption of any rectangular hyetograph significantly underestimates design hydrograph results. Furthermore, Singh [12] concluded that rainfall patterns with temporal variability result in higher peak discharges than one with constant rainfall intensity. For the same return period, different definitions of the temporal rainfall distribution yield different peak discharge values. Some of these methods that are used to define the temporal rainfall distribution can, from an engineering point of view, be regarded as conservative or not so conservative. For example, using the frequency storm method, one can obtain a design discharge that can be regarded as on the safe side (from the design perspective). On the other hand, the constant intensity method yields smaller peak discharge values. Different results can also be obtained using different Huff curves (e.g., 10%, 50%, or 90%). Moreover, if we do not fix the return period variable, numerous combinations are possible to define the design hyetograph. Thus, there are alternative approaches possible, such as the so-called optional design hyetograph [41]. However, this approach was not tested in this study.

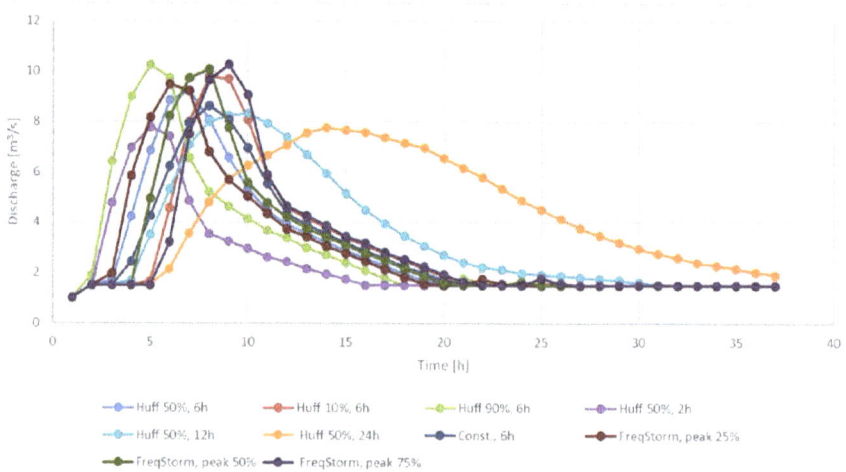

Figure 4. Comparison of outflow hydrographs for 10 selected scenarios for a 10-year return period.

In the next step, we compared maximum flood extents, maximum floodplain velocities, and floodplain volumes calculated from the hydraulic model simulations. Table 3 shows a comparison of these values for the 10 selected scenarios for the 10-year return period. One can notice that design rainfall selection yields more than a 35% difference in the maximum floodplain extent values (Table 3). The minimum extent of the flood was obtained using a scenario with a short rainfall duration of 2 h, and a 50% Huff curve, resulting in minimum hydrograph peak discharge. The maximum flood extent was obtained with the application of a scenario that represents a 90% Huff curve and a rainfall duration of 6 h, resulting in maximum hydrograph peak discharge. We also analysed which land-use types were flooded for these 10 scenarios, because flood damage depends on the flooded land-use types and property values (Table 4). For this purpose, a land use map of Slovenia was used [42]. The results show that the largest changes were associated with meadowland use type, which also covers the largest percent of the flooded area (Table 4). In the case of built areas, the largest extension of flooded areas (6.2×10^3 m^2) was calculated for the 90% Huff curve (6-h rainfall duration) scenario, whereas the smallest flooding extent on the built areas (3.5×10^3 m^2) was calculated for the 50% Huff curve (2 h rainfall duration) scenario. This also means that flood damage would be the largest for the previously mentioned scenario (Huff 90%, 6 h). For the 10-year return period, no flow was obtained for cross-sections 3 and 4, which are located on the 2D flow areas that are shown in Figure 3. This means that these areas were not flooded. Four times higher maximum water velocities were obtained for the scenario with a constant rainfall intensity of 6 h than for the scenario with a short rainfall duration of 2 h and the 50% Huff curve for cross-section 1 on the right bank of the flooded area. The water velocity has an important impact both on the stability of human body, and vehicles in the floodwater [27–30]. Similarly, also, floodplain volumes for different scenarios differ for an order of magnitude (more than 10 times) (Table 3), which indicates that the design rainfall definition has a significant impact on the simulated floodwater dynamics. Moreover, we have also compared the maximum water depths for defined scenarios for the 10-year return period (Table 5). It can be seen that the 90% Huff curve (6 h of rainfall duration) scenario yielded maximum water depth on the right and left floodplain areas (cross-sections 1 and 2) and the 50% Huff curve (2 h of rainfall duration) scenario resulted in minimum water depths (Table 5).

Table 3. Comparison among maximum floodplain extents, maximum floodplain velocities, and floodplain volumes for 10 selected scenarios for the 10-year return period. Bold values indicate maximum values in each column.

Scenario	Maximum Flood Water Extent (10^3 m^2)	Volume of Water Flowing through Cross-Section 1 (10^3 m^3)	Maximum Velocities at Cross-Section 1 (m/s)	Volume of Water Flowing through Cross-Section 2 (10^3 m^3)	Maximum Velocities at Cross-Section 2 (m/s)
Huff 50%, 6 h	89.3	4.2	0.21	2.4	0.17
Huff 10%, 6 h	97.1	4.7	0.34	3.1	0.19
Huff 90%, 6 h	**101.1**	**6.7**	0.20	2.5	0.19
Huff 50%, 2 h	65.4	0.5	0.09	1.1	0.15
Huff 50%, 12 h	80.7	3.1	0.14	2.5	0.15
Huff 50%, 24 h	72.9	2.3	0.18	2.0	0.14
Const., 6 h	79.8	1.9	**0.46**	2.1	0.16
FreqStorm, peak 25%	92.9	4.0	0.20	2.4	**0.20**
FreqStorm, peak 50%	99.5	5.5	0.41	2.8	0.19
FreqStorm, peak 75%	100.7	5.8	0.45	**3.5**	0.19

Table 4. Area (10^3 m^2) of flooded land use types for 10 scenarios for the 10-year return period.

Land Use/Scenario	Huff 50%, 6 h	Huff 10%, 6 h	Huff 90%, 6 h	Huff 50%, 2 h	Huff 50%, 12 h	Huff 50%, 24 h	Const., 6 h	FreqStorm, Peak 25%	FreqStorm, Peak 50%	FreqStorm, Peak 75%
Field	15.7	17.1	17.6	11.2	13.9	12.5	14.0	16.4	17.4	17.5
Meadow	55.2	60.6	63.4	38.3	49.3	43.8	48.5	57.7	62.3	63.3
Trees and bushes	1.4	1.5	1.6	1.0	1.3	1.2	1.3	1.4	1.6	1.6
Uncultivated agriculture land	0.0	0.0	0.0	0.0	0.0	0.0	0.0	0.0	0.0	0.0
Forest	0.6	0.8	0.9	0.0	0.6	0.4	0.4	0.7	0.9	0.9
Built areas	5.2	5.8	6.2	3.5	4.3	3.7	4.4	5.4	6.0	6.1
Water	11.3	11.3	11.3	11.3	11.3	11.3	11.3	11.3	11.3	11.3

Table 5. Comparison of maximum water depths for cross-sections 1 and 2 for different scenarios for the 10-year return period.

Scenario	Maximum Water Depth at Cross-Section 1 (m)	Maximum Water Depth at Cross-Section 2 (m)
Huff 50%, 6 h	0.25	0.49
Huff 10%, 6 h	0.30	0.56
IIuff 90%, 6 h	0.33	0.59
Huff 50%, 2 h	0.11	0.26
Huff 50%, 12 h	0.22	0.47
Huff 50%, 24 h	0.18	0.39
Const., 6 h	0.20	0.41
FreqStorm, peak 25%	0.28	0.51
FreqStorm, peak 50%	0.32	0.57
FreqStorm, peak 75%	0.33	0.59

3.2. 100-Year Return Period Event

We have also applied all 10 scenarios for the 100-year return period. Figure 5 shows a comparison of outflow hydrographs for the considered scenarios for the 100-year return period. Compared to the 10-year event (Figure 4), higher peak discharge values were obtained for all of the scenarios, as expected (Figure 5). Similarly, as for the 10-year return period, the maximum peak discharge value was obtained for the scenario using the frequency storm method, with a peak intensity position at 75% of the rainfall duration. On the other hand, the smallest peak discharge value was calculated for scenario based on the 50% Huff curve and 24 h of rainfall duration. Table 6 shows a comparison among the maximum floodplain water extents for investigated cases. The scenario based on the frequency storm method and a peak position at 75% yielded a floodplain extent that was about twice as large as the scenario that used the 50% Huff curve and 24 h of rainfall duration (Table 6). This means that the difference in the peak discharge value for a factor of 1.4 can result in a floodplain water extent that is more than twice as large (Figure 5 and Table 6). Figure 6 shows a comparison between the scenarios that caused the minimum and maximum floodplain extent for the 100-year return period. We also compared the volume of water flowing through cross-sections 1–4 (Figure 3), and the maximum water velocities through these cross-sections (Tables 6 and 7). The maximum floodplain water velocities exceed 1.2 m/s, and the differences among the maximum water velocities were more than 10 times for some of the scenarios (Table 7). Similar conclusions can also be made for the volume of water flowing through the different floodplain cross-sections (Tables 6 and 7). Table 8 shows which land use types were flooded during all of the considered scenarios for the 100-year return period. Similarly, as for the 10-year return period, the largest percentage of the flooded area was meadows. For the built areas, the largest extension of flooded areas (22.4×10^3 m^2) was calculated for the FreqStorm, peak 75% scenario, whereas the smallest flooding extent over the built areas (8.1×10^3 m^2) was calculated for the Huff 50%, 24-h scenario. Differences in the design rainfall resulted in a changed extension of the flooded built areas by a factor of 2.8. This poses huge uncertainty in predictions of the maximum flood extent (e.g., for a decision-maker). Further, the uncertainty in flood extent makes it difficult to assess potential flood damages (e.g., by using depth–damage curves), or plan future changes in land use in the flood hazard areas. Moreover, Table 9 shows a comparison between the maximum water depths for different scenarios for the 100-year return period. While the difference between the maximum and minimum water levels at the selected cross-sections seems small (in the range of 6–8 cm), it is well known that small changes in the shallow overflooding depth in urban areas can considerably increase the direct and indirect damage on buildings and urban infrastructure [43]. A review of flood damage studies revealed that that the variation in flood damage to properties could not be explained by inundation depth alone, and should be combined with other factors [44], such as water flow velocity. However, the results of our study show that the water velocities at the selected flood plain cross-sections can vary by a factor of 10.

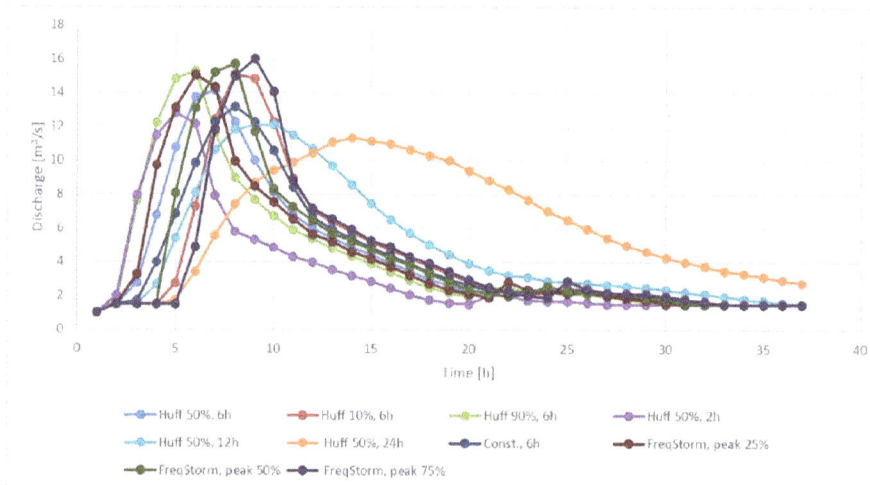

Figure 5. Comparison of outflow hydrographs for the 10 selected scenarios for the 100-year return period.

Table 6. Comparison among maximum floodplain extents, maximum floodplain velocities, and floodplain volumes (cross-sections 1 and 2) for the 10 selected scenarios for the 100-year return period. Bold values indicate the maximum values in each column.

Scenario	Maximum Flood Water Extent (10^3 m^2)	Volume of Water Flowing through Cross-Section 1 (10^3 m^3)	Maximum Velocities at Cross-Section 1 (m/s)	Volume of Water Flowing through Cross-Section 2 (10^3 m^3)	Maximum Velocities at Cross-Section 2 (m/s)
Huff 50%, 6 h	203.7	23.8	0.25	6.1	0.26
Huff 10%, 6 h	243.6	25.9	0.27	6.9	**0.29**
Huff 90%, 6 h	245.2	27.2	0.26	6.7	0.26
Huff 50%, 2 h	144.5	15.7	0.31	4.5	0.24
Huff 50%, 12 h	149.1	28.3	**0.69**	5.9	0.22
Huff 50%, 24 h	134.7	**38.2**	0.18	7.0	0.20
Const., 6 h	165.3	21.5	0.64	5.0	0.24
FreqStorm, peak 25%	237.4	27.1	0.26	5.3	0.28
FreqStorm, peak 50%	271.8	29.7	0.27	6.8	0.27
FreqStorm, peak 75%	**279.9**	28.3	0.28	**7.2**	0.28

Table 7. Comparison among maximum floodplain velocities and floodplain volumes (cross-sections 3 and 4) for different scenarios for the 100-year return period. Bold values indicate the maximum values in each column.

Scenario	Volume of Water Flowing through Cross-Section 3 (10^3 m^3)	Maximum Velocities at Cross-Section 3 (m/s)	Volume of Water Flowing through Cross-Section 4 (10^3 m^3)	Maximum Velocities at Cross-Section 4 (m/s)
Huff 50%, 6 h	8.0	0.54	2.6	0.16
Huff 10%, 6 h	11.9	0.64	5.0	0.44
Huff 90%, 6 h	11.4	0.49	4.8	1.38
Huff 50%, 2 h	2.7	0.75	0.6	0.13
Huff 50%, 12 h	5.2	0.93	1.2	0.13
Huff 50%, 24 h	3.8	0.32	0.8	0.12
Const., 6 h	5.3	0.84	1.1	0.14
FreqStorm, peak 25%	10.6	0.67	4.1	0.44
FreqStorm, peak 50%	13.7	0.41	6.5	0.46
FreqStorm, peak 75%	14.8	0.53	7.1	0.45

Table 8. Area (10^3 m^2) of flooded land use types for the 10 selected scenarios for the 100-year return period.

Land Use/Scenario	Huff 50%, 6 h	Huff 10%, 6 h	Huff 90%, 6 h	Huff 50%, 2 h	Huff 50%, 12 h	Huff 50%, 24 h	Const., 6 h	FreqStorm, Peak 25%	FreqStorm, Peak 50%	FreqStorm, Peak 75%
Field	41.5	52.7	52.5	21.7	22.7	20.3	29.1	50.2	59.8	62.1
Greenhouse	0.0	0.1	0.1	0.0	0.0	0.0	0.0	0.3	0.3	0.8
Orchard	4.4	10.2	10.2	0.0	0.0	0.0	0.0	9.6	10.7	10.8
Meadow	112.6	128.7	130.2	91.6	94.8	88.4	99.2	126.3	140.7	144.2
Trees and bushes	4.1	5.9	5.9	2.4	2.4	2.2	2.7	5.5	6.9	7.2
Uncultivated agriculture land	16.2	18.2	18.5	7.1	7.6	3.1	11.6	18.1	19.0	19.1
Forest	1.4	1.5	1.6	1.3	1.3	1.2	1.3	1.5	1.9	2.0
Built areas	12.2	14.9	15.1	9.1	9.1	8.1	10.1	14.6	21.2	22.4
Water	11.3	11.3	11.3	11.3	11.3	11.3	11.3	11.3	11.3	11.3

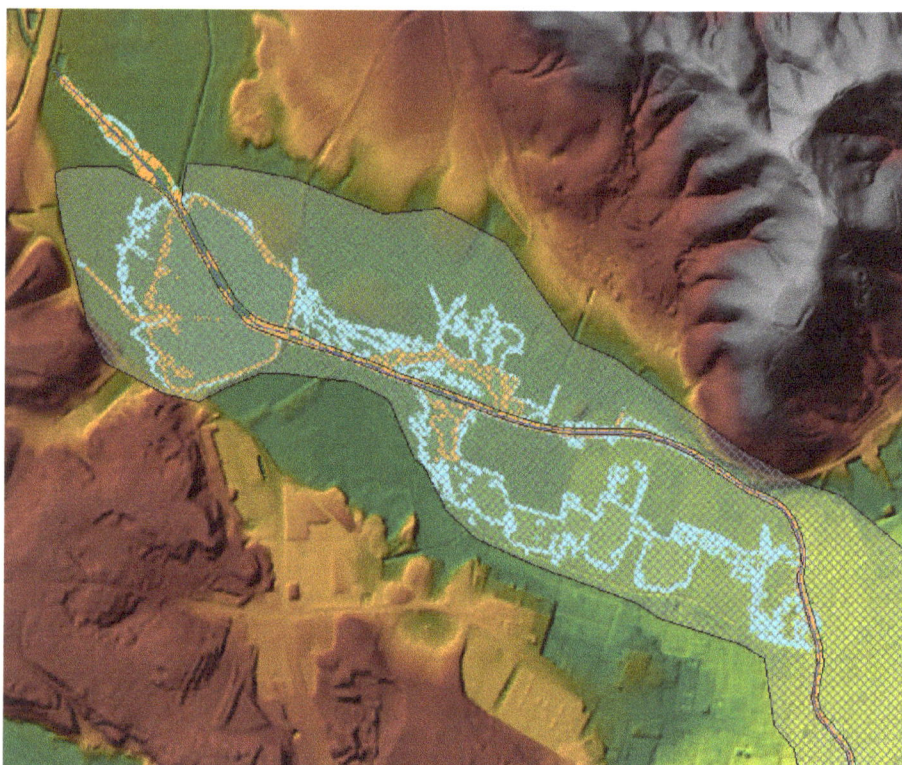

Figure 6. Comparison between the maximum floodplain extent for scenarios (vi) and (x) indicated with orange and light blue, respectively, for the 100-year return period.

Table 9. Comparison of maximum water depths for cross-sections 1 and 2 for different scenarios for the 100-year return period.

Scenario	Maximum Water Depth at Cross-Section 1 (m)	Maximum Water Depth at Cross-Section 2 (m)
Huff 50%, 6 h	0.52	0.80
Huff 10%, 6 h	0.57	0.84
Huff 90%, 6 h	0.57	0.85
Huff 50%, 2 h	0.46	0.75
Huff 50%, 12 h	0.47	0.75
Huff 50%, 24 h	0.45	0.73
Const., 6 h	0.49	0.77
FreqStorm, peak 25%	0.57	0.84
FreqStorm, peak 50%	0.60	0.86
FreqStorm, peak 75%	0.60	0.86

4. Conclusions

This study presents combined hydrological and hydraulic modelling results for the Glinščica Stream catchment in Slovenia, which can be regarded as a small-scale catchment (less than 20 km^2) that is ungauged in terms of discharges. This means that approaches suitable for ungauged catchments are the only option in order to derive design hydrographs, and more specifically design peak discharge values. This study evaluates 10 different design rainfall events (scenarios) that were used as input to the hydrological model. Both 10 and 100-year return period events were analysed. By using calibrated and validated hydrological models, the inputs for the hydraulic model were determined. Thus, the main aim was to evaluate the influence of the design rainfall selection in terms of the rainfall duration and temporal rainfall distribution defined using Huff curves on the hydraulic modelling results (e.g., shape of the outflow hydrograph, peak discharge values, floodplain water extents, maximum floodplain water velocities, and maximum water depths).

The 10 selected and considered scenarios in the study show that the maximum peak discharge value using different design hyetographs and rainfall durations can be 1.4 times larger than the minimum peak discharge value. At the same time, the maximum floodplain extent can be two times larger than the minimum flood extent, and the maximum floodplain water velocity can be 10 times larger than the minimum floodplain velocity scenarios. This means that design rainfall definition can significantly influence the hydraulic modelling results.

Thus, we recommend that the selection of the design rainfall event should be selected with care, and with the consideration of the typical temporal rainfall distribution of the region, which can be described using the Huff curves. Moreover, in order to select the crucial rainfall duration, an analysis of the past flood events could be useful, with the aim of identifying rainfall characteristics that can result in an extreme flood event, such as duration. In combination with the catchment time of concentration, this could be used to select the rainfall duration.

Acknowledgments: The results of the study are part of the Slovenian national research project J2-7322: "Modelling hydrologic response of nonhomogeneous catchments" and research Programme P2-0180: "Water Science and Technology, and Geotechnical Engineering: Tools and Methods for Process Analyses and Simulations, and Development of Technologies" that are financed by the Slovenian Research Agency (ARRS). We wish to thank the Slovenian Environment Agency (ARSO) for data provision.

Author Contributions: All authors drafted the manuscript and determined the aims of the research; N. Bezak carried out the hydrological and hydraulic calculations; All authors contributed to the manuscript writing and revision.

Conflicts of Interest: The authors declare no conflict of interest.

References

1. Zorn, M.; Komac, B. Damage caused by natural disasters in Slovenia and globally between 1995 and 2010. *Acta Geogr. Slov.* **2011**, *51*, 7–30. [CrossRef]
2. Blöschl, G.; Hall, J.; Parajka, J.; Perdigao, R.A.P.; Merz, B.; Arheimer, B.; Aronica, G.T.; Bilibashi, A.; Bonacci, O.; Borga, M.; et al. Changing climate shifts timing of European floods. *Science* **2017**, *357*, 588–590. [CrossRef] [PubMed]
3. Šraj, M.; Menih, M.; Bezak, N. Climate variability impact assessment on the flood risk in Slovenia. *Phys. Geogr.* **2016**, *37*, 73–87. [CrossRef]
4. Blöschl, G. Recent advances in flood hydrology—Contributions to implementing the Flood Directive. *Acta Hydrotech.* **2016**, *29*, 13–22.
5. Bezak, N.; Brilly, M.; Šraj, M. Comparison between the peaks-over-threshold method and the annual maximum method for flood frequency analysis. *Hydrol. Sci. J.* **2014**, *59*, 959–977. [CrossRef]
6. Šraj, M.; Bezak, N.; Brilly, M. Bivariate flood frequency analysis using the copula function: A case study of the Litija station on the Sava River. *Hydrol. Process.* **2015**, *29*, 225–238. [CrossRef]
7. Blöschl, G.; Sivapalan, M.; Wagener, T.; Viglione, A.; Savenije, H. *Runoff Prediction in Ungauged Basins: Synthesis across Processes, Places and Scales*, 1st ed.; Cambridge University Press: Cambridge, UK, 2013; 490p.
8. Bezak, N.; Šraj, M.; Mikoš, M. Design rainfall in engineering applications with focus on the design discharge. In *Engineering and Mathematical Topics in Rainfall*; InTech Press: London, UK, 2018; ISBN 978-953-51-5562-1.
9. Grimaldi, S.; Petroselli, A.; Serinaldi, F. A continuous simulation model for design-hydrograph estimation in small and ungauged watersheds. *Hydrol. Sci. J.* **2012**, *57*, 1035–1051. [CrossRef]
10. El-Jabi, N.; Sarraf, S. Effect of maximum rainfall position on rainfall-runoff relationship. *J. Hydraul. Eng.* **1991**, *117*, 681–685. [CrossRef]

11. Ball, J.E. The influence of storm temporal patterns on catchment response. *J. Hydrol.* **1994**, *158*, 285–303. [CrossRef]

12. Singh, V.P. Effect of spatial and temporal variability in rainfall and watershed characteristics on stream flow hydrograph. *Hydrol. Process.* **1997**, *11*, 1649–1669. [CrossRef]

13. Danil, E.I.; Michas, S.N.; Lazaridis, L.S. Hydrologic modeling for the determination of design discharges in ungauged basins. *Glob. NEST J.* **2005**, *7*, 296–305.

14. Šraj, M.; Dirnbek, L.; Brilly, M. The influence of effective rainfall on modeled runoff hydrograph. *J. Hydrol. Hydromech.* **2010**, *58*, 3–14. [CrossRef]

15. Sikoroska, A.E.; Seibert, J. Appropriate temporal resolution of precipitation data for discharge modelling in pre-alpine catchments. *Hydrol. Sci. J.* **2017**, *63*. [CrossRef]

16. Sikoroska, A.E.; Viviroli, D.; Seibert, J. Effective precipitation duration for runoff peaks based on catchment modelling. *J. Hydrol.* **2017**, *556*, 510–522. [CrossRef]

17. Alfieri, L.; Laio, F.; Claps, P. A simulation experiment for optimal design hyetograph selection. *Hydrol. Process.* **2008**, *22*, 813–820. [CrossRef]

18. Huff, F.A. Time distribution of rainfall in heavy storms. *Water Resour. Res.* **1967**, *3*, 1007–1019. [CrossRef]

19. Bonta, J.V. Development and utility of Huff curves for disaggregating precipitation amounts. *Appl. Eng. Agric.* **2004**, *20*, 641–653. [CrossRef]

20. Azli, M.; Rao, R. Development of Huff curves for Peninsular Malaysia. *J. Hydrol.* **2010**, *388*, 77–84. [CrossRef]

21. Dolšak, D.; Bezak, N.; Šraj, M. Temporal characteristics of rainfall events under three climate types in Slovenia. *J. Hydrol.* **2016**, *541*, 1395–1405. [CrossRef]

22. Huff, F. *Time Distributions of Heavy Rainstorms in Illinois*; Illinois State Water Survey, State of Illinois Department of Energy and Natural Resources: Champaign, IL, USA, 1990; Circular 173.

23. Savage, J.T.S.; Pianosi, F.; Bates, P.; Freer, J.; Wagener, T. Quantifying the importance of spatial resolution and other factors through global sensitivity analysis of a flood inundation model. *Water Resour. Res.* **2016**, *52*, 9146–9163. [CrossRef]

24. Hall, J.W.; Tarantola, S.; Bates, P.D.; Horritt, M.S. Distributed sensitivity analysis of flood inundation model calibration. *J. Hydraul. Eng.* **2005**, *131*, 117–126. [CrossRef]

25. Pestotnik, S.; Hojnik, T.; Šraj, M. Analysis of the possibility of using the distributed two-dimensional model Flo-2D for hydrological modeling. *Acta Hydrotech.* **2012**, *25*, 85–103.

26. Pappenberger, F.; Matgen, P.; Beven, K.J.; Henry, J.B.; Pfister, L.; de Fraipont, P. Influence of uncertain boundary conditions and model structure on flood inundation predictions. *Adv. Water Resour.* **2006**, *29*, 1430–1449. [CrossRef]

27. Shu, C.W.; Xia, J.Q.; Falconer, R.A.; Lin, B.L. Incipient velocity for partially submerged vehicles in floodwaters. *J. Hydraul. Res.* **2011**, *49*, 709–717. [CrossRef]

28. Xia, J.Q.; Falconer, R.A.; Wang, Y.J.; Xiao, X.W. New criterion for the stability of a human body in floodwaters. *J. Hydraul. Res.* **2014**, *52*, 93–104. [CrossRef]

29. Milanesi, L.; Pilotti, M.; Ranzi, R. A conceptual model of people's vulnerability to floods. *Water Resour. Res.* **2015**, *51*, 182–197. [CrossRef]

30. Milanesi, L.; Pilotti, M.; Bacchi, B. Using web-based observations to identify thresholds of a person's stability in a flow. *Water Resour. Res.* **2016**, *52*, 7793–7805. [CrossRef]

31. Kreibich, H.; Piroth, K.; Seifert, I.; Maiwald, H.; Kunert, U.; Schwarz, J.; Merz, B.; Thieken, A.H. Is flow velocity a significant parameter in flood damage modelling? *Nat. Hazards Earth Syst. Sci.* **2009**, *9*, 1679–1692. [CrossRef]

32. Hagemeier-Klose, M.; Wagner, K. Evaluation of flood hazard maps in print and web mapping services as information tools in flood risk communication. *Nat. Hazards Earth Syst. Sci.* **2009**, *9*, 563–574. [CrossRef]

33. Tacnet, J.-M.; Batton-Hubert, M.; Dezert, J. Information fusion for natural hazards in mountains. In *Advances and Applications of DSmT for Information Fusion*; Smarandache, F., Dezert, J., Eds.; American Research Press: Rehoboth, DE, USA, 2009; Chapter 23, Volume III, pp. 565–659.

34. Tacnet, J.-M.; Batton-Hubert, M.; Dezert, J.; Richard, D. Decision Support Tools for Natural Hazards Management under Uncertainty. In Proceedings of the 12th Congress Interpraevent, Grenoble, France, 23–26 April 2012; pp. 597–608. Available online: http://www.interpraevent.at/palm-cms/upload_files/Publikationen/Tagungsbeitraege/2012_1_597.pdf (accessed on 11 February 2018).

35. Bezak, N.; Šraj, M.; Rusjan, S.; Kogoj, M.; Vidmar, A.; Sečnik, M.; Brilly, M.; Mikoš, M. Comparison between two adjacent experimental torrential watersheds: Kuzlovec and Mačkov graben. *Acta Hydrotech.* **2013**, *26*, 85–97.

36. Šraj, M.; Bezak, N.; Rusjan, S.; Mikoš, M. Review of hydrological studies contributing to the advancement of hydrological sciences in Slovenia. *Acta Hydrotech.* **2016**, *29*, 47–71.

37. Brilly, M.; Rusjan, S.; Vidmar, A. Monitoring the impact of urbanisation on the Glinscica stream. *Phys. Chem. Earth Part A/B/C* **2006**, *31*, 1089–1096. [CrossRef]

38. Rusjan, S.; Fazarinc, R.; Mikoš, M. River rehabilitation of urban watercourses on the example of the Glinščica River in Ljubljana. *Acta Hydrotech.* **2003**, *21*, 1–22.

39. HEC-HMS 3.4. Reference Manual. Available online: http://www.hec.usace.army.mil/software/hec-hms/documentation/HEC-HMS_Users_Manual_3.4.pdf (accessed on 11 February 2018).

40. HEC-RAS 5.0. Reference Manual. Available online: http://www.hec.usace.army.mil/software/hec-ras/documentation/HEC-RAS%205.0%20Reference%20Manual.pdf (accessed on 11 February 2018).

41. Nguyen, T.A.; Grossi, G.; Ranzi, R. Design Storm for Mixed Urban and Agricultural Drainage Systems in the Northern Delta in Vietnam. *J. Irrig. Drain. Eng.* **2016**, *142*, 4015051. [CrossRef]

42. GERK. Land Use Map of Slovenia. Available online: http://rkg.gov.si/GERK/ (accessed on 11 February 2018).

43. Penning-Rowsell, E.; Johnson, C.; Tunstall, S.; Tapsell, S.; Morris, J.; Chatterton, J.B.; Green, C. *The Benefits of Flood and Coastal Risk Management: A Manual of Assessment Techniques*; Middlesex University Press: London, UK, 2005.

44. Hammond, M.J.; Chen, A.S.; Djordjević, S.; Butler, D.; Mark, O. Urban flood impact assessment: A state-of-the-art review. *Urban Water J.* **2015**, *12*, 14–29. [CrossRef]

geosciences

MDPI

Article

Simulating the Influence of Buildings on Flood Inundation in Urban Areas

Riccardo Beretta [1], Giovanni Ravazzani [1,*], Carlo Maiorano [2] and Marco Mancini [1]

[1]   Politecnico di Milano, Piazza Leonardo da Vinci, 32, 20133 Milano, Italy;
      riccardo4.beretta@mail.polimi.it (R.B.); marco.mancini@polimi.it (M.M.)
[2]   Modellistica e Monitoraggio Idrologico S.r.l, Via Ariberto 1, 20123 Milano, Italy; carlo.maiorano@mmidro.it
*   Correspondence: giovanni.ravazzani@polimi.it; Tel.: +39-02-2399-6231

Received: 19 January 2018; Accepted: 19 February 2018; Published: 24 February 2018

Abstract: Two-dimensional hydraulic modeling is fundamental to simulate flood events in urban area. Key factors to reach optimal results are detailed information about domain geometry and utility of hydrodynamic models to integrate the full or simplified Saint Venant equations in complex geometry. However, in some cases, detailed topographic datasets that represent the domain geometry are not available, so approximations—such as diffusive wave equation—is introduced whilst representing urban area with an adjusted roughness coefficient. In the present paper, different methods to represent buildings and approximation of the Saint Venant equations are tested by performing experiments on a scale physical model of urban district in laboratory. Simplified methods are tested for simulation of a real flood event which occurred in 2013 in the city of Olbia, Italy. Results show that accuracy of simulating flow depth with a detailed geometry is comparable to the one achieved with an adjusted roughness coefficient.

Keywords: urban topography; flood modeling; Saint Venant equations; laboratory experiment; buildings; roughness coefficient

1. Introduction

Flood events are one of the most dangerous natural phenomena connected to human activities, with possible consequences on people's safety and economic losses [1].

The flood hazard affecting densely populated areas is increasing in recent times, due to the intensification of extreme meteorological events and poorly managed urban development [2]. Two-dimensional flood inundation modeling is a pivotal component of flood risk assessment and management. It is therefore not surprising that over the last few decades significant efforts have been devoted to the development of increasingly complex algorithms to simulate the flow of water in streams and floodplains [3–10]. In areas with mild slope terrain, one-dimensional models may produce misleading results and two-dimensional (2D) models are recommended also for their ability to capture preferential flow directions caused by the presence of buildings [11,12].

The correct representation of buildings in a 2D model is a fundamental factor to reach good flood simulation results in urban areas. When detailed geometry is available, the individual shapes of buildings can be incorporated into the calculations. For large scale modeling or when detailed geometry information are not available, flow obstructions may be represented as areas with higher roughness coefficient (roughness approach). This accounts for the increased resistance induced by the presence of buildings in the urban area. Moreover, low accuracy of available topographic data justifies introducing some simplifications to the Saint Venant equations, also known as shallow water equations (SWE), that describe fluid dynamics [13–15]. In most practical applications of flood simulation, the diffusive wave simplification is preferred to the full solution (dynamic wave), since the local and convective acceleration terms are small in comparison to the bed slope [16].

In the last few decades, many aspects of urban flooding have been investigated through experimental studies [17,18], and few of these investigations have focused on how to properly represent buildings within inundation models [19,20].

The main objective of this paper is to verify the accuracy of the roughness approach against full buildings incorporation in flood simulation. Further analysis was dedicated to test diffusive model against full SVE. The novelty of this research is that three different methods to represent buildings are tested by performing a number of laboratory experiments carried out with a simplified urban district physical model, and reconstructing results with a hydraulic mathematical model considering both the solution of the full SWE and the diffusive simplification. Simplified methods are tested for simulation of a real flood event that hit the city of Olbia, Italy, on November 2013.

2. Materials and Methods

2.1. Experimental Setup

Experiments were performed on a physical scale model at the Fantoli Hydraulic Laboratory at the Politecnico di Milano (Figure 1a). The model was implemented for verifying the hydraulic performance of the dam body of an on-stream detention basin designed for flood risk reduction of the Fosso di Pratolungo river, a small tributary of the river Aniene, in the Lazio region, Italy. Flow into the physical model is regulated by two triangular Thompson weirs, often used in laboratory experiments for their high sensitivity to low flow rates. The maximum flow rate achievable, is 110 L/s. The ratio between the physical model lengths and the prototype is 1:25. Specifically for this work, we considered the channel reach and floodplain downstream of the dam artifact (427 cm × 223 cm), where six bricks (12 cm × 25 cm) were placed for the representation of a small urban area with regular simple geometry (Figure 1c). An impermeable foam layer was attached to the lower side of the bricks in order to follow the irregularities of floodplain reproduced in the physical model (Figure 1b). A removable bridge was placed within the channel (Figure 1c) and the channel end was blocked with a board (Figure 1a) in order to promote floodplain inundation.

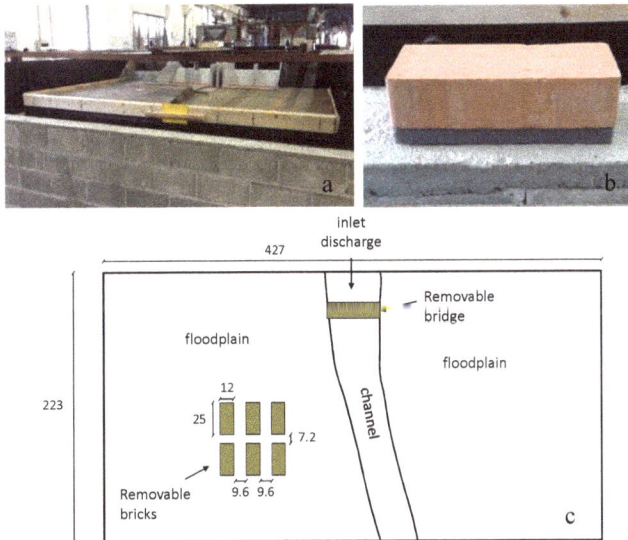

Figure 1. (a) Physical scale model of the dam body and the downstream floodplain; (b) brick with a layer of foam lining the lower side used to represent the urban area; (c) layout of the experimental setup denoting inlet discharge, channel, floodplain, and removable bridge and bricks (dimension in cm).

Water level and velocity were measured using a portable high precision nonius hydrometric rod and a micro-current-meter, respectively, in the points shown in Figure 2. The hydrometric rod was used to measure floodplain and channel bed elevation with a spatial resolution of 1 cm, leading to the digital elevation model shown in Figure 2.

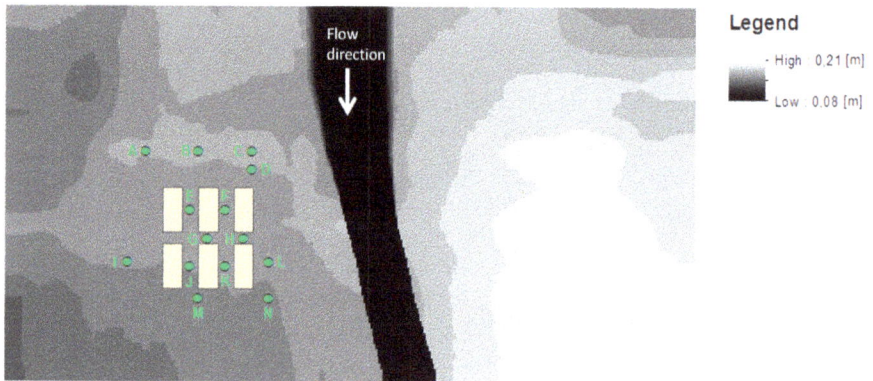

Figure 2. Digital Elevation Model of the floodplain with the six bricks used to represent a simplified urban district and locations of points where measures were acquired, marked with letters from A to N.

2.2. Mathematical Hydraulic Modelling

In order to simulate the flood inundation, the Hec-Ras model was employed [21]. As from release 5.0, Hec-Ras is designed to simulate one-dimensional, two-dimensional, and combined one/two-dimensional unsteady flow through a full network of open channels, floodplains, and alluvial fans. For the purpose of this work, flood inundation was simulated with unsteady two-dimensional solution of the full SWE and the simplified diffusive equation. When a steady state was required, this was reached by setting as input a constant discharge hydrograph long enough to reach the steady condition.

Several methods have been proposed to set the friction coefficient when roughness approach is employed to simulate flow obstacles such as the equivalent friction slope method [22,23] or similar methods [24]. In this work, we chose to use roughness Manning coefficient reported in the Hec-Ras 2D manual [25] as this is what is likely done for practical engineering applications. Further analysis to verify possible improvement when using different approaches to set roughness coefficient is ongoing.

Buildings were modeled in three different ways:

1. Method 1: incorporation of buildings using the detailed digital elevation model (DEM) with 1 cm spatial resolution.
2. Method 2: buildings are replaced by a flat area with high roughness (Manning coefficient = 10).
3. Method 3: all urban area is replaced by a flat area with high roughness (Manning coefficient = 0.15).

The mean relative absolute error (MRAE) was computed as goodness of fit index

$$MRAE = \frac{\sum_{i=1}^{n} |X_{obs,i} - X_{mod,i}| / X_{obs,i}}{n} \tag{1}$$

where X_{obs} and X_{mod} are the observed and modeled values, respectively, n is the number of points compared.

2.3. Hydrologic Model

Flood hydrograph of the six streams flowing to Olbia during the 2013 flood were simulated with the FEST model (flash-flood event-based spatially distributed rainfall-runoff transformation) [26–29]. FEST is a distributed, raster-based hydrologic model developed focusing on flash-flood event simulation. As a distributed model, FEST can manage spatial distribution of meteorological forcings, and heterogeneity in hill slope and drainage network morphology (slope, roughness, etc.) and land use.

The FEST model has three principal components. In the first component, the flow path network is automatically derived from the digital elevation model using a least-cost path algorithm [30]. In the second component, the surface runoff is computed for each elementary cell using the SCS-CN method [31,32]. The third component performs the runoff routing throughout the hill slope and the river network through a diffusion wave scheme based on the Muskingum–Cunge method in its non-linear form with the time variable celerity [33]. Spatial resolution of input maps was 10 m.

2.4. The 2013 Flood in Olbia

Olbia is a flood-prone city located in Sardinia, Italy, that developed in an alluvial plain bounded on the West side by a steep mountains chain and on the East side by the Tyrrhenian Sea. Six creeks cross this area with drainage area ranging from 0.5 km^2 to 38.4 km^2 (Figure 3). A steep slope in the upper part and a mild slope in the valley where city mostly expanded characterize them.

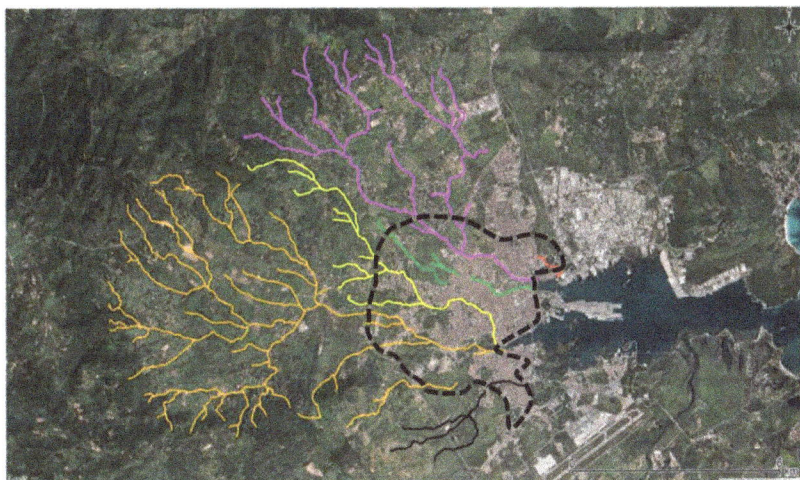

Figure 3. Creeks draining to the Olbia city center. Dashed line marks the study area where flood simulation has been conducted.

The peculiar morphology of the territory together with the urbanization pressure have contributed to transform a flood prone area into a high flood risk territory demonstrated by catastrophic floods that hit Olbia in 1970, 2013, and 2015.

Specifically, on 18 November 2013, the island of Sardinia (Italy) was affected by a meteorological event, named Cleopatra, characterized by extreme rainfall intensity (rain rate exceeded 120 mm/h in some localities), and amount (more than 450 mm of cumulated rainfall in 15 h) that sets the maximum return period of precipitation well above 200 years. Continuous rain over two days resulted in the overflowing of the rivers in the north-eastern part of Sardinia. Olbia was one of the affected cities of the island, with discharge values that reached the 25-year return period. Images and videos of the flood can be seen on the page dedicated by BBC to the Cleopatra cyclone affecting Sardinia (http://www.bbc.com/news/world-europe-24996292).

After the flood, the technical office of the Municipality carried out a survey of the flooded areas in the urban center of Olbia.

3. Results

3.1. Simulation of Flow Depth and Velocity of the Laboratory Experiments

The first phase of experiments was dedicated to calibrate the Manning roughness coefficient of the channel and the floodplains. In this phase, bricks and bridge were removed from the physical model. Normal depth was set as boundary condition on domain border. The domain was implemented with a square mesh with 1 cm spatial resolution.

Measurements of the free-surface profile in the channel and over the floodplain for various flow rates (15, 18, and 21 L/s) were compared to values computed with Hec-Ras with different roughness coefficient. The roughness coefficient value that minimized the difference between measured and computed water profile was $0.0166\,\mathrm{s}\,\mathrm{m}^{-1/3}$, which is in consistent with the expected value for concrete.

In the second phase, the bricks and bridge were positioned in the model, and experiments were performed considering a constant flow rate of 22.6 L/s, discharged through the dam bottom spillway of the physical model. Flow depth and velocity measured values were compared to mathematical model simulation results obtained with the solution of the diffusive equation and considering the three methods for representing buildings described in Section 2.2. Simulated water levels and velocities are shown in Figures 4 and 5, respectively.

In Tables 1 and 2 observed and simulated water level and velocity, respectively, the 14 points monitored are reported. MRAE and standard deviation values computed between observed and simulated water levels and velocities are reported in Table 3. Errors related to water levels are lower than values calculated for water velocity; this is probably also justified by the relatively higher uncertainty intrinsically involved in the velocity measuring in very low water depth (indeed, in point I, it was not possible to get velocity measurement).

As a general comment, the three methods tested are all equivalent in simulating water depths, while Method 3 is not able to correctly capture water velocities within the area approximated with a homogeneous roughness coefficient. In fact, velocity computed inside the buildings (points E, F, G, H, J, and K) do not have a physical meaning when Method 3 is used. Method 3 is intended for considering effect of urban area on flood inundating surrounding places and not to investigate flow dynamics inside the urban area.

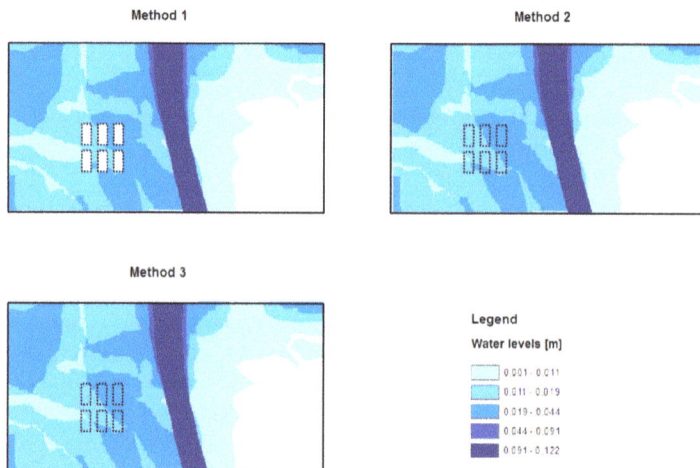

Figure 4. Water levels (m) on the laboratory model simulated with the three methods. The dashed lines denote brick locations.

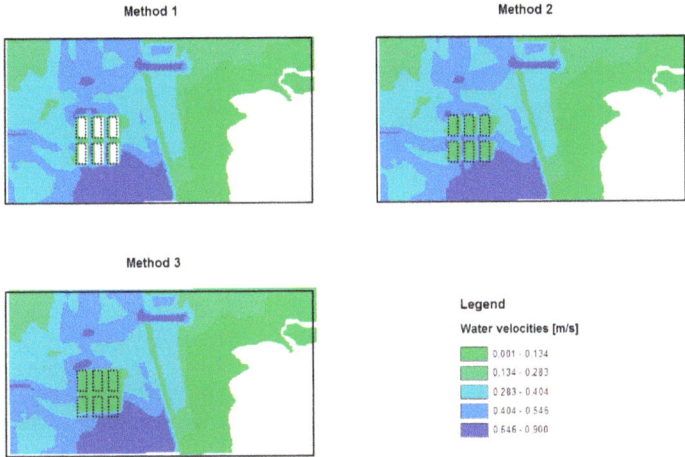

Figure 5. Water velocities on the laboratory model simulated with the three methods. Dashed lines denote brick locations.

Table 1. Observed and simulated water levels with Methods 1, 2, and 3.

Point	Observed Water Level (m)	Simulated Water Level (m)		
		Method 1	Method 2	Method 3
A	0.014	0.009	0.009	0.009
B	0.022	0.014	0.013	0.014
C	0.022	0.016	0.016	0.016
D	0.044	0.024	0.023	0.024
E	0.015	0.019	0.019	0.019
F	0.021	0.021	0.021	0.022
G	0.016	0.018	0.019	0.019
H	0.024	0.022	0.021	0.022
I	0.011	0.008	0.009	0.009
J	0.010	0.014	0.014	0.014
K	0.017	0.018	0.018	0.018
L	0.040	0.023	0.022	0.022
M	0.019	0.019	0.019	0.019
N	0.044	0.027	0.026	0.026

Table 2. Observed and simulated water velocities with Methods 1, 2, and 3.

Point	Observed Water Velocity (m/s)	Simulated Water Velocity (m/s)		
		Method 1	Method 2	Method 3
A	0.505	0.345	0.331	0.351
B	0.416	0.392	0.370	0.369
C	0.862	0.340	0.355	0.341
D	0.590	0.373	0.368	0.349
E	0.359	0.400	0.350	0.072
F	0.497	0.400	0.330	0.058
G	0.267	0.103	0.442	0.078
H	0.267	0.117	0.450	0.072
I	NA	0.412	0.459	0.440
J	0.382	0.490	0.530	0.086
K	0.566	0.550	0.550	0.092
L	0.673	0.526	0.515	0.546
M	0.244	0.390	0.416	0.432
N	0.659	0.700	0.680	0.687

Note: NA = measure was not possible.

Table 3. Mean relative absolute error, with standard deviation in brackets, for water level and velocity simulation with the three different methods.

	Method 1	Method 2	Method 3
Level	0.248 (0.157)	0.257 (0.161)	0.252 (0.154)
Velocity	0.309 (0.214)	0.347 (0.244)	0.551 (0.288)

In Method 2 and 3, as buildings that are considered impervious in Method 1 are represented only with a different roughness coefficient, water is free to flow in the area occupied by buildings, although with velocities close to zero. While this does not represent an issue when performing constant flow rate simulations, it could introduce error in inundation volume when simulating unsteady flow transient condition with discharge changing over time.

For this reason, further simulations for Methods 1 and 2 were performed, considering two different triangular symmetric hydrographs characterized by the same peak discharge, and two different durations, 1 and 10 min (denoted H1 and H10, respectively), so to consider two hydrographs with different volume (Figure 6). Moreover, water flow was also simulated running full SWE, to evaluate possible differences against simplified diffusive scheme. As the laboratory model was not intended for reproducing unsteady hydrograph, in this test we could only compare simulated data. The goodness of fit indexes are to be interpreted as the deviation of Method 2 with respect to the Method 1 simulation. Water depth and velocity in points F, G, and J are considered in this analysis at 20, 40, and 60 s for H1, and 3, 6, and 10 min for H10.

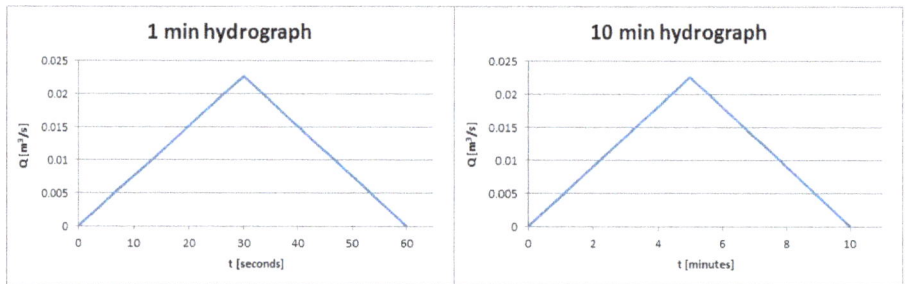

Figure 6. Hydrographs characterized by the same peak discharge, and two different durations, 1 and 10 min, used to run unsteady simulations.

The results shown in Table 4 confirm small differences between Methods 1 and 2 concerning water levels, especially using diffusive model, with a maximum deviation of 0.11 when full SWE are solved with H1. This confirms that the roughness approach and diffusive solution of SWE are good enough to simulate water depths even with unsteady flow hydrographs.

On the other hand, by analyzing water velocities, greater differences between Method 1 and 2 are reported for both the simulations, especially for the diffusive model.

Table 4. Mean relative absolute deviation of Method 2 respect to Method 1 in reconstructing water levels and velocities with 1 min (H1) and 10 min (H10) duration hydrographs.

	Water Levels		Water Velocities	
Simulation	Diffusive	Dynamic	Diffusive	Dynamic
H1	0.04	0.11	0.64	0.27
H10	0.04	0.09	0.71	0.22

Finally, an advantage of the diffusive solution is that it is faster than full SWE; in particular, the simplified simulation took about 20% time less than the full solution on a laptop computer with 1.5 GHz CPU clock speed and 2 GB ram.

3.2. Simulation of Olbia Flood Inundation

Results presented in the previous section show that Methods 1 and 2 provide similar accuracy in simulating water depth and velocity in urban area. This is relevant when one flood event has to be simulated in an urban area for which detailed buildings geometry is not available. In order to validate these findings, the 2013 flood that occurred in Olbia was simulated with the assumptions of Method 2, that are 2D diffusive solution model and buildings represented with high roughness cells (Manning coefficient = 10). Method 1 could not be applied as available DEM does not include building geometry, only terrain elevation is provided. Method 3 was not applied as it is assumed to be more suited to flood simulation over larger areas with many urbanized zones in them. In the case of the Olbia flood, we are interested in reconstructing detailed inundation in one single urban area.

Hydrographs of the six streams flowing to Olbia reconstructed by the FEST hydrological model were used as forcing input of the hydraulic model. The domain was implemented with a square mesh with 15 m spatial resolution deriving information from an available LIDAR survey with 1 m spatial resolution.

By comparing the flood extent simulated by the hydraulic model and the one surveyed after the flood (Figures 7 and 8), a good agreement can be observed. The small discrepancy is probably due to the contribution of the subsurface urban drainage flow that is not considered in the mathematical model, and uncertainties in hydrological reconstruction of stream flood hydrographs. This confirms that the use of a simplified equation and approximation of buildings as high friction cells does not introduce significant error in practical applications when flood area must be assessed, even in a complex area such as the one considered in this analysis.

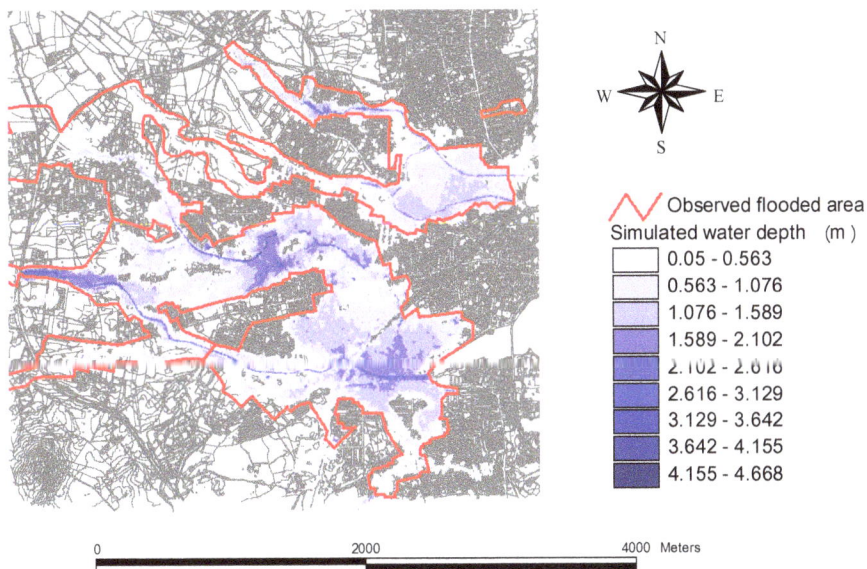

Figure 7. Water depth (m) simulated by the 2D hydraulic model compared to the observed flood extent.

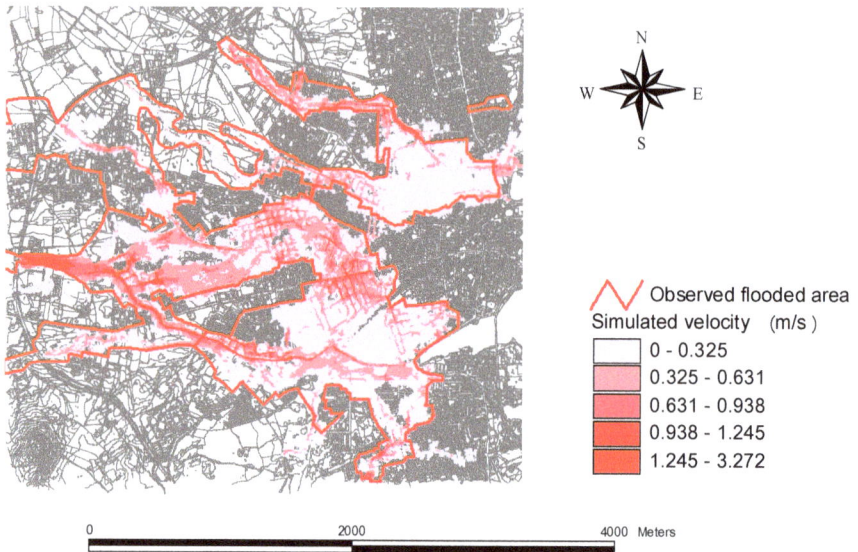

Figure 8. Flow velocity (m/s) simulated by the 2D hydraulic model compared to the observed flood extent.

4. Conclusions

In this paper, three different methods to simulate the influence of buildings on flood inundation have been tested against measurement undertaken on a simplified urban district model in laboratory. The first method considered detailed information about topography of area containing the buildings. According to the further two methods tested, single buildings or the entire urban district are represented with a high friction area.

Results show that adoption of 2D diffusive model and roughness parameter method to represent buildings is a good option to model water heights, even when unsteady discharge with rapid change in time is considered.

On the other hand, water velocities are significantly better reconstructed by methods that consider the effect of single buildings, like in Methods 1 and 2. Results of unsteady analysis show that solution of the full SWE is less sensitive to the method considered for representing buildings respect to diffusive simplification.

Application to the simulation of a real flood that occurred in Olbia (Italy) demonstrated that use of 2D diffusive model and setting high friction instead of detailed building geometry is an effective method to assess flood inundation extent.

Acknowledgments: We thank the three anonymous reviewers that provided comments that helped to improve this manuscript.

Author Contributions: Riccardo Beretta, Giovanni Ravazzani, and Marco Mancini conceived and designed the experiments; Riccardo Beretta performed the experiments; Riccardo Beretta, Giovanni Ravazzani, and Carlo Maiorano analyzed the data; Riccardo Beretta and Giovanni Ravazzani wrote the paper.

Conflicts of Interest: The authors declare no conflict of interest.

References

1. Berz, G.; Kron, W.; Loster, T.; Rauch, E.; Schimetschek, J.; Schmieder, J.; Siebert, A.; Smolka, A.; Wirtz, A. World map of natural hazards—A global view of the distribution and intensity of significant exposures. *Nat. Hazards* **2001**, *23*, 443–465. [CrossRef]
2. Milly, P.C.D.; Wetherald, R.T.; Dunne, K.A.; Delworth, T.L. Increasing risk of great floods in a changing climate. *Nature* **2002**, *415*, 514–517. [CrossRef] [PubMed]
3. De Almeida, G.A.M.; Bates, P.; Freer, J.; Souvignet, M. Improving the stability of a simple formulation of the shallow water equations for 2-D flood modeling. *Water Resour. Res.* **2012**, *48*, W05528. [CrossRef]
4. Peraire, J.; Zienkiewicz, O.C.; Morgan, K. Shallow water problems: A general explicit formulation. *Int. J. Numer. Methods Eng.* **1986**, *22*, 547–574. [CrossRef]
5. Toro, E.F.; Garcia-Navarro, P. Godunov-type methods for freesurface shallow flows: A review. *J. Hydraul. Res.* **2007**, *45*, 736–751. [CrossRef]
6. Guinot, V.; Frazão, S. Flux and source term discretization in two-dimensional shallow water models with porosity on unstructured grids. *Int. J. Numer. Method Fluids* **2006**, *50*, 309–345. [CrossRef]
7. LeVeque, R.J.; George, D.L. High-resolution finite volume methods for the shallow water equations with bathymetry and dry states. In *Advanced Numerical Models for Simulating Tsunami Waves and Runup*; World Scientific: Singapore, 2008. Available online: http://www.amath.washington.edu/rjl/pubs/catalina04 (accessed on 10 January 2018).
8. Sanders, B.; Schubert, J.E.; Gallegos, H.A. Integral formulation of shallow-water equations with anisotropic porosity for urban flood modeling. *J. Hydrol.* **2008**, *362*, 19–38. [CrossRef]
9. Li, S.; Duffy, C.J. Fully coupled approach to modeling shallow water flow, sediment transport, and bed evolution in rivers. *Water Resour. Res.* **2011**, *47*, W03508. [CrossRef]
10. Tsakiris, G. Flood risk assessment: Concepts, modelling, applications. *Nat. Hazard Earth Syst. Sci.* **2014**, *14*, 1361–1369. [CrossRef]
11. Alcrudo, F. Mathematical Modelling Techniques for Flood Propagation in Urban Areas. IMPACT Project Technical Report. Available online: http://www.impact-project.net/AnnexII_DetailedTechnicalReports/AnnexII_PartB_WP3/Modelling_techniques_for_urban_flooding.pdf (accessed on 10 January 2018).
12. Liu, L.; Liu, Y.; Wang, X.; Yu, D.; Liu, K.; Huang, H.; Hu, G. Developing an effective 2-D urban flood inundation model for city emergency management based on cellular automata. *Nat. Hazards Earth Syst. Sci.* **2015**, *15*, 381–391. [CrossRef]
13. Dottori, F.; Todini, E. Testing a simple 2D hydraulic model in an urban flood experiment. *Hydrol. Proces.* **2013**, *27*, 1301–1320. [CrossRef]
14. McMillan, H.K.; Brasington, J. Reduced complexity strategies for modelling urban floodplain inundation. *Geomorphology* **2007**, *90*, 226–243. [CrossRef]
15. Hunter, N.M.; Bates, P.D.; Horritt, M.S.; Wilson, M.D. Simple spatially distributed models for predicting flood inundation: A review. *Geomorphology* **2007**, *90*, 208–225. [CrossRef]
16. Prestininzi, P. Suitability of the diffusive model for dam break simulation: Application to a CADAM experiment. *J. Hydrol.* **2008**, *361*, 172–185. [CrossRef]
17. Lopes, P.; Leandro, J.; Carvalho, R.F.; Páscoa, P.; Martins, R. Numerical and experimental investigation of a gully under surcharge conditions. *Urban Water J.* **2015**, *12*, 468–476. [CrossRef]
18. Bazin, P.-H.; Nakagawa, H.; Kawaike, K.; Paquier, A.; Mignot, E. Modeling flow exchanges between a street and an underground drainage pipe during urban floods. *J. Hydraul. Eng.* **2014**, *140*, 4014051. [CrossRef]
19. Zhou, Q.; Yu, W.; Chen, A.S.; Jiang, C.; Fu, G. Experimental assessment of building blockage effects in a simplified urban district. *Procedia Eng.* **2016**, *154*, 844–852. [CrossRef]
20. Testa, G.; Zuccala, D.; Alcrudo, F.; Mulet, J.; Soares-Frazão, S. Flash flood flow experiment in a simplified urban district. *J. Hydraul. Res.* **2007**, *45*, 37–44. [CrossRef]
21. US Army Corps of Engineers—Hydrologic Engineering Center. Hydraulic Reference Manual. Available online: http://www.hec.usace.army.mil/software/hec-ras/documentation/HEC-RAS%205.0%20Reference%20Manual.pdf (accessed on 10 January 2018).
22. Kim, J.; Ivanov, V.Y.; Katopodes, N.D. Hydraulic resistance to overland flow on surfaces with partially submerged vegetation. *Water Resour. Res.* **2012**, *48*, W10540.

23. Sammarco, P.; Di Risio, M. Effects of moored boats on the gradually varied free-surface profiles of river flows. *J. Waterw. Port Coast. Ocean Eng.* **2016**, *143*, 04016020. [CrossRef]

24. Yamashita, K.; Suppasri, A.; Oishi, Y.; Imamura, F. Development of a tsunami inundation analysis model for urban areas using a porous body model. *Geosciences* **2018**, *8*, 12. [CrossRef]

25. US Army Corps of Engineers—Hydrologic Engineering Center. HEC-RAS 5.0 River Analysis System 2D Modeling User's Manual. Available online: http://www.hec.usace.army.mil/software/hec-ras/documentation/HEC-RAS%205.0%202D%20Modeling%20Users%20Manual.pdf (accessed on 10 January 2018).

26. Ravazzani, G.; Amengual, A.; Ceppi, A.; Homar, V.; Romero, R.; Lombardi, G.; Mancini, M. Potentialities of ensemble strategies for flood forecasting over the Milano urban area. *J. Hydrol.* **2016**, *539*, 237–253. [CrossRef]

27. Ravazzani, G.; Bocchiola, D.; Groppelli, B.; Soncini, A.; Rulli, M.C.; Colombo, F.; Mancini, M.; Rosso, R. Continuous stream flow simulation for index flood estimation in an Alpine basin of Northern Italy. *Hydrol. Sci. J.* **2015**, *60*, 1013–1025. [CrossRef]

28. Ravazzani, G.; Gianoli, P.; Meucci, S.; Mancini, M. Assessing downstream impacts of detention basins in urbanized river basins using a distributed hydrological model. *Water Resour. Manag.* **2014**, *28*, 1033–1044. [CrossRef]

29. Ravazzani, G.; Gianoli, P.; Meucci, S.; Mancini, M. Indirect estimation of design flood in urbanized river basins using a distributed hydrological model. *J. Hydrol. Eng.* **2014**, *19*, 235–242. [CrossRef]

30. Ehlschlaeger, C.R. Using the AT search algorithm to develop hydrologic models from digital elevation data. In Proceedings of the International Geographic Information System (IGIS) Symposium, Baltimore, MD, USA, 18–19 March 1989.

31. Soil Conservation Service. *National Engineering Handbook, Hydrology, Section 4*; US Department of Agriculture: Washington, DC, USA, 1986.

32. Miliani, F.; Ravazzani, G.; Mancini, M. Adaptation of precipitation index for the estimation of Antecedent Moisture Condition (AMC) in large mountainous basins. *J. Hydrol. Eng.* **2011**, *16*, 218–227. [CrossRef]

33. Ponce, V.M.; Chaganti, P.V. Variable—Parameter Muskingum—Cunge method revisited. *J. Hydrol.* **1994**, *162*, 433–439. [CrossRef]

geosciences

MDPI

Technical Note

Examples of Application of $^{GA}SAKe$ for Predicting the Occurrence of Rainfall-Induced Landslides in Southern Italy

Oreste Terranova [1], Stefano Luigi Gariano [2,*], Pasquale Iaquinta [1], Valeria Lupiano [1], Valeria Rago [1] and Giulio Iovine [1]

[1] Istituto di Ricerca per la Protezione Idrogeologica (IRPI), Consiglio Nazionale delle Ricerche (CNR), via Cavour 6, 87036 Rende (CS), Italy; oreste.terranova@irpi.cnr.it (O.T.); pasquale.iaquinta@irpi.cnr.it (P.I.); valeria.lupiano@irpi.cnr.it (V.L.); valeria.rago@irpi.cnr.it (V.R.); giulio.iovine@irpi.cnr.it (G.I.)

[2] Istituto di Ricerca per la Protezione Idrogeologica (IRPI), Consiglio Nazionale delle Ricerche (CNR), via Madonna Alta 126, 06128 Perugia, Italy

* Correspondence: gariano@irpi.cnr.it; Tel.: +39-075-5014-424

Received: 12 January 2018; Accepted: 22 February 2018; Published: 24 February 2018

Abstract: $^{GA}SAKe$ is an empirical-hydrological model aimed at forecasting the time of occurrence of landslides. Activations can be predicted of either single landslides or sets of slope movements of the same type in a homogeneous environment. The model requires a rainfall series and a set of dates of landslide activation as input data. Calibration is performed through genetic algorithms, and allows for determining a family of optimal kernels to weight antecedent rainfall properly. As output, the mobility function highlights critical conditions of slope stability. Based on suitable calibration and validation samples of activation dates, the model represents a useful tool to be integrated in early-warning systems for geo-hydrological risk mitigation purposes. In the present paper, examples of application to three rock slides in Calabria and to cases of soil slips in Campania are discussed. Calibration and validation are discussed, based on independent datasets. Obtained results are either excellent for two of the Calabrian rock slides or just promising for the remaining case studies. The best performances of the model take advantage of an accurate knowledge of the activation history of the landslides, and a proper hydrological characterization of the sites. For such cases, $^{GA}SAKe$ could be usefully employed within early-warning systems for geo-hydrological risk mitigation and Civil Protection purposes. Finally, a new release of the model is presently under test: its innovative features are briefly presented.

Keywords: landslide forecasting; threshold; hydrological model; genetic algorithms; Calabria; Campania

1. Introduction

Rainfall-induced landslides often cause significant economic loss and casualties in Calabria, as in most part of the Italian Peninsula [1,2]. Therefore, their prediction assumes a crucial role in geo-hydrological risk mitigation. The timing of activation of such phenomena is usually predicted by means of either empirical (e.g., [3] and references therein) or physically-based [4–6] approaches. To relate rainfall to time of slope instability, the extent of the landslide and the physical characteristics of both rainfall and involved materials (soil/rock) must be considered—e.g., in terms of intensity, duration, amount, and of infiltration capacity. Unfortunately, because of the high cost of field investigations, the parameters required by the physically-based approach are known only for a very limited number of case studies. Therefore, to model the triggering conditions of slope movements—either shallow or deep-seated—a threshold-based modelling approach can be employed [7,8]. Empirical thresholds can be expressed in terms of curves, delimiting the portion of Cartesian planes containing rainfall or hydrological conditions related to known activations. Examples of definition and application of empirical rainfall thresholds in Southern Italy are provided by [9–13]. In hydrological models, kernels (or filter functions) are commonly employed to express the influence of rainfall on runoff and groundwater dynamics: they are usually defined in terms of a simple, continuous analytical function. The base time (t_b) of a given kernel expresses the temporal extent in which rainfall seems to have a significant effect on slope stability. The shape and the base time of the kernel are related to magnitude of the landslide and to hydro-geological complexity of the site under investigation. Finally, the predictive tool (named mobility function) can be obtained through the convolution integral between the kernel and the rainfall series [14–16].

2. Materials and Methods

$^{GA}SAKe$ (Genetic-Algorithm-based Self Adaptive Kernel) is an empirical-hydrological model for predicting the timing of activation of slope movements of different types [17]. A linear and steady slope-stability response to rainfall and a classic threshold scheme are assumed in the model: the exceedance of the threshold determines the triggering of the landslide [18]. Though inspired by the *FLaIR* (Forecasting Landslides Induced by Rainfall) model [19,20] (already applied in several case studies, see [21–23]), it differs for several features and is suitable to handle complex cases: in fact, the model adopts a discrete kernel, instead of a continuous one, and is based on a genetic-algorithm procedure that allows for an effective, automated, self-adaptive calibration. The model can be applied either to single landslides, characterized by several historical activations, or to a set of similar slope movements in a homogeneous geomorphological context. Case studies may be of any depth, from shallow to deep-seated. The inputs of the model are: (i) the rainfall series and (ii) the set of known dates of landslide activation. Regarding rainfall data, $^{GA}SAKe$ needs as input a single rainfall series at a given time scale (e.g., daily, hourly, sub-hourly rainfall); the series can be derived from one only rain gauge or by combining rainfall data from different gauges into a "synthetic" series (e.g., by means of geostatistical techniques). The temporal extent of the rainfall series must start before the first landslide activation date and terminate after the last date (e.g., at the beginning and at the end of hydrological years). The mobility function—i.e., the output of the model as obtained through calibration—highlights the most critical conditions for the stability of the considered case study: its values depend on both rainfall amounts and shape/base time of the kernel. Triggering conditions occur when the value of the mobility function exceeds a given threshold.

More in detail, calibration of the model is based on genetic algorithms ("GA", in the following;[24,25]), which allow to obtain families of optimal, discretized solutions (kernels) that maximize the fitness function. At the beginning of a given optimization experiment, a family of kernels is randomly generated. The application to such kernels of a sequence of genetic operators—namely *selection*, *crossover* and *mutation*—constrained by prefixed probabilities, allows to generate a new population of candidate solutions (genotypes), to be tested as kernels in the successive GA iteration. A mobility function (phenotype) is then obtained by convolution between each kernel and rainfall, and its performance is evaluated by applying a suitable fitness function. Thanks to inherent proprieties of GA (cf., Fundamental Theorem [24]), better individuals (i.e., characterized by higher fitness values) can be obtained over time through iterations.

In $^{GA}SAKe$ the evaluation of the phenotypes (and, therefore, of the related genotypes) is primarily based on the *fitness values*, Φ, defined on the relative position (*rank*, k_i) of the peaks of the mobility function corresponding to the L dates of landslide activation [17]. It is:

$$\Phi = \frac{\sum_{i=1}^{L} k_i^{-1}}{\sum_{i=1}^{L} i^{-1}}. \tag{1}$$

When two (or more) mobility functions share the same value of fitness, a further distinction can be made based on the safety margin, Δz_{cr}, defined as:

$$\Delta z_{cr} = \left(z_{j-min} - z_{cr} \right) / z_{j-min}, \tag{2}$$

in which z_{j-min} is the height of the lowest peak of the mobility function that corresponds to one of the activation dates, and z_{cr} is the critical height, i.e., the height of the highest peak of the mobility function located just below z_{j-min}. Kernels with similar Φ can therefore be distinguished, and those more useful for predictive purposes selected, being characterized by larger separations among the peaks of the mobility function either corresponding to dates of activation or not.

Over several GA iterations, the mobility function is forced toward a shape characterized by peaks coinciding with the dates of landslide occurrence. An optimal kernel leads to a mobility function having the highest peaks in correspondence to such dates; further peaks may also be present, but characterized by lower values.

The best Q kernels obtained in the calibration phase can be utilized to synthetize "average" kernels to be employed for validation, by considering further dates of landslide activation (and appropriate rainfall series). Up to date, a set of Q = 100 optimal kernels has been employed during tests and first applications of the model to several case studies in Southern Italy. Once validated, the model can finally be applied—e.g., in support of an early-warning system—to estimate the timing of future landslide activations in the same study area, by employing measured or forecasted rainfall. In Figure 1, the calibration and validation procedures are schematically shown. For the sake of brevity, a complete description of the model is not reported here; however, a more detailed overview of its main features and of the calibration/validation procedures can be found in [3].

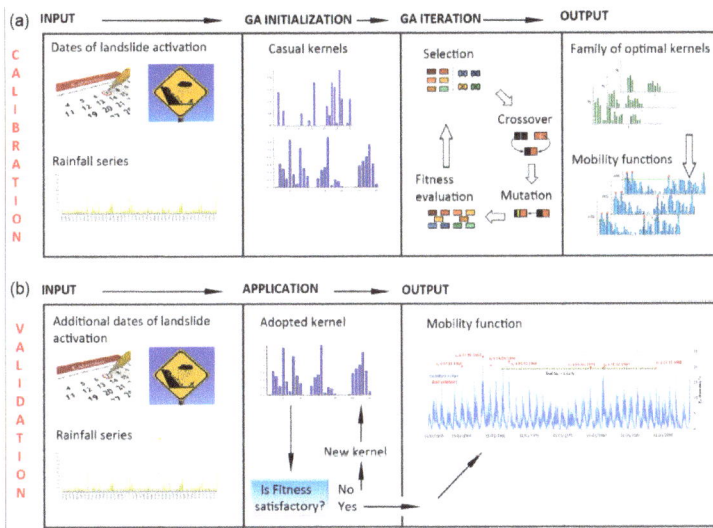

Figure 1. Schematic sketch of: (**a**) the calibration and (**b**) the validation procedures of ^{GA}SAKe.

3. Case Studies

The model was applied to different case studies in Southern Italy (Figure 2), and precisely to 3 rock slides in northern Calabria and to a set of shallow landslides (soil slip-debris flows) in Campania. Table 1 lists the average rainfall recorded near the considered case studies, while Tables 2 and 3 list the main characteristics of the considered slope movements.

Table 1. Average monthly and annual rainfall (MR, in mm) and number of rainy days (MRD) at rain gauges near the considered case studies. Key: (AC) Acri rain gauge (750 m a.s.l.); (MU) Montalto Uffugo rain gauge (468 m a.s.l.); (GR) Gragnano rain gauge (185 m a.s.l.).

Rain Gauge	Variable	September	October	November	December	January	February	March	April	May	June	July	August	Year
AC	MR	57.9	105.7	130.5	160.1	141.4	120.8	102.4	73.6	54.8	24.4	16.2	24.7	1012.4
	MRD	6	9	10	13	13	11	10	9	6	3	2	3	96
MU	MR	70.4	125.1	187.9	220.8	198.1	160.3	132.8	98.9	64.6	27.8	18.3	28.6	1333.6
	MRD	7	11	13	14	14	13	13	11	8	5	3	4	114
GR	MR	90.1	144.8	202.1	209.3	181.1	152.8	119.0	106.4	62.7	31.8	23.2	36.4	1359.7
	MRD	6	9	12	10	11	10	9	9	7	4	3	3	94

Table 2. Main characteristics of the Calabrian case studies. In Italics, the dates of activation used for validation. Employed data series range from the beginning of the calibration periods to the end of the validation period (cf. last column). No missing data characterize the considered rain series.

Landslide (Involved Lithotype)	Landslide Type	Dimensions	Activation Dates	Optimization Periods (Rain Gauge, Distance from Landslides)
Acri—Serra di Buda (Palaeozoic metamorphic and intrusive rocks)	rock slide	length: 550 m width: 400 m depth: 45–50 m	(1) 20.11.1937 (2) 29.12.1937 (3) 12.1944–01.1945 (4) 01.12.1980 (5) 28.11.1984 (6) *01.04.1985*	calibration period: 01.01.1932–31.01.1985 validation period: 01.02.1985–01.06.1985 (Acri, 1.2 km)
San Benedetto Ullano—San Rocco (Palaeozoic metamorphic rocks)	rock slide	length: 550 m width: 300 m depth: 15–35 m	(1) 28.01.2009 (2) 31.01.2010 (3) *15.03.2013*	calibration period: 01.01.1970–30.04.2010 validation period: 01.05.2010–30.04.2013 (Montalto Uffugo, 3.5 km)
San Fili—Uncino (Miocene sedimentary rocks overlaying Palaeozoic metamorphic rocks)	rock slide	length: 650 m width: 200 m depth: 25–30 m	(1) 16.01.1960 (2) 01.11.1962–14.04.1963 (3) 15.04.1964 (4) 14.12.1966 (5) 13.02.1979 (6) *12.1980*	calibration period: 01.09.1959–31.08.1980 validation period: 01.09.1980–31.03.1981 (Montalto Uffugo, 8 km)

Table 3. Main characteristics of the considered Campanian case studies. Key: (M) multiple activation; (S) single activation. In Italics, the date of activation used for validation. Employed data series range from the beginning of the calibration periods to the end of the validation period (cf. last column). The Gragnano rain gauge was employed for the Campanian cases. Missing values (about 2% of the whole set) were taken from the Castellammare and Tramonti-Chiunzi gauges.

Landslide (Involved Lithotype)	Landslide Type	Average Dimensions	Activation Dates (Type)/Affected Site	Optimization Periods (Rain Gauge, Average Distance from Landslides)
Sorrento Peninsula (Pleistocene volcanic and volcanoclastic deposits overlaying Mesozoic limestone)	soil slip	source area: 100–20000 m^2 source depth: 0.5–4 m	(1) 17.02.1963 (M)/Gragnano, Pimonte, Castellammare. (2) 23.11.1966 (S)/Vico Equense (Scrajo), Arola, Ticciano. (3) 15–24.03.1969 (M)/Cava de' Tirreni, Agerola, Scrajo Seiano. (4) 02.01.1971 (S)/Gragnano. (5) 21.01.1971 (S)/Gragnano. (6) 04.11.1980 (S)/Vico Equense (Scrajo). (7) 14.11.1982 (S)/Pozzano. (8) 22.02.1986 (M)/Palma Campania, Castellammare, Vico Equense. (9) 23.02.1987 (S)/Gragnano, Castellammare. (10) 23.11.1991 (S)/Pozzano. *(11) 10.01.1997 (M)/Pozzano, Castellammare, Nocera, Pagani, Amalfitana Coast.*	calibration: 17.01.1963–10.12.1996 validation: 11.12.1996–10.02.1997 (Gragnano, Castellammare, and Tramonti-Chiunzi, 4.5 km)

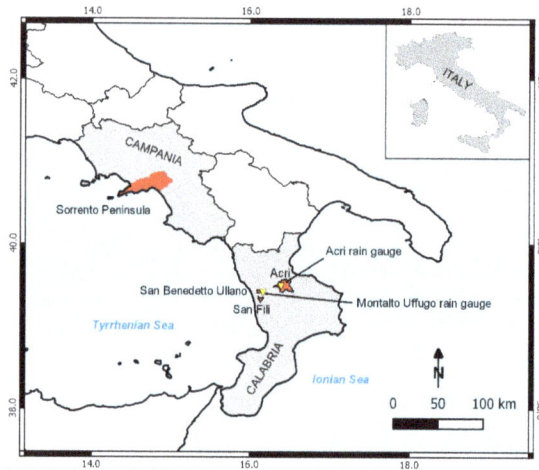

Figure 2. Location of the case studies (marked in red) in Calabria and Campania regions (marked in grey). Yellow triangles indicate Acri and Montalto Uffugo rain gauges.

3.1. Calabrian Case Studies

Calabria is an accretionary wedge made of a series of Jurassic to Early Cretaceous ophiolite-bearing tectonic units, plus overlying Hercynian and pre-Hercynian basement nappes [26]. Because of its geodynamic history [27–30], the lithological units that make up the Calabrian Arc are commonly characterized by pervasive fracture systems that favored the development of severe weathering processes. The combined effect of tectonic disturbance, differentiated uplift, erosive processes, and chemical-physical alteration influenced physiographic setting of the region, allowing for widespread slope movements of various types and extensions. In addition to earthquakes, meteoric events represent the main triggering factor of landslides in Calabria [31]. These latter pose serious risk conditions in much of the region [32–39]. The climate is Mediterranean (Csa, according to [40]). The Tyrrhenian sector is rainier than the Ionian one (1200–2000 mm vs. 500 mm); nevertheless, the most severe storms occur more frequently on the Ionian side of the region [41]. According to [42], heavy and frequent winter rainfall, caused by cold fronts mainly approaching from NW, and autumn rains, determined by cold air masses from NE, affect the region. In spring, rains show lower intensities than in autumn, while strong convective storms are common at the end of summer. Average yearly rainfall varies between 1000 and 2000 mm/y in mountainous and internal areas, and between 600 and 900 mm/y in coastal areas, with a mean regional value of about 1150 mm/y. Over 70% of the yearly precipitation occurs from October to March, with negligible monthly values from June to September. Concerning the effects of climate change in Calabria, a reduction of annual and winter amounts and an increase of summer rainfall has been recorded in the past decades (e.g., [43,44]).

3.1.1. The Acri—Serra di Buda Landslide

The village of Acri (Figure 3) is located on the Sila Massif, on the right flank of the Crati Graben, a tectonic depression belonging to the Calabrian-Sicilian Rift Zone [27]. The area is marked by a couple of regional, recent/active systems, made of N–S trending normal faults and of WNW–ESE trending strike-slip faults [45,46]. Along the N–S trending system, Palaeozoic metamorphic and igneous rocks (gneiss and granite, commonly weathered) of the Sila Massif, to the East, give place to Late Miocene-Quaternary sediments (mainly conglomerate, sand and clay) of the graben, to the West [47].

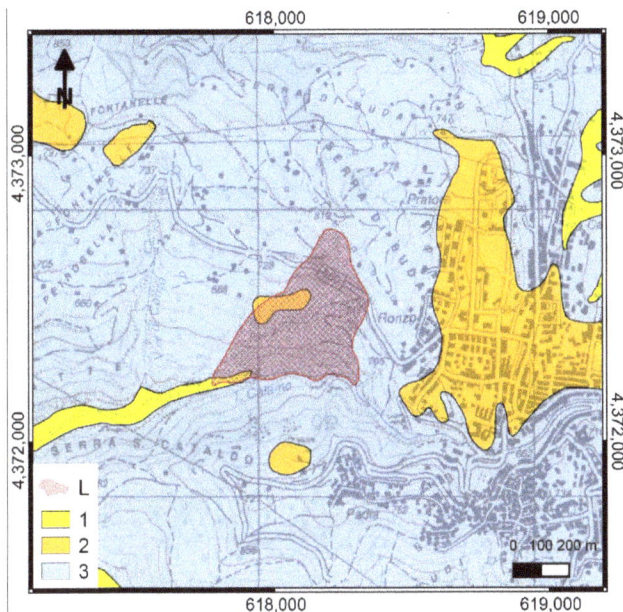

Figure 3. Lithological sketch of the Serra di Buda landslide at Acri. Key: (1) Alluvium, colluvium and residual soil; (2) conglomerate and sandstone; (3) Igneous and medium-high grade metamorphic rock. The landslide (L) is marked by a red hatched polygon. Topographic base map after 1:25,000 IGMI sheets; lithological map after [47], Geological map of Calabria (scale 1:25,000), mod.

The translational rock slide of Serra di Buda is a portion of a Sackung [48] that threatens the surroundings of the village of Acri. The landslide involves weathered gneissic and granitic rocks. In the last 100 years, it suffered from several reactivations and the Civil Authorities had to close the state road to traffic on several occasions, thus causing serious connection problems with the main urban centers of the area (mainly located in the nearby valley). In the present study, only a subset of the known dates of activation are considered: based on hydrological analyses, [49] showed either scarce or no direct correlation between some of the historical dates of mobilization and rainfall amounts. For such dates, further analyses are needed to better evaluate the possible causes of activation [50].

3.1.2. The San Benedetto Ullano—San Rocco Landslide

The village of San Benedetto Ullano (Figure 4) is located on the left flank of the Crati Graben, at the base of the Coastal Chain. The area is marked by the San Fili-San Marco Argentano normal fault, a N–S trending active structure ca. 30 km long. Along the fault, the metamorphic rocks (mainly gneiss, schist and phyllite) of the Coastal Chain, to the West, give place to Pliocene-Quaternary sediments (conglomerate, sand and clay) and subordinate Miocene outcrops (conglomerate and evaporate rocks), to the East [47]. Note that, at the rain gauge of Montalto Uffugo (in the surroundings of San Benedetto Ullano), the highest Calabrian amounts of mean annual rainfall are recorded.

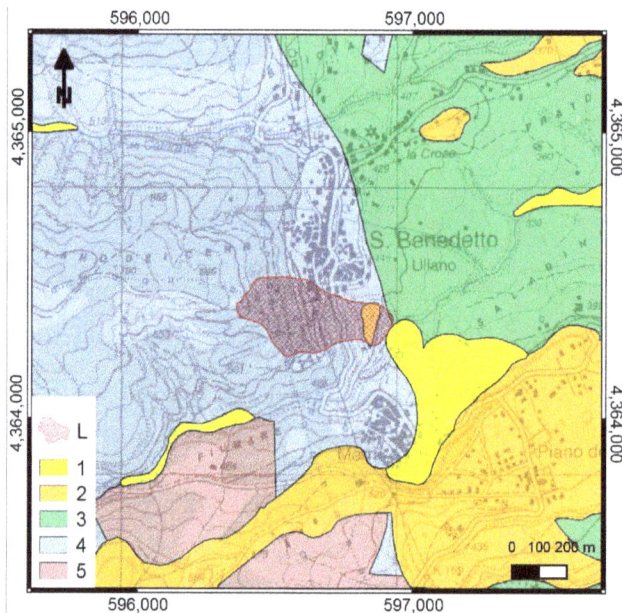

Figure 4. Lithological sketch of the San Rocco landslide at San Benedetto Ullano. Key: (1) Alluvium, colluvium, and residual soil; (2) conglomerate and sandstone; (3) Clay and clayey flysch; (4) Igneous and medium-high grade metamorphic rock; (5) Argillite and low-grade metamorphic rock. The landslide (L) is marked by a red hatched polygon. Topographic base map after 1:25,000 IGMI sheets; lithological map after [47], Geological map of Calabria (scale 1:25,000), mod.

The San Rocco translational rock slide developed mainly in gneissic rocks at the southern margin of the village, between the historical center and the cemetery [14,51,52]. Between January 2009 and March 2013, a series of 3 major re-activations was observed, with serious damage to the provincial road; the Cemetery and several buildings at the southern margin of the village were also affected by widespread opening of fissures and fractures.

3.1.3. The San Fili—Uncino Landslide

The village of San Fili (Figure 5) is located on the left flank of the Crati graben. In the area, N–S trending normal faults mark the transition between Palaeozoic weathered metamorphic rocks of the Coastal Chain (migmatitic gneiss and biotitic schist), mantled by a late Miocene sedimentary cover of conglomerate, arenite and marly clay, to the West, and Pliocene–Quaternary sediments (sand and clay) of the graben, to the East [47]. The village lies in the intermediate sector between two faults, cut by a NE–SW trending connection fault delimiting Miocene sediments, to the North, from gneissic rocks, to the South.

The Uncino rock slide developed at the western margin of San Fili, involving Miocene sediments (clay and subordinate sandstone) overlaying Palaeozoic metamorphic rocks (gneiss and biotitic schist). In historical time, the slope movements repeatedly affected the village, damaging the railway and the local road network, as well as some buildings: the most ancient known activation dates to the beginning of the twentieth century [53]; from 1960 to 1990, the regional railroad was frequently damaged or even interrupted [54]. Cumulated antecedent rains corresponding to known activations of the Uncino landslide were analyzed by considering the records of the Montalto Uffugo rain gauge [17].

Based on trends of antecedent rains in the 30–180 days before activations, the activation dates to be used for hydrological analyses could be selected.

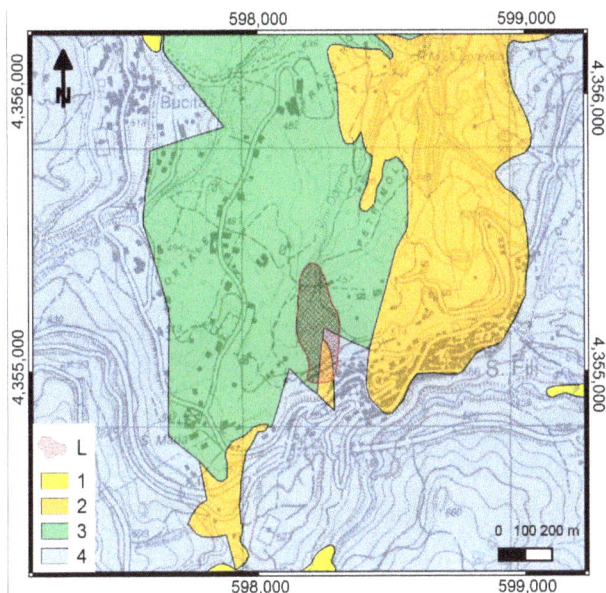

Figure 5. Lithological sketch of the Uncino landslide at San Fili. Key: (1) Alluvium, colluvium, and residual soil; (2) conglomerate and sandstone; (3) Clay and clayey flysch; (4) Igneous and medium-high grade metamorphic rock. The landslide (L) is marked by a red hatched polygon. Topographic base map after 1:25,000 IGMI sheets; lithological map after [47], Geological map of Calabria (scale 1:25,000), mod.

3.2. Campanian Ccase Studies

The Sorrento Peninsula (Figure 6) is in western Campania, Southern Italy, within the frame of the Neogene Apennine Chain [55]. In the area, Mesozoic limestone mainly crops out, covered by Miocene flysch, Pleistocene volcanic deposits (pyroclastic fall, ignimbrite), and Pleistocene detrital–alluvial deposits [56]. The carbonate bedrock constitutes a monocline, gently dipping towards WNW, mantled by sedimentary and volcanoclastic deposits, with thicknesses ranging from a few decimeters to tens of meters.

Hot, dry summers and moderately cold and rainy winters characterize the study area; its climate is Mediterranean (Csa, according to [40]). Average annual rainfall varies from 900 mm, west of Sorrento, to 1500 mm at Mt. Faito; moving inland to the East, it reaches 1600 mm at Mt. Cerreto and 1700 mm at the Chiunzi pass [57]. On average, annual totals are concentrated on about 95 rainy days. During the driest 6 months (from April to September), only 30% of the annual rainfall is recorded in about 30 rainy days. During the three wettest months (November, October, and December), a similar amount is recorded in about 34 rainy days [58]. In the area, convective rainstorms commonly occur at the beginning of the rainy season (from September to October). In autumn–winter, either high intensity or long duration rainfall are usually recorded, while uniformly distributed rains generally occur in spring. Both at sea level and higher elevations, the southern slopes of the Peninsula are characterized by smaller average annual maxima of daily rainfall with respect to the northern slopes [59]. Severe storms frequently trigger shallow landslides in the volcanoclastic cover of the Peninsula. Landslide sources show a wide range of planforms (i.e., elongated, triangular, grape-like and compound), extensions (from 100 m up to 20,000 m^2), and depths (from 0.5 to 4 m). Soil slips usually propagate seaward as either debris flows or avalanches, often increasing their original volume by entraining material along the tracks, and eventually causing casualties and serious damage to urbanized areas and transportation facilities [60–62].

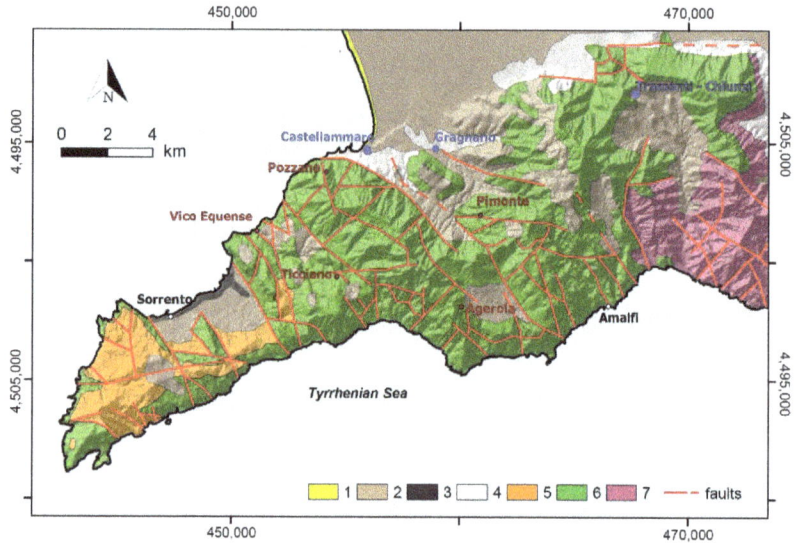

Figure 6. Lithological sketch of the Sorrento Peninsula. Key: (1) beach deposit (Holocene); (2) pyroclastic fall deposit (late Pleistocene–Holocene); (3) Campanian ignimbrite (late Pleistocene); (4) detrital alluvial deposit (Pleistocene); (5) flysch deposit (Miocene); (6) limestone (Mesozoic); (7) dolomitic limestone (Mesozoic). Red lines mark the main faults; red circles and red labels, the sites affected by shallow landslide activations; blue circles, the rain gauges; white circles, the main localities (after [17], mod.).

In the second half of the 20th century, shallow landslides activated on several occasions nearby Castellammare di Stabia: in Table 3, the major events recorded between Vico Equense and Gragnano are listed, with details on types of events and affected sites. All such events occurred between November and March, a period characterized by a medium to low suction range and included in the rainy season (October to April), according to [63]. Rainfall responsible for landslide occurrences in the Sorrento Peninsula were extracted from the records of the nearest gauges (i.e., Gragnano, Castellammare, and Tramonti-Chiunzi; Figure 6). The trends of antecedent rains look quite different. Based on hydrological consideration, the dates of activation were selected for calibration and validation of the model.

4. Results and Discussion

For each case study, the model was optimized by considering a subset of calibration dates, and then validated against further dates of activation. The calibration and validation periods were defined based on rainfall data availability, seasonality, and quality of information on activation dates (Tables 2 and 3). More in detail, optimal kernels were obtained in calibration by averaging the best 100 filter functions. Validation was then performed by applying such average kernels to the remaining activation dates (i.e., a temporal validation was performed). The optimal kernels and mobility functions for the considered case studies are shown in Figures 7–10. Calibration and validation results are synthetized in Table 4.

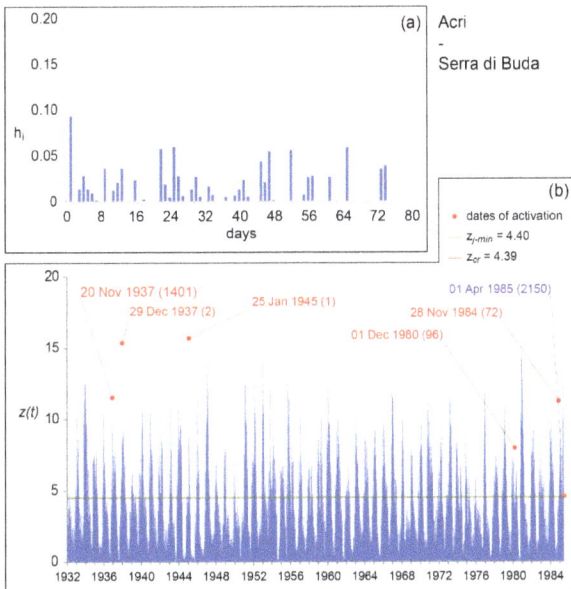

Figure 7. Acri—Serra di Buda case study: (**a**) optimal kernel; (**b**) mobility function obtained by applying the optimal kernel to the entire set of available activation dates (cf. [50]).

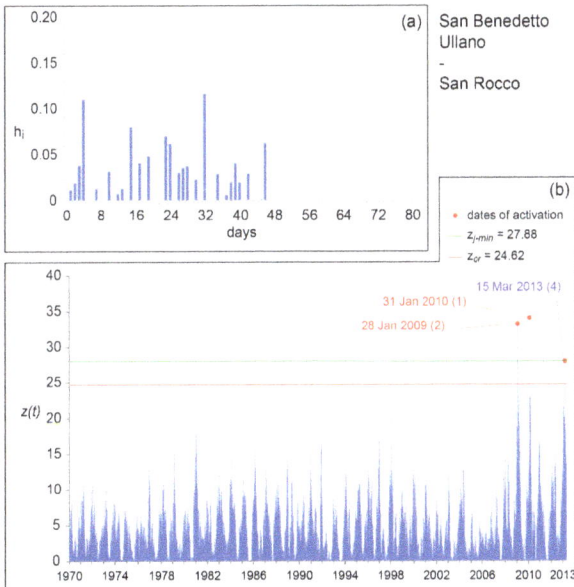

Figure 8. San Benedetto Ullano—San Rocco case study: (**a**) optimal kernel; (**b**) mobility function obtained by applying the optimal kernel to the entire set of available activation dates (cf. [54,64]).

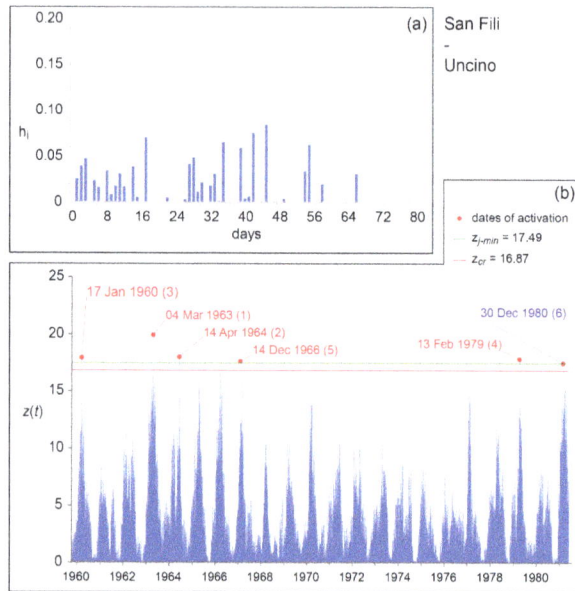

Figure 9. San Fili—Uncino case study: (**a**) optimal kernel; (**b**) mobility function obtained by applying the optimal kernel to the entire set of available activation dates (cf. [17]).

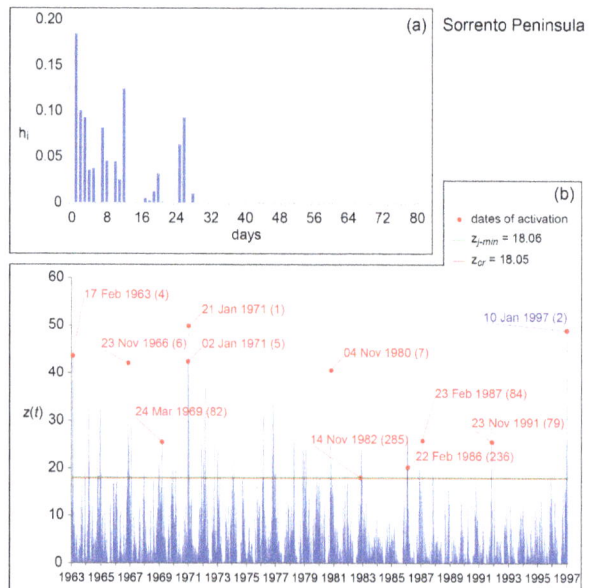

Figure 10. Sorrento Peninsula case study: (**a**) optimal kernel; (**b**) mobility function obtained by applying the optimal kernel to the entire set of available activation dates (cf. [17]).

Table 4. Model results for the considered case studies. For each case, the following details are listed: landslide type; number of activation dates employed for calibration and validation; base time of the optimal kernel; maximum fitness obtained in calibration and in validation.

Case Study	Landslide Type	Activation Dates (Calibration + Validation)	t_b (Days)	Φ_c (Calibration)	Φ_v (Validation)
Acri —Serra di Buda	rock slide	5 + 1	74	82.8%	62.2%
San Benedetto Ullano—San Rocco	rock slide	2 + 1	46	100%	96.2%
San Fili—Uncino	rock slide	5 + 1	66	100%	100%
Sorrento Peninsula	soil slip	10 + 1	28	80.6%	76.3%

The base times of the optimal kernels for the considered case studies are quite consistent with landslide magnitudes, being t_b of the Sorrento Peninsula about a half of the Calabrian cases. Base times also reflect permeability and extent of underground water patterns of the drainage basins feeding the landslide sites. In the case of the shallow landslides of the Sorrento Peninsula, shorter base times and higher influence of most recent rainfall can be observed with respect to the Calabrian rock slides, evidencing a faster slope stability response to rainfall as potential triggering factor.

As for the case studies considered in this paper, calibration and validation results show fitness values ranging from 81% to 100 % and from 62% to 100%, respectively. The best model performances were obtained for the case studies of San Fili—Uncino and San Benedetto Ullano—San Rocco, while less satisfactorily results characterize the Sorrento Peninsula and Acri—Serra di Buda cases. It should be noticed that, for the San Rocco landslide, the fitness obtained in validation is affected by a false alarm predicted by the model just the day before the activation of the phenomenon (as known from local archives).

Generally speaking, case studies characterized by fewer activation dates are simpler to model. On the other hand, model performances are known to be hampered by quality or completeness of input data (e.g., missing dates of activation; dates related to other triggering factors; unsuitability of the rain gauge network) [65]. Furthermore, available rain gauges are usually far from the sites of interest, and located at either different elevations or aspects. Uncertainty in rainfall estimation has a strong impact on the identification of the triggering rainfall [66,67]. Still, information regarding activation dates is commonly incomplete or poorly accurate [10]. As a whole, such uncertainties may strongly affect the calibration and the validation of the models.

In this study, the best results obtained were presumably favored by:

- accurate recording of damaging events along the railway track, for the Uncino rock slide;
- accurate monitoring of landslide activations and damaging events at the margin of the village, for the San Rocco rock slide;
- good representativeness of the rainfall series recorded at the Montalto Uffugo rain gauge, for both the Uncino and the San Rocco rock slides.

As for the weakest results, they may be explained by considering the following issues:

- for the Sorrento Peninsula case study, significant heterogeneities in slope materials and differences in extent of shallow landslides reasonably affected model performances. In addition, dates of landslide activation may be missing, especially for soils slips triggered in remote areas. Rainfall events responsible for shallow landslide activations are usually short and spatially limited (e.g., convective storms), and can barely be recorded by the rain gauge network. Consequently, the representativeness of the rain series cannot be guaranteed for such type of meteoric events.

- for the Acri case study, some of the considered activations may refer to secondary portions of the rock slide. For some dates, historical archives are not, in fact, detailed enough to permit an accurate understanding of the mobilized volumes. Consequently, movements affecting only portions of the rock slide—or even other nearby landslides developed in the same Sackung—may be erroneously attributed to the investigated phenomenon.

5. Conclusions and Perspectives

In the present paper, examples of application of $^{GA}SAKe$ have been presented, with reference to soil slips and rock slides in Southern Italy. For all cases, both calibration and validation of the model were performed using independent datasets. Obtained results sound either excellent (cf. San Fili—Uncino and San Benedetto Ullano—San Rocco cases) or just promising (Sorrento Peninsula, and Acri—Serra di Buda cases).

For the Uncino and the San Rocco landslides, the activation dates are correctly predicted by the model, evidently thanks to an accurate knowledge of the activation history of the landslide, and a proper hydrological characterization of the site. For such case, $^{GA}SAKe$ could be applied to predict the timing of activation of future landslide activations in the same areas. Thanks to kernels obtained in calibration, the model could be usefully employed within early-warning systems for geo-hydrological risk mitigation and Civil Protection purposes, also based on predicted rainfall.

Conversely, for the Serra di Buda and the Sorrento Peninsula cases, weaker model performances appear to be influenced by an inaccurate knowledge of the activation dates and/or rainfall series—as generally occurs with hydrological models. Nevertheless, such results still sound encouraging, provided that quality and completeness of data were improved. For past activations, further investigations may allow to better define the history of landslide mobilizations in the considered areas; rainfall data recorded at the surface by the gauge network may profitably be combined with weather radar prospections (cf. e.g., [68,69]). Moreover, it would be desirable that regional geological and hydrographic services were strengthened to guarantee adequate activities of surveying and monitoring of the geo-hydrological events, by implementing and updating suitable data bases on triggered slope movements (including details such as location, type, type of involved materials, dates of activation, damage). As for the rains, the networks may be improved by adding further gauges and/or radar installations.

It is worth observing that, due to its empirical/hydrological nature, $^{GA}SAKe$ needs a representative sample of data for calibration and validation. Such phases are both crucial for properly tuning any prediction model, particularly when it is aimed at being implemented into an early-warning system. In some of the cases analyzed in this paper, the number of available activation dates was not large enough to permit the selection of different sub-samples, thus only one date of activation (i.e., the last) was considered for all validations. Nevertheless, such type of approach can be viewed in terms of an "operative" application of the model in an operational warning system (i.e., used to predict the "next" activation, based on the known history of landslide movements in a given area).

With respect to other empirical/hydrological models, the main advantages implemented in $^{GA}SAKe$ are: (i) the automatic calibration of the model by means of genetic algorithms; (ii) the high degrees of freedom of the kernel (in fact, a fixed mathematical function is not required). The adoption of Genetic Algorithms allows for a thorough exploration of the "solution space", and guarantees that best solutions to any problem can be obtained (following Holland's Fundamental Theorem [24]). Moreover, the initial shape of the kernel can be suggested from the operator or even left at random. During model iterations, the shape of the kernels evolves (as their length), thanks to genetic operators, and can reach any type of final configuration. The above features are deemed of outmost value, as they allow to limit subjectivity in the application of a model [70].

If analyzed in detail, the kernels obtained with $^{GA}SAKe$ may appear difficult to understand in physical terms. Nevertheless, in most real cases, slope instability needs to be explained by a complex interplay of several groundwater paths. These latter commonly result in kernels with quite irregular patterns, and cannot be simulated by simplified analytical functions. According to [71], kernels characterized by complex patterns (and many parameters) are commonly needed to simulate groundwater dynamics. Being characterized by higher complexity, resulting kernels do not necessarily imply greater predictive uncertainties [72,73].

A new release of the model is presently being tested against a set of case studies of different types and extent, selected in diverse geological contexts. Among the improvements, some concern the Genetic Algorithms utilized for the optimization: in particular, a set of selection criteria (roulette, ranking, stochastic tournament, deterministic tournament) will be available shortly, also considering different combinations of ordering criteria (by including the safety margin, the base time, and the first-order momentum to the main fitness of the kernel). Obtained kernels can be post-processed and analyzed in terms of either control points or analytical functions, to allow for deeper hydro-geological ruminations. The final regression of kernels into analytical functions may allow a better understanding of complex groundwater behaviors, and hence landslide responses to antecedent rainfall. Finally, some computational improvements concern smaller requirements of storage capacity and higher execution speed, thanks to parallel optimization that allows for reducing elaborations by ca. 1/50, which will allow improvements in the model performances for an effective implementation in early-warning systems.

Acknowledgments: Authors are grateful to Alessio De Rango, Donato D'Ambrosio, William Spataro, and Rocco Rongo (Department of Mathematics and Informatics, University of Calabria, Italy) for discussions and support in the development and testing phases of the model. Authors acknowledge the Editor and three anonymous Referees for their comments and suggestions that allowed improvements to the original version of the manuscript.

Author Contributions: All the authors contributed to the design and implementation of the research, to the analysis of the results and to the preparation of the manuscript.

Conflicts of Interest: The authors declare no conflict of interest.

References

1. Guzzetti, F. Landslides fatalities and the evaluation of landslide risk in Italy. *Eng. Geol.* **2000**, *58*, 89–107. [CrossRef]

2. Iovine, G.; Gariano, S.L.; Terranova, O. Alcune riflessioni sull'esposizione al rischio da frane superficiali alla luce dei recenti eventi in Italia meridionale. *Geologi Calabria* **2009**, *10*, 4–31. (In Italian)

3. Peruccacci, S.; Brunetti, M.T.; Gariano, S.L.; Melillo, M.; Rossi, M.; Guzzetti, F. Rainfall thresholds for possible landslide occurrence in Italy. *Geomorphology* **2017**, *290*, 39–57. [CrossRef]

4. Greco, R.; Comegna, R.; Damiano, E.; Guida, A.; Olivares, L.; Picarelli, L. Hydrological modelling of a slope covered with shallow pyroclastic deposits from field monitoring data. *Hydrol. Earth Syst. Sci.* **2013**, *17*, 4001–4013. [CrossRef]

5. Peres, D.J.; Cancelliere, A. Derivation and evaluation of landslide-triggering thresholds by a Monte Carlo approach. *Hydrol. Earth Syst. Sci.* **2014**, *18*, 4913–4931. [CrossRef]

6. Alvioli, M.; Baum, R.L. Parallelization of the TRIGRS model for rainfall-induced landslides using the message passing interface. *Environ. Modell. Softw.* **2016**, *81*, 122–135. [CrossRef]

7. Rossi, M.; Peruccacci, S.; Brunetti, M.T.; Marchesini, I.; Luciani, S.; Ardizzone, F.; Balducci, V.; Bianchi, C.; Cardinali, M.; Fiorucci, F.M. SANF: National warning system for rainfall-induced landslides in Italy. In *Landslides and Engineered Slopes: Protecting Society through Improved Understanding*; Taylor & Francis Group: London, UK, 2012; pp. 1895–1899.

8. De Luca, D.; Versace, P. Diversity of rainfall thresholds for early warning of hydro-geological disasters. *Adv. Geosci.* **2017**, *44*, 53–60. [CrossRef]

9. Vennari, C.; Gariano, S.L.; Antronico, L.; Brunetti, M.T.; Iovine, G.; Peruccacci, S.; Terranova, O.; Guzzetti, F. Rainfall thresholds for shallow landslide occurrence in Calabria, southern Italy. *Nat. Hazards Earth Syst. Sci.* **2014**, *14*, 317–330. [CrossRef]

10. Gariano, S.L.; Brunetti, M.T.; Iovine, G.; Melillo, M.; Peruccacci, S.; Terranova, O.; Vennari, C.; Guzzetti, F. Calibration and validation of rainfall thresholds for shallow landslide forecasting in Sicily, southern Italy. *Geomorphology* **2015**, *228*, 653–665. [CrossRef]

11. Pisano, L.; Vennari, C.; Vessia, G.; Trabace, M.; Amoruso, G.; Loiacono, P.; Parise, M. Data collection for reconstructing empirical rainfall thresholds for shallow landslides: Challenges and improvements in the Daunia Sub-Apennine (Southern Italy). *Rend. Online Soc. Geol. Ital.* **2015**, *35*, 236–239. [CrossRef]

12. Melillo, M.; Brunetti, M.T.; Peruccacci, S.; Gariano, S.L.; Guzzetti, F. Rainfall thresholds for the possible landslide occurrence in Sicily (Southern Italy) based on the automatic reconstruction of rainfall events. *Landslides* **2016**, *13*, 165–172. [CrossRef]

13. Piciullo, L.; Gariano, S.L.; Melillo, M.; Brunetti, M.T.; Peruccacci, S.; Guzzetti, F.; Calvello, M. Definition and performance of a threshold-based regional early warning model for rainfall-induced landslides. *Landslides* **2017**, *14*, 995–1008. [CrossRef]

14. Iovine, G.; Iaquinta, P.; Terranova, O. Emergency management of landslide risk during Autumn-Winter 2008/2009 in Calabria (Italy). The example of San Benedetto Ullano. In Proceedings of the 18th World IMACS Congress and MODSIM09 International Congress on Modelling and Simulation, Cairns, Australia, 13–17 July 2009; pp. 2686–2693.

15. Iovine, G.; Petrucci, O.; Rizzo, V.; Tansi, C. The March 7th 2005 Cavallerizzo (Cerzeto) landslide in Calabria-Southern Italy. In Proceeding of the 10th IAEG Congress: Engineering Geology for Tomorrow's Cities, Nottingham, UK, 6–10 September 2006.

16. Terranova, O.; Lollino, P.; Gariano, S.L.; Iaquinta, P.; Iovine, G. Un sistema integrato di sorveglianza per la mitigazione del rischio da frana. *Geologi Calabria* **2010**, *11*, 6–28. (In Italian)

17. Terranova, O.; Gariano, S.L.; Iaquinta, P.; Iovine, G. GASAKe: Forecasting landslide activations by a genetic-algorithms-based hydrological model. *Geosci. Model Dev.* **2015**, *8*, 1955–1978. [CrossRef]

18. Terranova, O.; Iaquinta, P.; Gariano, S.L.; Greco, R.; Iovine, G. CMSAKe: A hydrological model to forecasting landslide activations. In *Landslide Science and Practice*; Springer: Berlin, Germany, 2013; pp. 73–79.

19. Sirangelo, B.; Versace, P. A real time forecasting for landslides triggered by rainfall. *Meccanica* **1996**, *31*, 1–13. [CrossRef]

20. Capparelli, G.; Versace, P. FLaIR and SUSHI: Two mathematical models for early warning of landslides induced by rainfall. *Landslides* **2011**, *8*, 67–79. [CrossRef]

21. Peres, D.J.; Cancelliere, A. Defining rainfall thresholds for early warning of rainfall-triggered landslides: The case of North-East Sicily. In *Landslide Science and Practice*; Springer: Berlin, Germany, 2013; pp. 257–263.

22. Capparelli, G.; Giorgio, M.; Greco, R. Shallow landslides risk mitigation by early warning: The Sarno case. In *Landslide Science and Practice*; Springer: Berlin, Germany, 2013; pp. 767–772.

23. Greco, R.; Giorgio, M.; Capparelli, G.; Versace, P. Early warning of rainfall-induced landslides based on empirical mobility function predictor. *Eng. Geol.* **2013**, *153*, 68–79. [CrossRef]

24. Holland, J.H. *Adaptation in Natural and Artificial Systems: An Introductory Analysis with Applications to Biology, Control and Artificial Intelligence*; MIT Press: Cambridge, MA, USA, 1992; ISBN 0262082136.

25. D'Ambrosio, D.; Spataro, W.; Rongo, R.; Iovine, G. Genetic algorithms, optimization, and evolutionary modeling. In *Treatise on Geomorphology. Quantitative Modeling of Geomorphology*; Shroder, J., Baas, A.C.W., Eds.; Academic Press: San Diego, CA, USA, 2013; pp. 74–97.

26. Amodio-Morelli, L.; Bonardi, G.; Colonna, V.; Dietrich, D.; Giunta, G.; Ippolito, F.; Liguori, V.; Lorenzoni, S.; Paglionico, A.; Perrone, V. L'arco calabro-peloritano nell'orogene appenninico-maghrebide. *Mem. Soc. Geol. Ital.* **1976**, *17*, 1–60. (In Italian)

27. Monaco, C.; Tortorici, L. Active faulting in the Calabrian Arc and eastern Sicily. *J. Geodyn.* **2000**, *29*, 407–424. [CrossRef]

28. Tortorici, L.; Monaco, C.; Tansi, C.; Cocina, O. Recent and active tectonics in the Calabrian Arc (southern Italy). *Tectonophysics* **1995**, *243*, 37–55. [CrossRef]

29. Van Dijk, J.P.; Bello, M.; Brancaleoni, G.P.; Cantarella, G.; Costa, V.; Frixa, A.; Golfetto, F.; Merlini, S.; Riva, M.; Torricelli, S.; Toscano, C.; Zerilli, A. A regional structural model for the northern sector of the Calabrian Arc (southern Italy). *Tectonophysics* **2000**, *324*, 267–320. [CrossRef]

30. Tansi, C.; Muto, F.; Critelli, S.; Iovine, G. Neogene-Quaternary strike-slip tectonics in the central Calabrian Arc (southern Italy). *J. Geodyn.* **2007**, *43*, 393–414. [CrossRef]

31. Sorriso-Valvo, G.M. Mass movements and slope evolution in Calabria. Proceedings of 4th International Conference and Field Workshop on Landslides, Tokyo, Japan, 23–31 August 1985; pp. 23–30.

32. Carrara, A.; Catalano, E.; Sorriso-Valvo, G.M.; Reali, C.; Merenda, L.; Rizzo, V. Landslide morphometry and typology in two zones, Calabria, Italy. *Bull. Eng. Geol. Environ.* **1977**, *16*, 8–13.

33. Carrara, A.; Merenda, L.; Nicoletti, P.G.; Sorriso-Valvo, G.M. Slope instability in Calabria, Italy. In Proceedings of the Polish-Italian Seminar on Superficial Mass Movement in Mountain Regions, Szymbark, Poland, 17–19 May 1979; pp. 47–62.

34. Crescenzi, E.; Grassi, D.; Iovine, G.; Merenda, L.; Miceli, F.; Sdao, F. Fenomeni di instabilità franosa nei centri abitati calabri: Esempi rappresentativi. *Geol. Appl. Hydrogeo.* **1996**, *31*, 203–226. (In Italian)

35. Iovine, G.; Merenda, L. Nota illustrativa alla "Carta delle frane e della mobilizzazione diastrofica dal 1973 ad oggi nel bacino del Torrente Straface (Alto Jonio; Calabria)". *Geol. Appl. Hydrogeol.* **1996**, *31*, 107–128. (In Italian)

36. Iovine, G.; Petrucci, O. Effetti sui versanti e nel fondovalle indotti da un evento pluviale eccezionale nel bacino di una fiumara calabra (T. Pagliara). *Boll. Soc. Geol. Ital.* **1998**, *117*, 821–840. (In Italian)

37. Iovine, G.; Parise, M.; Tansi, C. Slope movements and tectonics in North-Eastern Calabria (Southern Italy). In Proceedings of the 7th International Symposium on Landslides (ISL'96): Landslides Glissements de Terrain, Trondheim, 17–21 June 1996; pp. 785–790.

38. Ferrari, E.; Iovine, G.; Petrucci, O. Evaluating landslide hazard through geomorphologic, hydrologic and historical analyses in north-eastern Calabria (southern Italy). In Proceedings of the EGS Plinius Conference on Mediterranean Storms, Maratea, Italia, 14–16 October 1999; pp. 425–438.

39. Tansi, C.; Iovine, G.; Folino-Gallo, M. Tettonica attiva e recente; e manifestazioni gravitative profonde; lungo il bordo orientale del graben del Fiume Crati (Calabria settentrionale). *Boll. Soc. Geol. Ital.* **2005**, *124*, 563–578. (In Italian)

40. Köppen, W.P. *Climatologia con un Estudio de los Climas de la Tierra*; Fondo de Cultura Economica: Ciudad de Mexico City, Mexico, 1948; p. 479.

41. Terranova, O. Caratteristiche degli eventi pluviometrici a scala giornaliera in Calabria. In Proceeding of the XXIX Convegno di Idraulica e Costruzioni Idrauliche, Trento, Italy, 7–10 September 2004; pp. 343–350. (In Italian)

42. Terranova, O.; Gariano, S.L. Rainstorms able to induce flash floods in a Mediterranean-climate region (Calabria, southern Italy). *Nat. Hazard Earth Syst. Sci.* **2014**, *14*, 2423–2434. [CrossRef]

43. Ferrari, E.; Terranova, O. Non-parametric detection of trends and change point years in monthly and annual rainfalls. In Proceedings of the 1st Italian-Russian Workshop on New Trend in Hydrology, Rende (CS), Italy, 24–26 September 2002; pp. 177–188. (In Italian)

44. Terranova, O.; Gariano, S.L. Regional investigation on seasonality of erosivity in the Mediterranean environment. *Environ. Earth Sci.* **2015**, *73*, 311–324. [CrossRef]

45. Iovine, G.; Tansi, C.; Folino-Gallo, M. Strutture da accomodamento tettono-gravitativo nell'evoluzione tardiva dei sistemi di catena: Il caso di studio di Acri (Calabria settentrionale). *Boll. Soc. Geol. Ital.* **2004**, *123*, 39–51. (In Italian)

46. Tansi, C.; Talarico, A.; Iovine, G.; Folino Gallo, M.; Falcone, G. Interpretation of radon anomalies in seismotectonic and tectonic-gravitational setting of the south-eastern Crati Graben (Northern Calabria, Italy). *Tectonophysics* **2005**, *396*, 181–193. [CrossRef]

47. *Carta Geologica della Calabria*; CASMEZ: Ercolano, Napoli, Italia, 1967. (In Italian)

48. Sorriso-Valvo, G.M. *1:250,000 Scale Map of the Large Landslides and of the Deep-Seated Gravitational Slope Deformations of Calabria*; Selca: Firenze, Italia, 1996.

49. Terranova, O.; Antronico, L.; Gullà, G. Landslide triggering scenarios in homogeneous geological contexts: The area surrounding Acri (Calabria, Italy). *Geomorphology* **2007**, *87*, 250–267. [CrossRef]

50. Gariano, S.L.; Terranova, O.G.; Greco, R.; Iaquinta, P.; Iovine, G. Forecasting the timing of activation of rainfall-induced landslides. An application of GA-SAKe to the Acri case study (Calabria, Southern Italy). *Geophys. Res. Abstr.* **2013**, *15*, EGU2013-678.

51. Iovine, G.; Lollino, P.; Gariano, S.L.; Terranova, O. Coupling limit equilibrium analyses and real-time monitoring to refine a landslide surveillance system in Calabria (southern Italy). *Nat. Hazard Earth Syst. Sci.* **2010**, *10*, 2341–2354. [CrossRef]

52. Capparelli, G.; Iaquinta, P.; Iovine, G.; Terranova, O.G.; Versace, P. Modelling the rainfall-induced mobilization of a large slope movement in northern Calabria. *Nat. Hazards* **2012**, *61*, 247–256. [CrossRef]

53. Sorriso-Valvo, G.M.; Antronico, L.; Catalano, E.; Gullà, G.; Tansi, C.; Dramis, F.; Ferrucci, F.; Fantucci, R. *The Temporal Stability and Activity of Landslides in Europe with Respect to Climatic Change (TESLEC)*; Final Report; Istituto di Ricerca per la Protezione Idrogeologica (IRPI): Turin, Italy, 1996.

54. Iovine, G.; De Rango, A.; Gariano, S.L.; Terranova, O. Forecasting landslide activations by means of GA-SAKe. An example of application to three case studies in Calabria (Southern Italy). *Geophys. Res. Abstr.* **2016**, *18*, 4645.

55. Ippolito, F.; D'Argenio, B.; Pescatore, T.; Scandone, P. Structural–stratigraphic units and tectonic framework of Southern Apennines. In *Geology of Italy*; Squyres, C., Ed.; Earth Sciences Society of the Libyan Arab Republic: Tripoli, Libya, 1975; pp. 317–328.

56. Di Crescenzo, G.; Santo, A. Analisi morfologica delle frane da scorrimento-colata rapida in depositi piroclastici della Penisola Sorrentina (Campania). *Geogr. Fis. Dinam. Quat.* **1999**, *22*, 57–72. (In Italian)

57. Ducci, D.; Tranfaglia, G. L'impatto dei cambiamenti climatici sulle risorse idriche sotterranee in Campania. *Boll Ordine Geol. Campania* **2005**, *1–4*, 13–21. (In Italian)

58. *Servizio Idrografico. Annali Idrologici: Parte I*; Compartimento di Napoli, Istituto poligrafico e Zecca dello Stato: Rome, Italy, 1948–1999.

59. Rossi, F.; Villani, P. *Valutazione Delle Piene in Campania*; CNR-GNDCI publications No. 1470; Grafica Metelliana: Cava de' Tirreni, Italia, 1994. (In Italian)

60. Mele, R.; Del Prete, S. Lo studio della franosità storica come utile strumento per la valutazione della pericolosità da frane. Un esempio nell'area di Gragnano (Campania). *Boll. Soc. Geol. Ital.* **1999**, *118*, 91–111. (In Italian)

61. Calcaterra, D.; Santo, A. The January 10, 1997 Pozzano landslide, Sorrento Peninsula, Italy. *Eng. Geol.* **2004**, *75*, 181–200.

62. Di Crescenzo, G.; Santo, A. Debris slides-rapid earth flows in the carbonate massifs of the Campania region (Southern Italy): Morphological and morphometric data for evaluating triggering susceptibility. *Geomorphology* **2005**, *66*, 255–276.

63. Cascini, L.; Sorbino, G.; Cuomo, S.; Ferlisi, S. Seasonal effects of rainfall on the shallow pyroclastic deposits of the Campania region (southern Italy). *Landslides* **2014**, *11*, 779–792. [CrossRef]

64. Terranova, O.; Greco, V.R.; Gariano, S.L.; Pascale, S.; Rago, V.; Caloiero, P.; Iovine, G. Monitoring and modelling for landslide risk mitigation and reduction. The case study of San Benedetto Ullano (Northern Calabria-Italy). *Geophys. Res. Abstr.* **2016**, *18*, 4708.

65. Peres, D.J.; Cancelliere, A.; Greco, R.; Bogaard, T.A. Influence of uncertain identification of triggering rainfall on the assessment of landslide early warning thresholds. *Nat. Hazards Earth Syst. Sci. Discuss.* **2017**. [CrossRef]

66. Nikolopoulos, E.I.; Crema, S.; Marchi, L.; Marra, F.; Guzzetti, F.; Borga, M. Impact of uncertainty in rainfall estimation on the identification of rainfall thresholds for debris flow occurrence. *Geomorphology* **2014**, *221*, 286–297. [CrossRef]

67. Marra, F.; Destro, E.; Nikolopoulos, E.I.; Zoccatelli, D.; Creutin, J.D.; Guzzetti, F.; Borga, M. Impact of rainfall spatial aggregation on the identification of debris flow occurrence thresholds. *Hydrol. Earth Syst. Sci.* **2017**, *21*, 4525–4532. [CrossRef]

68. Gabriele, S.; Terranova, O.; Pascale, S.; Rago, V.; Chiaravalloti, F.; Sabatino, P.; Brocca, L.; Laviola, S.; Baldini, L.; Federico, S. RAMSES: A nowcasting system for mitigating geo-hydrological risk along the railway. *Geophys. Res. Abstr.* **2016**, *18*, 8462.

69. Rago, V.; Chiaravalloti, F.; Chiodo, G.; Gabriele, S.; Lupiano, V.; Nicastro, R.; Pellegrino, A.D.; Procopio, A.; Siviglia, S.; Terranova, O.G.; Iovine, G. Geomorphic effects caused by heavy rainfall in southern Calabria (Italy) on 30 October–1 November 2015. *J. Maps* **2017**, *13*, 836–843. [CrossRef]

70. D'Ambrosio, D.; Spataro, W.; Rongo, R.; Iovine, G. Genetic algorithms, optimization, and evolutionary modeling. In *Treatise on Geomorphology, Volume 2, Quantitative Modeling of Geomorphology*; Baas, A., Ed.; Academic Press: San Diego, CA, USA, 2013; pp. 74–97.

71. Pinault, J.-L.; Plagnes, V.; Aquilina, L. Inverse modeling of the hydrological and the hydrochemical behavior of hydrosystems: Characterization of karst system functioning. *Water Resour. Res.* **2001**, *37*, 2191–2204. [CrossRef]

72. Fienen, M.N.; Doherty, J.E.; Hunt, R.J.; Reeves, H. *Using Prediction Uncertainty Analysis to Design Hydrologic Monitoring Networks: Example Applications from the Great Lakes Water Availability Pilot Project*; Scientific Investigations Report 2010-5159, U.S. Geological Survey: Reston, VA, USA, 2010.

73. Long, A.J. RRAWFLOW: Rainfall-response aquifer and watershed flow model (v1.15). *Geosci. Model Dev.* **2015**, *8*, 865–880. [CrossRef]

geosciences

MDPI

Article

SPI Trend Analysis of New Zealand Applying the ITA Technique

Tommaso Caloiero

National Research Council—Institute for Agricultural and Forest Systems in Mediterranean (CNR-ISAFOM), Via Cavour 4/6, 87036 Rende, Cosenza, Italy; tommaso.caloiero@isafom.cnr.it; Tel.: +39-0984-841-464

Received: 28 February 2018; Accepted: 13 March 2018; Published: 15 March 2018

Abstract: A natural temporary imbalance of water availability, consisting of persistent lower-than-average or higher-than-average precipitation, can cause extreme dry and wet conditions that adversely impact agricultural yields, water resources, infrastructure, and human systems. In this study, dry and wet periods in New Zealand were expressed using the Standardized Precipitation Index (SPI). First, both the short term (3 and 6 months) and the long term (12 and 24 months) SPI were estimated, and then, possible trends in the SPI values were detected by means of a new graphical technique, the Innovative Trend Analysis (ITA), which allows the trend identification of the low, medium, and high values of a series. Results show that, in every area currently subject to drought, an increase in this phenomenon can be expected. Specifically, the results of this paper highlight that agricultural regions on the eastern side of the South Island, as well as the north-eastern regions of the North Island, are the most consistently vulnerable areas. In fact, in these regions, the trend analysis mainly showed a general reduction in all the values of the SPI: that is, a tendency toward heavier droughts and weaker wet periods.

Keywords: drought; SPI; trend; New Zealand

---

1. Introduction

Recently, the adverse impacts of climate change have become the focus of considerable international attention due to the increase in phenomena such as flood, heat waves, forest fires, and droughts [1,2]. Among these damaging climate events, drought phenomena play a significant role in socio-economic and health terms, even though their impact on populations depends on the vulnerable elements [3]. Moreover, understanding drought phenomena is paramount for the appropriate planning and management of water resources [4]. For example, different drought events have been detected during the last decades [5–7], and drought is expected to become more frequent in the 21st century in some seasons and areas [8] following precipitation and/or evapotranspiration variability [9].

In recent years, several researchers have analyzed drought events in several parts of the world [10–17], even though drought phenomena are difficult to detect and to monitor due to their complex nature. Usually, drought severity is evaluated by means of drought indices since they facilitate communication of climate anomalies to diverse user audiences; they also allow scientists to assess quantitatively climate anomalies in terms of their intensity, duration, frequency, recurrence probability, and spatial extent [3,18]. In the past few decades, numerous indices were proposed for identifying and monitoring drought events. Some of these indices refer to meteorological drought (scarcity of precipitation) and are based on the analysis of the rainfall information only. Thus, different categories of drought can be investigated by choosing appropriate temporal scales addressing different categories of users. Other indices, however, are more suitable to describe hydrological drought (scarcity in surface and subsurface water supplies), agricultural drought (water shortage compared to the typical needs crops irrigation), and socio-economic drought (referred to the global water consumption).

Specifically, meteorological drought consists of temporary lower-than-average precipitation and results in diminished water resources availability [19], which impact on economic activities, human lives, and the environment [20]. The most well-known index for analyzing the meteorological drought is undoubtedly the Standardized Precipitation Index (SPI) proposed by McKee et al. [21], which has been extensively applied in different countries [22–28]. This drought index can be considered one of the most robust and effective drought indices, as it can be evaluated for different time scales and allows the analysis of different drought categories [29]. Moreover, the evaluation of the SPI requires only precipitation data, making it easier to calculate more than complex indices, and allows for the comparison of drought conditions in different regions and for different time periods [30–33]. Due to its intrinsic probabilistic nature, the SPI is the ideal candidate for carrying out drought risk analysis [34,35]. With this aim, several authors focused on the SPI trend [36–38]. These studies are mainly based on non-parametric tests, which are better suited to deal with non-normally distributed hydrometeorology data than the parametric methods. Recently, Şen [39] proposed the Innovative Trend Analysis (ITA) technique, which allows a graphical trend evaluation of the low, medium, and high values in the data. The ITA technique was widely applied to the trend detection of several hydrological variables. Haktanir and Citakoglu [40] analyzed the annual maximum rainfall series by means of the ITA method. Kisi and Ay [41] studied some water quality parameters registered at five Turkish stations by means of the ITA and the MK. Şen [42] and Ay and Kisi [43] applied the ITA to Turkish temperature data. The ITA technique was also used to analyze the trends of heat waves [44], monthly pan evaporations [45], and streamflow data [46].

Since agriculture is one of the largest sectors of the tradable economy, a period of drought in New Zealand can have significant ecological, social, and economic impacts [47]. In fact, New Zealand experiences rainfall deficits and short duration of dry spells that are not as unusual as isolated drought events at regional level. For example, the widespread drought event that affected New Zealand from late 2007 to the end of autumn 2008 caused damages of about 2.8 billion New Zealand dollars [48]. The 2013 drought in New Zealand was estimated to have caused GDP (Gross Domestic Product) to fall by 0.6% [49]. Regional scenarios of drought in New Zealand evidenced an increase in drought trends during this century in all the areas presently subject to drought [50]. Furthermore, based on the latest climate and impact modelling, more droughts can be expected in the future in some locations such as the agricultural regions on the Eastern coast and particularly the Canterbury Plains, as well as Northland [51].

In this article, drought events in several regions of New Zealand have been studied by applying the SPI at various time scales (3, 6, 12, and 24 months) starting from a database of 294 monthly rainfall series in the period 1951–2010. In particular, this work aims to identify the most drought-prone regions of New Zealand by analyzing its evolution through the identification of the SPI trend at different timescales by means of the Innovative Trend Analysis (ITA), which allows the trend identification of the low, medium, and high values of a series.

2. Methodology

2.1. Standardized Precipitation Index

In this study, dry and wet periods were evaluated using the SPI at different time scale (3, 6, 12 and 24 months). In fact, while the 3- and 6-month SPI describe droughts that affect plant life and farming, the 12- and 24-month SPI influence the way how water supplies/reserves are managed [52,53]. Angelidis et al. [54] offered a meticulous description of the method to compute the SPI.

In order to calculate the index, for each time scale, an appropriate probability density function (PDF) must be fitted to the frequency distribution of the cumulated precipitation. In particular, a gamma function is considered. The shape and the scale parameters must be estimated for each month of the year and for each time aggregation, for example by using the approximation of Thom [55].

Since the gamma distribution is undefined for a rainfall amount $x = 0$, in order to take into account the zero values that occur in a sample set, a modified cumulative distribution function (CDF) must be considered.

$$H(x) = q + (1-q)G(x) \tag{1}$$

with $G(x)$ the CDF and q the probability of zero precipitation, given by the ratio between the number of zero in the rainfall series (m) and the number of observations (n).

Finally, the CDF is changed into the standard normal distribution by using, for example, the approximate conversion provided by Abramowitz and Stegun [56].

$$z = \text{SPI} = -\left(t - \frac{c_0 + c_1 t + c_2 t^2}{1 + d_1 t + d_2 t^2 + d_3 t^3}\right), t = \sqrt{\ln\left(\frac{1}{(H(x))^2}\right)} \text{ for } 0 < H(x) < 0.5, \tag{2}$$

$$z = \text{SPI} = +\left(t - \frac{c_0 + c_1 t + c_2 t^2}{1 + d_1 t + d_2 t^2 + d_3 t^3}\right), t = \sqrt{\ln\left(\frac{1}{(1 - H(x))^2}\right)} \text{ for } 0.5 < H(x) < 1, \tag{3}$$

with c_0, c_1, c_2, d_1, d_2, and d_3 as mathematical constants.

Table 1 reports the climatic classification according to the SPI provided by the National Drought Mitigation Center. This index is now habitually used in the classification of wet periods, even though the original classification provided by McKee et al. [21] was limited to drought periods only.

Table 1. Climate classification according to the Standardized Precipitation Index (SPI) values.

SPI Value	Class	Probability (%)
SPI $\geq$ 2.00	Extremely wet	2.3
1.5 $\leq$ SPI < 2.00	Severely wet	4.4
SPI < 1.50	Moderately wet	9.2
SPI < 1.00	Mildly wet	34.1
$-1.00 \leq$ SPI < 0.00	Mild drought	34.1
$-1.50 \leq$ SPI < -1.00	Moderate drought	9.2
$-2.00 \leq$ SPI < -1.50	Severe drought	4.4
SPI < -2.00	Extreme drought	2.3

2.2. Innovative Trend Analysis (ITA)

The ITA method has been first proposed by Şen [39]: unlike the MK test or other methods, the ITA's greatest advantage is the fact that it does not require any assumptions (serial correlation, non-normality, sample number, and so on). First, the time series is divided into two equal parts, which are separately sorted in ascending order. Then, the first and the second half of the time series are located on the X-axis and on the Y-axis, respectively. In Figure 1, a graphical representation of the innovative method on a Cartesian coordinate system is shown. If the data are collected on the 1:1 ideal line (45° line), there is no trend in the time series. If data are located on the upper triangular area of the ideal line, an increasing trend in the time series exists, while if data are accumulated in the lower triangular area of the 1:1 line, there is a decreasing trend in the time series [39,42]. Thus, trends of low, medium, and high values of any hydro-meteorological or hydro-climatic time series can be clearly identified through this method.

For example, the results shown in Figure 1 identify a positive trend of the lowest values and a negative trend of the highest ones, while no trends can be detect for the medium values that lies near the 1:1 ideal line.

In this work, in order to easily and better identify the possible trend of the severe dry and wet conditions, two vertical bands have been added in Figure 1: a red band corresponding to the severe drought limit (SPI = -1.5) and a blue band corresponding to the severe wet limit (SPI = 1.5).

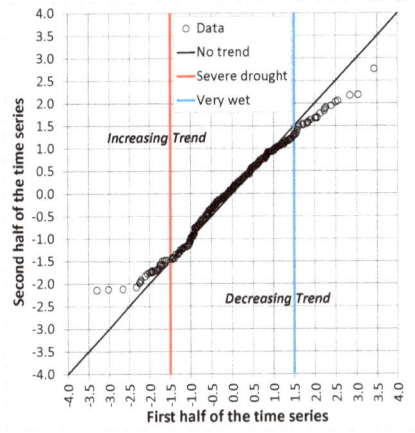

Figure 1. Example of the Innovative Trend Analysis (ITA) proposed by Şen [39].

3. Study Area

New Zealand, with an elongated shape of 1930 km and a maximum width of 400 km and a surface area of about 270,000 km^2, is located in the southern hemisphere 2500 km east from Australia. Its altitude distribution is mainly mountainous; in fact, 76% of its territory presents an elevation higher than 600 m a.s.l. with a maximum altitude of 3724 m (Figure 2).

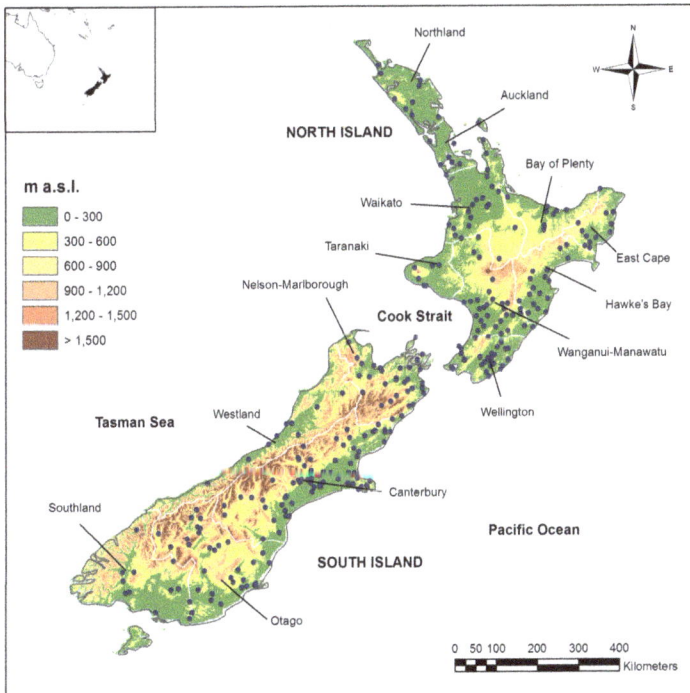

Figure 2. Location of the 14 regions of New Zealand and of the selected 294 rain gauge stations on a Digital Elevation Model (DEM) of New Zealand.

New Zealand's climate is mainly influenced by the following physical factors [57]:

a. The water masses surrounding the country, which results in cool summers and moderately cold winters, and the cold air masses from Antarctica, which cause snow and frost in some areas of the country;

b. The mountains crossing both the islands, which constitute a barrier against air flows coming in from south-south west, cause significant differences in the rainfall amounts, also within short distances;

c. The proximity of the Australia's eastern land/sea boundary, which is characterized by a low-pressure region of cyclonic circulation toward the Tasman Sea.

For a more thorough description of the New Zealand's climate, interested readers can refer to the National Institute of Water and Atmosphere Research (NIWA) of New Zealand, which published a detailed analysis of New Zealand's climate [58].

In order to carry out an SPI analysis at different time scales, the New Zealand National Climate database of the National Institute of Water and Atmospheric Research (NIWA) has been selected. Several studies on New Zealand's climate [59–65] made extensive use of this database, which presents high-quality data and complete, or near complete, records for the period 1951–2010. Until 2010, the New Zealand National Climate database consisted of measurements collected at 3011 stations, with a density of one station per 89 km^2. In particular, for the present analysis, the monthly rainfall data were extracted and used after performing record error checks and metadata analyses for inhomogeneities detection. Following these procedures, a number of station series, which presented either low quality records or few years of observation for statistical purposes (<50 available years of data), were discarded. The series ending before 2010 and those showing more than 5% of lacking data were also discarded. Thus, the final selection included 294 series longer than 50 years with a density of one station per 913 km^2 (Figure 2).

4. Results and Discussion

Following the NIWA, which provides climate maps at regional scale, for every region of New Zealand shown in Figure 2, an average SPI series has been evaluated for each time scale, and a trend analysis has been performed through the application of the ITA approach. In particular, the SPI has been evaluated for each rain gauge and for each time scale and then a simple arithmetic average of the obtained SPI values has been computed for each region.

Before applying the ITA, the number of months showing severe or extreme dry and wet conditions was evaluated. As a result, different conditions have been detected considering the different time scales.

As to what concerns the 3-month SPI (Figure 3a) in nine out of 14 regions the number of months showing severe or extreme wet conditions is higher than the ones showing severe or extreme dry conditions. In particular, in the South Island, in the Canterbury and Southland areas, only in 15 and 18 months dry conditions have been detected while, instead, 32 and 27 months evidenced wet conditions, respectively. Differently from these regions, in the Hawke's Bay area (North Island), the months that showed dry conditions (35 months) clearly outperform the months (24) in which wet conditions have been detected.

The 6-month SPI (Figure 3b) showed a different behavior with respect to the 3-month SPI. In fact, in eight regions, the number of months showing severe or extreme dry conditions is higher than the one presenting wet conditions. Dry conditions have been detected especially in the North Island and, specifically, in the Auckland (27 months against 16), Bay of Plenty (36 months against 29), East Cape (37 months against 31), Wanganui-Manawatu (28 months against 13), and Hawke's Bay (31 months against 19) areas. In the South Island, only the Nelson-Marlborough region showed dry conditions in 26 months (17 wet months), while in the other regions, wet conditions have been evidenced.

Concerning the 12-month SPI (Figure 3c) an equal number of regions showed prevailing dry or wet conditions. In particular, in the North Island, in six out of nine regions, the number of months showing dry conditions is higher than the ones showing wet conditions. Conversely, in the South Island, four regions showed wet conditions, and only one region (Nelson-Marlborough) evidenced dry conditions. Specifically, in the North Island, relevant results have been obtained in the Bay of Plenty and in the East Cape regions, with 38 and 40 months showing severe or extreme dry conditions, respectively, while, in the South Island, the Westland (45 months against 24) and the Southland (38 months against 22) regions marked dry conditions have been detected.

Finally, the 24-month SPI (Figure 3d) showed a clear difference between the two islands. In fact, in the North Island, severe or extreme dry conditions have been detected in all the regions, while severe or extreme wet conditions have been detected in all the regions of the South Island. In particular, the driest conditions in the North Island have been detected in the Taranaki (73 months against 34) and in the Bay of Plenty (61 months against 37) regions, while, in the South Island, the Southland (73 months against 34) and the Canterbury (61 months against 37) areas showed the highest number of months with wet conditions.

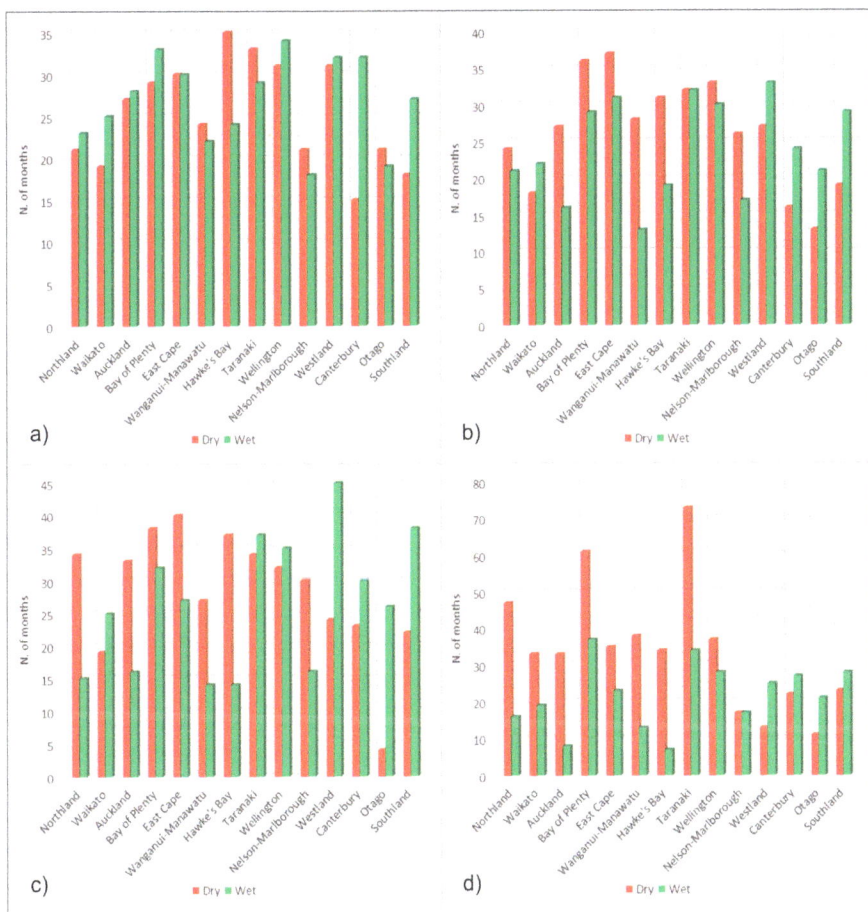

Figure 3. Number of months showing severe or extreme dry and wet conditions for the 3-month (**a**); 6-month (**b**); 12-month (**c**); and 24-month SPI (**d**).

With the aim to detect possible trends in the 3-, 6-, 12-, and 24-month SPI values, for each region the ITA method was applied to the monthly series of the index. The ITA method allowed to evidence the tendencies of both low and high SPI values, thus including values referred to wet conditions. As a result of the ITA approach, Figures 4–7 show the results obtained at regional level for the 3-, 6-, 12- and 24-month SPI, respectively. All the SPI series were divided into two 30-year sub-series: from 1951 to 1980, and from 1981 to 2010.

Generally, the main result obtained for the 3-month SPI values was a negative trend of the highest values of the index, which is related to weaker wet periods (Figure 4). This tendency has been detected in nine out of 14 regions but with a different behavior of the lowest SPI values. In fact, in four regions of the North Island (Northland, Auckland, Bay of Planty and East Cape) and in the Canterbury region in the South Island, a negative trend of both the lowest and the highest values of the index has been detected, thus evidencing heavier droughts and weaker wet periods. At the same time, in the Waikato, Wellington, Southland, and Westland regions, a positive trend of the lowest values and negative trend of the highest ones has been evidenced, both indicating weaker droughts and wet periods. Differently from the previous regions, the Otago, Wanganui-Manawatu, and Taranaki areas evidenced a tendency through weaker droughts and heavier wet periods given by a positive trend of both the lowest and the highest SPI values. Finally, in the Nelson-Marlborough region, a negative trend of the lowest values (heavier droughts) and a positive trend of the highest values (heavier wet periods) have been detected. The results of the ITA methods on the Hawke's Bay region did not show a clear tendency, with the lowest and highest values falling close to the no trend line.

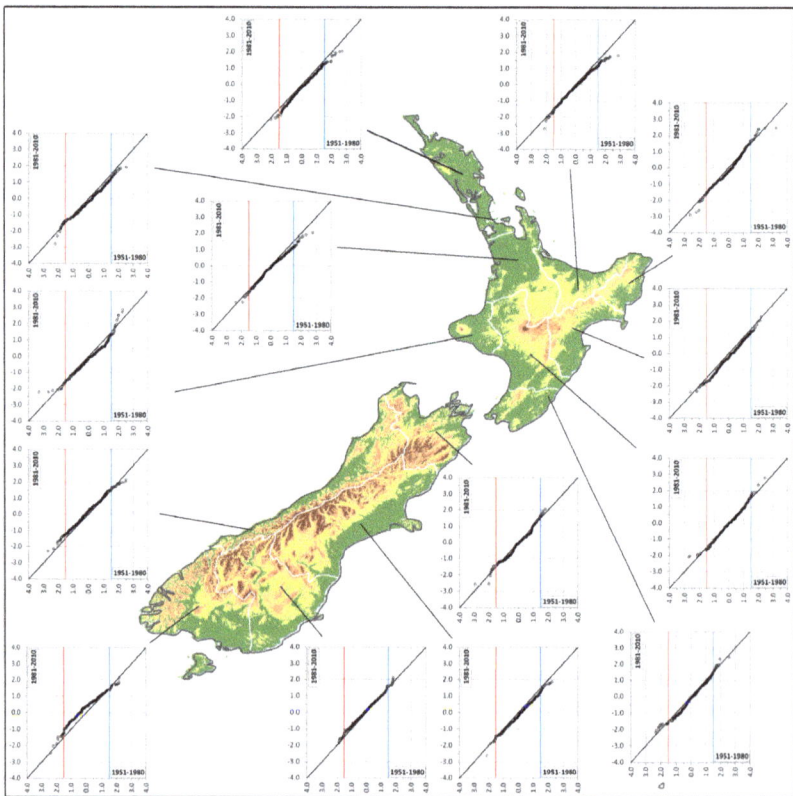

Figure 4. Regional results of the Innovative Trend Analysis (ITA) method applied to the 3-month SPI.

As regards the 6-month SPI, results confirm the ones obtained for the 3-month SPI, with a spreading negative trend of the highest values of the index (Figure 5). In fact, similar results to the 3-month SPI have been obtained in Northland, Auckland, Bay of Planty, Wanganui-Manawatu, Taranaki, Nelson-Marlborough, Westland, and Southland. Different from the results obtained for the 3-month SPI, in the Waikato and in the Hawke's Bay regions, a negative trend of both the lowest and the highest values of the SPI index has been detected. Moreover, in the Canterbury and Otago region, a positive trend of the lowest values (weaker droughts) and a negative trend of the highest values (weaker wet periods) have been evidenced. Finally, the Wellington and East Cape regions did not show a clear tendency.

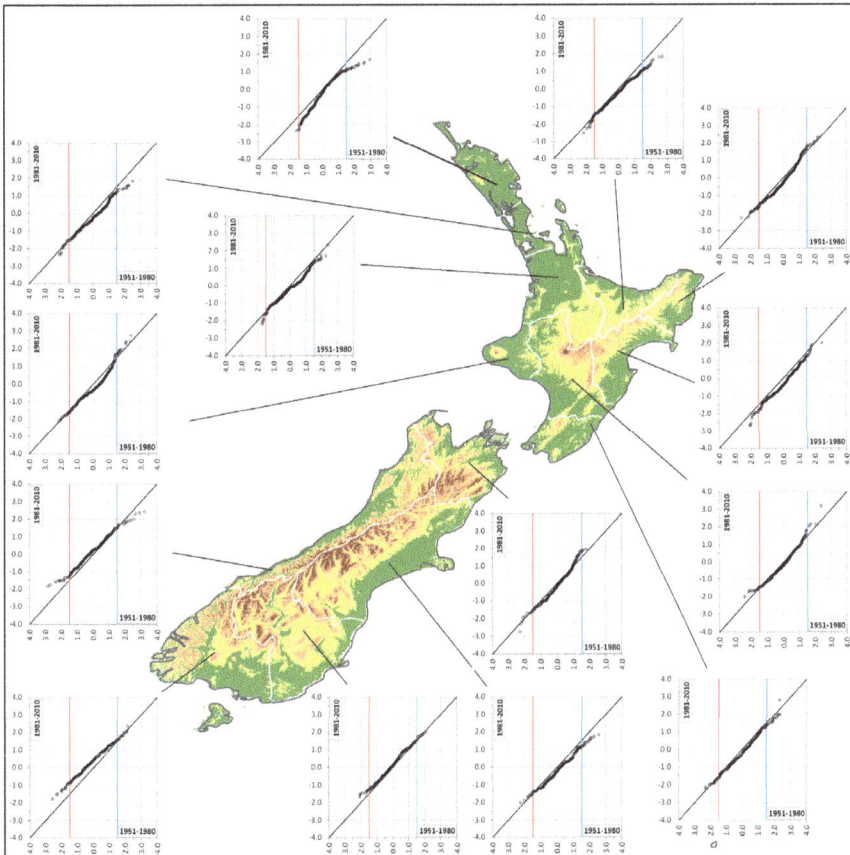

Figure 5. Regional results of the ITA method applied to the 6-month SPI.

Studies on the 3- and 6-month SPI trends, which impact vegetation and agricultural practices, are paramount for New Zealand because agriculture is one of the largest sectors of the economy. As a summary of the results of the trend analysis on these time scales, in the South Island, which is generally characterized by wet conditions, a negative trend of the highest SPI value has been detected in the majority of the regions, thus indicating a tendency toward weaker wet periods. In particular, in the agricultural area of the Canterbury, the 3-month SPI evidenced a decreasing trend of both the lowest and the highest values of the index, which lead to heavier droughts and weaker wet periods. In the North Island, where the majority of the regions showed a higher percentage of dry months than wet months, a general negative trend of both the lowest and the highest SPI values has been identified, with the exception of some region such as Wanganui-Manawatu and Taranaki.

Considering the long time scales, the 12-month SPI trend showed similar results than the 6-month SPI but with an increase in the negative trend (Figure 6). In fact, in the Northland, Waikato, Auckland, Bay of Planty, East Cape, Hawke's Bay, and Canterbury regions, a reduction of all the SPI values has been detected, thus confirming a clear tendency toward heavier droughts and weaker wet periods. On the contrary, the Wanganui-Manawatu, Nelson-Marlborough, and Otago regions showed trend results that indicate weaker droughts and heavier wet periods. An increase of the lowest values and a decrease in the highest values (weaker wet and dry periods) has been evidenced in the Wellington, Westland, and Southland regions. In the Taranaki area, no clear tendencies have been detected in the severe and extreme SPI values.

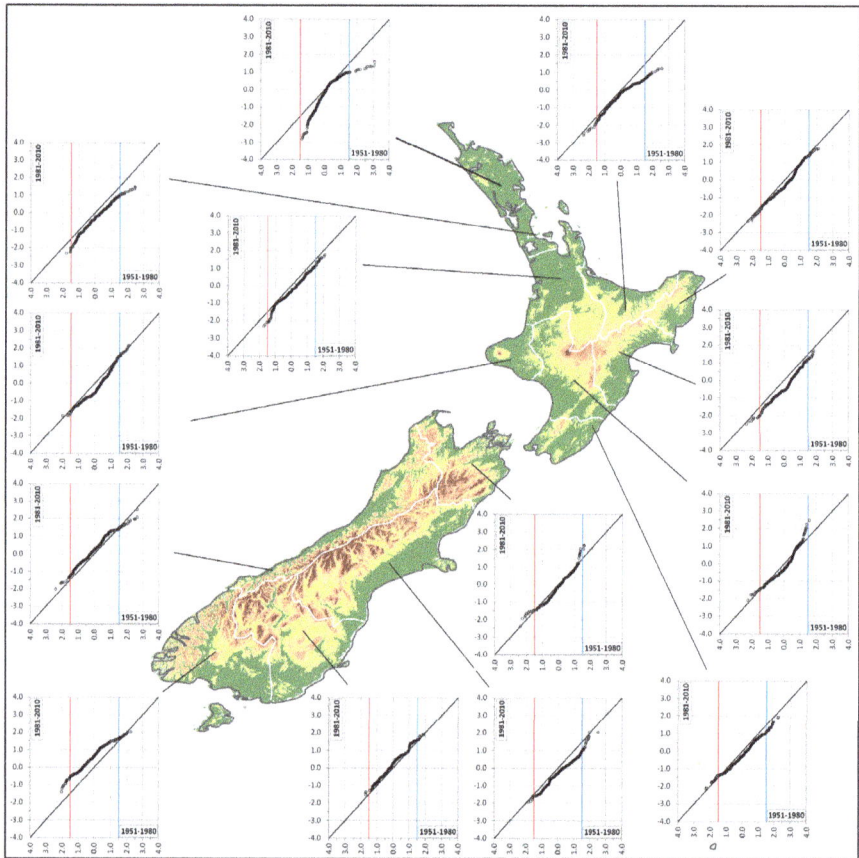

Figure 6. Regional results of the ITA method applied to the 12-month SPI.

As regards the 24-month SPI, in eight regions (Northland, Waikato, Auckland, Bay of Planty, East Cape, Wanganui-Manawatu, Hawke's Bay, and Canterbury), the results of the trend analysis with the ITA method evidenced a clear reduction of the lowest SPI values. Among these regions, the same result has been obtained for the highest values, with the exception of the Waikato and Auckland, in which the highest values showed a positive trend. By contrast, the Taranaki, Nelson-Marlborough, Otago Westland, and Southland regions showed an increase in the lowest values, with concomitant decrease in the highest values in the Taranaki and Otago areas and increase in the highest values in the Nelson-Marlborough, Westland, and Southland regions. As also evidenced for the 3-month SPI, the Wellington region did not show a clear tendency.

The 12- and 24-month SPI are a broad proxy for water resource management. Results of the trend analysis on these time scales confirm the ones obtained on a short time scale but with an increase in the number of regions where a negative trend has been detected. In fact, current global level assessments suggest that droughts are expected to both increase and decrease following future climate change depending upon geographic location [66]. Based on the latest climate and impact modelling, New Zealand can expect more droughts in the future in some locations [51].

As a result, this work has evidenced an increase in drought trend in all the areas that are presently subject to drought, supporting what has been evidenced in past studies [50,64]. In fact, the results of this paper confirm the geographic pattern of change found by Mullan et al. [50] and Caloiero [64], which mainly detected a drought increase in the future projections on the East Coast and no change in drought projections for the West Coast of the South Island. Specifically, as also evidenced by Clark et al. [51] and Caloiero [64], the results of this paper highlight that key agricultural regions on the Eastern side such as the Canterbury Plains are the most consistently vulnerable areas in the South Island, together with other regions in the North Island, including a key primary industry region like Waikato.

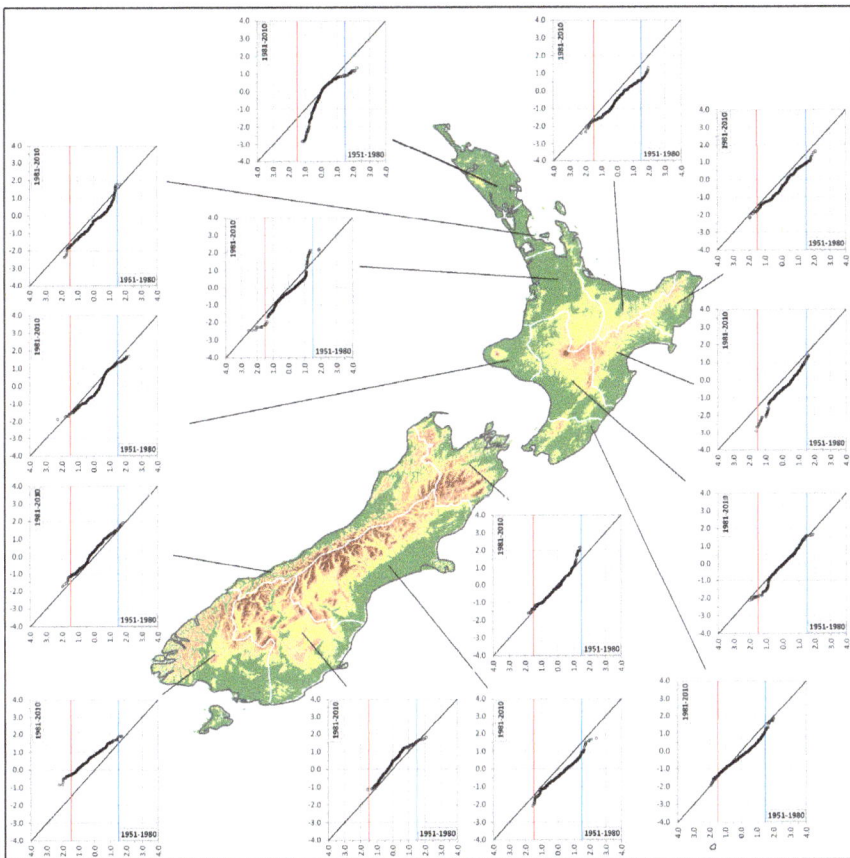

Figure 7. Regional results of the ITA method applied to the 24-month SPI.

5. Conclusions

Generally, the methods used for trend detection allow researchers to analyze general tendencies of climatic variables, but they do not make it possible to identify sub-trends. In this paper, for every region of New Zealand, an average SPI series has first been evaluated at various time scales (3, 6, 12, and 24 months). Then, for each region, the number of months showing severe or extreme dry and wet conditions has been detected. Finally, a graphical technique, based on the Innovative Trend Analysis (ITA) proposed by Şen [39], was applied to the SPI series. As a result, a different behavior emerged between the two islands. In fact, for the several time scale, in the majority of the regions of the North Island, the number of months showing dry conditions is higher than the ones showing wet conditions, while in the South Island, the number of months showing wet conditions is higher than the ones showing dry conditions. In particular, for the 24-month time scale, dry conditions have been detected in all the regions of the North Island, while wet conditions have been detected in the South Island ones. As to what concerns the trend analysis, on a short time scale, the results showed a negative trend of the highest SPI value in the South Island, thus indicating a tendency toward weaker wet periods. In particular, in the agricultural area of the Canterbury, the 3-month SPI evidenced a decreasing trend in both the lowest and the highest values of the index, which lead to heavier droughts and weaker wet periods. In the North Island, a general negative trend of both the lowest and the highest SPI values has been identified, thus evidencing heavier droughts and weaker wet periods. Results of the trend analysis on long time scales confirm the ones obtained on a short time scale but with an increase in the number of regions where a negative trend has been detected, highlighting that agricultural regions on the eastern side of the South Island, as well as the north-eastern regions of the North Island, are the most consistently vulnerable areas.

Acknowledgments: The author would like to thank the National Institute of Water and Atmosphere Research for providing access to the New Zealand meteorological data from the National Climate Database.

Conflicts of Interest: The author declares no conflict of interest.

References

1. Estrela, T.; Vargas, E. Drought management plans in the European Union. *Water Resour. Manag.* **2010**, *26*, 1537–1553. [CrossRef]
2. Kreibich, H.; Di Baldassarre, G.; Vorogushyn, S.; Aerts, J.C.J.H.; Apel, H.; Aronica, G.T.; Arnbjerg-Nielsen, K.; Bouwer, L.M.; Bubeck, P.; Caloiero, T.; et al. Adaptation to flood risk: Results of international paired flood event studies. *Earth's Future* **2017**, *5*, 953–965. [CrossRef]
3. Wilhite, D.A.; Hayes, M.J.; Svodoba, M.D. Drought monitoring and assessment in the US. In *Drought and Drought Mitigation in Europe*; Voght, J.V., Somma, F., Eds.; Kluwers: Dordrecht, The Netherlands, 2000; pp. 149–160.
4. Yevjevich, V.; Da Cunha, L.; Vlachos, E. *Coping with Droughts*; Water Resources Publications: Littleton, CO, USA, 1983.
5. Fink, A.H.; Brücher, T.; Krüger, A.; Leckebush, G.C.; Pinto, J.G.; Ulbrich, U. The 2003 European summer heatwaves and drought-synoptic diagnosis and impacts. *Weather* **2004**, *59*, 209–216. [CrossRef]
6. Lloyd-Huhes, B.; Saunders, M.A. A drought climatology for Europe. *Int. J. Climatol.* **2002**, *22*, 1571–1592. [CrossRef]
7. Zaidman, M.D.; Rees, H.G.; Young, A.R. Spatio-temporal development of streamflow droughts in north-west Europe. *Hydrol. Earth Syst. Sci.* **2012**, *5*, 733–751. [CrossRef]
8. Hannaford, J.; Lloyd-Hughes, B.; Keef, C.; Parry, S.; Prudhomme, C. Examining the large-scale spatial coherence of European drought using regional indicators of precipitation and streamflow deficit. *Hydrol. Process.* **2011**, *25*, 1146–1162. [CrossRef]
9. Intergovernmental Panel on Climate Change (IPCC). *Summary for Policymakers*; Fifth Assessment Report of the Intergovernmental Panel on Climate Change; Cambridge University Press: Cambridge, UK, 2013.
10. Buttafuoco, G.; Caloiero, T.; Ricca, N.; Guagliardi, I. Assessment of drought and its uncertainty in a southern Italy area (Calabria region). *Measurement* **2018**, *113*, 205–210. [CrossRef]

11. Caloiero, T.; Coscarelli, R.; Ferrari, E.; Sirangelo, B. Analysis of Dry Spells in Southern Italy (Calabria). *Water* **2015**, *7*, 3009–3023. [CrossRef]

12. Fang, K.; Gou, X.; Chen, F.; Davi, N.; Liu, C. Spatiotemporal drought variability for central and eastern Asia over the past seven centuries derived from tree-ring based reconstructions. *Quat. Int.* **2013**, *283*, 107–116. [CrossRef]

13. Feng, S.; Hu, Q.; Oglesby, R.J. Influence of Atlantic sea surface temperatures on persistent drought in North America. *Clim. Dyn.* **2011**, *37*, 569–586. [CrossRef]

14. Hua, T.; Wang, X.M.; Zhang, C.X.; Lang, L.L. Temporal and spatial variations in the Palmer Drought Severity Index over the past four centuries in arid, semiarid, and semihumid East Asia. *Chin. Sci. Bull.* **2013**, *58*, 4143–4152. [CrossRef]

15. Minetti, J.L.; Vargas, W.M.; Poblete, A.G.; de la Zerda, L.R.; Acuña, L.R. Regional droughts in southern South America. *Theor. Appl. Climatol.* **2010**, *102*, 403–415. [CrossRef]

16. Sirangelo, B.; Caloiero, T.; Coscarelli, R.; Ferrari, E. A stochastic model for the analysis of the temporal change of dry spells. *Stoch. Environ. Res. Risk Assess.* **2015**, *29*, 143–155. [CrossRef]

17. Sirangelo, B.; Caloiero, T.; Coscarelli, R.; Ferrari, E. Stochastic analysis of long dry spells in Calabria (Southern Italy). *Theor. Appl. Climatol.* **2017**, *127*, 711–724. [CrossRef]

18. Tsakiris, G.; Pangalou, D.; Vangelis, H. Regional drought assessment based on the Reconnaissance Drought Index (RDI). *Water Resour. Manag.* **2007**, *21*, 821–833. [CrossRef]

19. Tabari, H.; Abghari, H.; Hosseinzadeh Talaee, P. Temporal trends and spatial characteristics of drought and rainfall in arid and semi-arid regions of Iran. *Hydrol. Process.* **2012**, *26*, 3351–3361. [CrossRef]

20. Bayissa, Y.A.; Moges, S.A.; Xuan, Y.; Van Andel, S.J.; Maskey, S.; Solomatine, D.P.; Griensven, A.; Van Tadesse, T. Spatio-temporal assessment of meteorological drought under the influence of varying record length: The case of Upper Blue Nile Basin, Ethiopia. *Hydrol. Sci. J.* **2015**, *60*, 1927–1942. [CrossRef]

21. McKee, T.B.; Doesken, N.J.; Kleist, J. The relationship of drought frequency and duration to time scales. In Proceedings of the 8th Conference on Applied Climatology, Anaheim, CA, USA, 17–22 January 1993; pp. 179–184.

22. Khan, S.; Gabriel, H.F.; Rana, T. Standard precipitation index to track drought and assess impact of rainfall on watertables in irrigation areas. *Irrig. Drain. Syst.* **2008**, *22*, 159–177. [CrossRef]

23. Logan, K.E.; Brunsell, N.A.; Jones, A.R.; Feddema, J.J. Assessing spatiotemporal variability of drought in the US central plains. *J. Arid Environ.* **2010**, *74*, 247–255. [CrossRef]

24. Manatsa, D.; Mukwada, G.; Siziba, E.; Chinyanganya, T. Analysis of multidimensional aspects of agricultural droughts in Zimbabwe using the Standardized Precipitation Index (SPI). *Theor. Appl. Climatol.* **2010**, *102*, 287–305. [CrossRef]

25. Patel, N.R.; Yadav, K. Monitoring spatio-temporal pattern of drought stress using integrated drought index over Bundelkhand region, India. *Nat. Hazards* **2015**, *77*, 663–677. [CrossRef]

26. Raziei, T.; Saghafian, B.; Paulo, A.A.; Pereira, L.S.; Bordi, I. Spatial patterns and temporal variability of drought in Western Iran. *Water Resour. Manag.* **2009**, *23*, 439–455. [CrossRef]

27. Buttafuoco, G.; Caloiero, T. Drought events at different timescales in southern Italy (Calabria). *J. Maps* **2014**, *10*, 529–537. [CrossRef]

28. Zhai, L.; Feng, Q. Spatial and temporal pattern of precipitation and drought in Gansu Province Northwest China. *Nat. Hazards* **2009**, *49*, 1–24. [CrossRef]

29. Capra, A.; Scicolone, B. Spatiotemporal variability of drought on a short–medium time scale in the Calabria Region (Southern Italy). *Theor. Appl. Climatol.* **2012**, *3*, 471–488. [CrossRef]

30. Wu, H.; Hayes, M.J.; Wilhite, D.A.; Svoboda, M.D. The effect of the length of record on the standardized precipitation index calculation. *Int. J. Climatol.* **2005**, *25*, 505–520. [CrossRef]

31. Vicente-Serrano, S.M. Differences in spatial patterns of drought on different time sales: An analysis of the Iberian Peninsula. *Water Resour. Manag.* **2006**, *20*, 37–60. [CrossRef]

32. Buttafuoco, G.; Caloiero, T.; Coscarelli, R. Analyses of Drought Events in Calabria (Southern Italy) Using Standardized Precipitation Index. *Water Resour. Manag.* **2015**, *29*, 557–573. [CrossRef]

33. Caloiero, T.; Coscarelli, R.; Ferrari, E.; Sirangelo, B. An Analysis of the Occurrence Probabilities of Wet and Dry Periods through a Stochastic Monthly Rainfall Model. *Water* **2016**, *8*, 39. [CrossRef]

34. Guttman, N.B. Accepting the standardized precipitation index: A calculating algorithm. *J. Am. Water Resour. Assoc.* **1999**, *35*, 311–323. [CrossRef]

35. Cancelliere, A.; Di Mauro, G.; Bonaccorso, B.; Rossi, G. Drought forecasting using the Standardised Precipitation Index. *Water Resour. Manag.* **2007**, *21*, 801–819. [CrossRef]

36. Bordi, I.; Fraedrich, K.; Sutera, A. Observed drought and wetness trends in Europe: An update. *Hydrol. Earth Syst. Sci.* **2009**, *13*, 1519–1530. [CrossRef]

37. Golian, S.; Mazdiyasni, O.; AghaKouchak, A. Trends in meteorological and agricultural droughts in Iran. *Theor. Appl. Climatol.* **2015**, *119*, 679–688. [CrossRef]

38. Zhai, J.; Su, B.; Krysanova, V.; Vetter, T.; Gao, C.; Jiang, T. Spatial variation and trends in PDSI and SPI indices and their relation to streamflow in 10 large regions of China. *J. Clim.* **2010**, *23*, 649–663. [CrossRef]

39. Şen, Z. An innovative trend analysis methodology. *J. Hydrol. Eng.* **2012**, *17*, 1042–1046. [CrossRef]

40. Haktanir, T.; Citakoglu, H. Trend, independence, stationarity, and homogeneity tests on maximum rainfall series of standard durations recorded in Turkey. *J. Hydrol. Eng.* **2014**, *19*, 501–509. [CrossRef]

41. Kisi, O.; Ay, M. Comparison of Mann–Kendall and innovative trend method for water quality parameters of the Kizilirmak River, Turkey. *J. Hydrol.* **2014**, *513*, 362–375. [CrossRef]

42. Şen, Z. Trend identification simulation and application. *J. Hydrol. Eng.* **2014**, *19*, 635–642. [CrossRef]

43. Ay, M.; Kisi, O. Investigation of trend analysis of monthly total precipitation by an innovative method. *Theor. Appl. Climatol.* **2015**, *120*, 617–629. [CrossRef]

44. Martínez-Austria, P.F.; Bandala, E.R.; Patiño-Gómez, C. Temperature and heat wave trends in northwest Mexico. *Phys. Chem. Earth* **2015**, *91*, 20–26. [CrossRef]

45. Kisi, O. An innovative method for trend analysis of monthly pan evaporations. *J. Hydrol.* **2015**, *527*, 1123–1129. [CrossRef]

46. Tabari, H.; Willems, P. Investigation of streamflow variation using an innovative trend analysis approach in northwest Iran. In Proceedings of the 36th IAHR World Congress, The Hague, The Nederland, 28 June–3 July 2015.

47. Palmer, J.G.; Cook, E.R.; Turney, C.S.M.; Allen, K.; Fenwick, P.; Cook, B.; O'Donnell, A.J.; Lough, J.M.; Grierson, P.F.G.; Baker, P. Drought variability in the eastern Australia and New Zealand summer drought atlas (ANZDA, CE 1500–2012) modulated by the Interdecadal Pacific Oscillation. *Environ. Res. Lett.* **2015**, *10*, 124002. [CrossRef]

48. MAF. *Regional and National Impacts of the 2007–2008 Drought*; Butcher Partners Ltd.: Tai Tapu, New Zealand, 2009.

49. Kamber, G.; McDonald, C.; Price, G. *Drying Out: Investigating the Economic Effects of Drought in New Zealand*; Reserve Bank of New Zealand: Wellington, New Zealand, 2013.

50. Mullan, B.; Porteous, A.; Wratt, D.; Hollis, M. *Changes in Drought Risk with Climate Change*; National Institute of Water & Atmospheric Research: Wellington, New Zealand, 2015.

51. Clark, A.; Mullan, B.; Porteous, A. *Scenarios of Regional Drought under Climate Change*; National Institute of Water & Atmospheric Research: Wellington, New Zealand, 2011.

52. Edwards, D.; McKee, T. *Characteristics of 20th Century Drought in the United States at Multiple Scale*; Atmospheric Science Paper 634; Department of Atmospheric Science Colorado State University: Fort Collins, CO, USA, 1997.

53. Bonaccorso, B.; Bordi, I.; Cancelliere, A.; Rossi, G.; Sutera, A. Spatial variability of drought: An analysis of SPI in Sicily. *Water Resour. Manag.* **2003**, *17*, 273–296. [CrossRef]

54. Angelidis, P.; Maris, F.; Kotsovinos, N.; Hrissanthou, V. Computation of drought index SPI with Alternative Distribution Functions. *Water Resour. Manag.* **2012**, *26*, 2453–2473. [CrossRef]

55. Thom, H.C.S. A note on the gamma distribution. *Mon. Weather Rev.* **1958**, *86*, 117–122. [CrossRef]

56. Abramowitz, M.; Stegun, I.A. *Handbook of Mathematical Functions with Formulas, Graphs, and Mathematical Tables*; Dover Publications, INC.: New York, NY, USA, 1970.

57. Oliver, J.E. *Encyclopedia of World Climatology*; Springer: Amsterdam, The Netherlands, 2005.

58. NIWA. Overview of New Zealand Climate. Available online: http://www.niwa.co.nz/education-and-training/schools/resources/climate/overview (accessed on 26 February 2018).

59. Salinger, M.J.; Mullan, A.B. New Zealand climate: Temperature and precipitation variations and their links with atmospheric circulation 1930–1994. *Int. J. Climatol.* **1999**, *19*, 1049–1071. [CrossRef]

60. Griffiths, G.M.; Salinger, M.J.; Leleu, I. Trends in extreme daily rainfall across the South Pacific and relationship to the South Pacific Convergence Zone. *Int. J. Climatol.* **2003**, *23*, 847–869. [CrossRef]

61. Dravitzki, S.; McGregor, J. Extreme precipitation of the Waikato region, New Zealand. *Int. J. Climatol.* **2011**, *31*, 1803–1812. [CrossRef]

62. Caloiero, T. Analysis of daily rainfall concentration in New Zealand. *Nat. Hazards* **2014**, *72*, 389–404. [CrossRef]

63. Caloiero, T. Analysis of rainfall trend in New Zealand. *Environ. Earth. Sci.* **2015**, *73*, 6297–6310. [CrossRef]
64. Caloiero, T. Drought analysis in New Zealand using the standardized precipitation index. *Environ. Earth. Sci.* **2017**, *76*, 569. [CrossRef]
65. Caloiero, T. Trend of monthly temperature and daily extreme temperature during 1951–2012 in New Zealand. *Theor. Appl. Climatol.* **2017**, *129*, 111–117. [CrossRef]
66. Wang, G. Agricultural drought in a future climate: Results from 15 global climate models participating in the IPCC 4th assessment. *Clim. Dyn.* **2005**, *25*, 739–753. [CrossRef]

geosciences

MDPI

Article

Comparison of SCS and Green-Ampt Distributed Models for Flood Modelling in a Small Cultivated Catchment in Senegal

Christophe Bouvier *, Lamia Bouchenaki and Yves Tramblay

Institut de Recherche pour le Développement (IRD), HydroSciences Montpellier, UMR 5569 CNRS-IRD-UM, 34000 Montpellier, France; bouchenaki.lamia@gmail.com (L.B.); yves.tramblay@ird.fr (Y.T.)
* Correspondence: jean-christophe.bouvier@umontpellier.fr; Tel.: +33-467-149-073

Received: 29 January 2018; Accepted: 29 March 2018; Published: 4 April 2018

Abstract: The vulnerability to floods in Africa has increased over the last decades, together with a modification of land cover as urbanized areas are increasing, agricultural practices are changing, and deforestation is increasing. Rainfall-runoff models that properly represent land use change and hydrologic response should be useful for the development of water management and mitigation plans. Although some studies have applied rainfall-runoff models in West Africa for flood modelling, there is still a need to develop such models, while many data are available and have not still been used for modelling improvement. The Ndiba catchment (16.2 km^2), which is located in an agricultural area in south Senegal, is such catchment, where a lot of hydro-climatic data has been collected between 1983 and 1992. Twenty-eight flood events have been extracted and modelled by two event-based rainfall-runoff models that are based on the Soil Conservation Service (SCS) or the Green-Ampt (GA) models for runoff, both coupled with the distributed Lag and Route (LR) for routing. Both models were able to reproduce the flood events after calibration, but they had to account for that the infiltration processes are highly dependent on the tillage of the soils and the growing of the crops during the rainy season, which made the initialization of the event-based models difficult. The most influent parameters for both models (the maximal water storage capacity for SCS, the hydraulic conductivity at saturation for Green-Ampt) were mostly related to the development stage of the vegetation, described by a Normalized Difference Vegetation Index (NDVI) anomaly. The SCS model performed finally better than the Green-Ampt model, because Green-Ampt was very sensitive to the variability of the hydraulic conductivity at saturation. The variability of the parameters of the models highlights the complexity of this kind of cultivated catchment, with highly non stationary conditions. The models could be improved by a better knowledge of the tillage practices, and a better integration of these practices in the parameters predictors.

Keywords: flood modelling; Agricultural Small Catchment; SCS-CN; Green-Ampt; Senegal

1. Introduction

The vulnerability to floods in Africa has increased over the last decades, in terms of human fatalities and economic losses [1], while in most of African countries, flood warning and prevision systems are not existent [2]. Consequently, there is a need for a better knowledge of these events and to improve rainfall-runoff models that could be used in the development of prevision systems. However, up to now, only a few studies have applied rainfall-runoff models in West Africa for flood modelling (e.g., [3,4]) or the analysis of the hydrological processes in small catchments less than some tenth or hundreds of squared kilometers [5,6]. In West Africa, flood prediction usually derive from synthetic guidelines [7] or regional studies [8–10]. But, such works are now ancient and could be revisited or developed, while many data are available in West Africa and have not still been used for modelling improvement.

The small agricultural area Thysse Kaymor is one of these catchments that have not been the object of modelling studies at the catchment scale, despite that many hydro-climatic data have been collected. Five nested catchments have been monitored during the period 1982–1992, including rainfall, runoff, and water content and hydrodynamic properties of the soils. The role of the agriculture on the runoff was clearly shown by [11], who claimed that the runoff conditions are first high before the tillage, and then reduce along the rainy season due to the growth of the crops. They also found that the Normalized Difference Vegetation Index (NDVI) could be a good predictor of the runoff conditions. However, modelling was only performed at the plot scale [12], and not at the catchment scale.

The aim of this study is to assess the skill of two well-known event-based models—the Soil Conservation Service (SCS) model and the Green-Ampt (GA) model in reproducing the flood processes in a semi-arid agricultural catchment of Senegal. To understand the behavior of a watershed during a specific rainfall episode, event-based models are often preferred over continuous models since they require less input data [13] and they possibly reduce the complexity of the hydrological processes. However, the event-based models need to set the initial condition for each event, and to relate this condition to an external predictor [14,15]. Therefore, an event-based model must not only be assessed from its skill to reproduce the floods, but also from the goodness of the relationship between the initial condition of the model and the external predictor, which is linked with. The latter part is often neglected, whereas it is the most important. This dual calibration of an event-based model is addressed in this paper, focusing on the relationship of the initial condition and external predictor, such as antecedent precipitation, event-rainfall characteristics, or satellite-derived normalized difference vegetation index (NDVI) to represent the vegetation stage.

2. Study Area and Data Collection

This study was conducted using data and catchment description from hydrological campaign reports done between 1983 and 1990 by two institutions (Office de la Recherche Scientifique et Technique Outre-Mer (now IRD) and the Institut Sénégalais de Recherche Agricole) in the Thysse Kaymor area.

2.1. Ndiba Catchment

The Ndiba catchment (16.2 km^2) is situated in the rural community of Thysse Kaymor, in the Sine Saloum region in southern Senegal (Figure 1). The elevations range between 13 and 45 m, and the slope is about 1%. The Ndiba catchment is mostly (over 50%) cultivated (millet and groundnut), and the other parts are bush areas [16].

The Ndiba catchment, as most of the Thysse Kaymor's region, is characterized by sandy soils at the surface, but the proportion of clay increases with depth especially between 0.6 and 1.5 m depth [17]. The soil porosity was assumed to decrease with the depth between 0.31 and 0.25 cm$^3 \cdot$cm^{-3} [18]: these values were obtained from experimental measurements of both water content θ and head pressure h at depths 10, 20, 30, 40, 60, 80, 100, 120, 140, 160 cm, in two sites, with either artificial rainfall or Müntz experiment; when the soil saturation was not reached, the water content at saturation was extrapolated from the experimental points (h,θ). The hydraulic conductivities at saturation ranged between 10 and 50 mm$\cdot$h^{-1}, depending on the type of tillage (perpendicular or parallel to the slope) and the cumulative rainfall since the last tillage [19]; the results were obtained from 48 infiltration tests that were performed by a disk infiltrometer, each test for a different site within an 900 m^2 area; the potential storage of water in these areas was estimated to 170 mm in the first 170 cm. During the first rainfall events at the beginning of the wet season, before tillage, the runoff coefficients are usually high since the soils are almost bare due to animal grazing and soil crusting. Then, the runoff coefficient tends to decrease gradually as the amount of vegetation increase during the growing season [11,20].

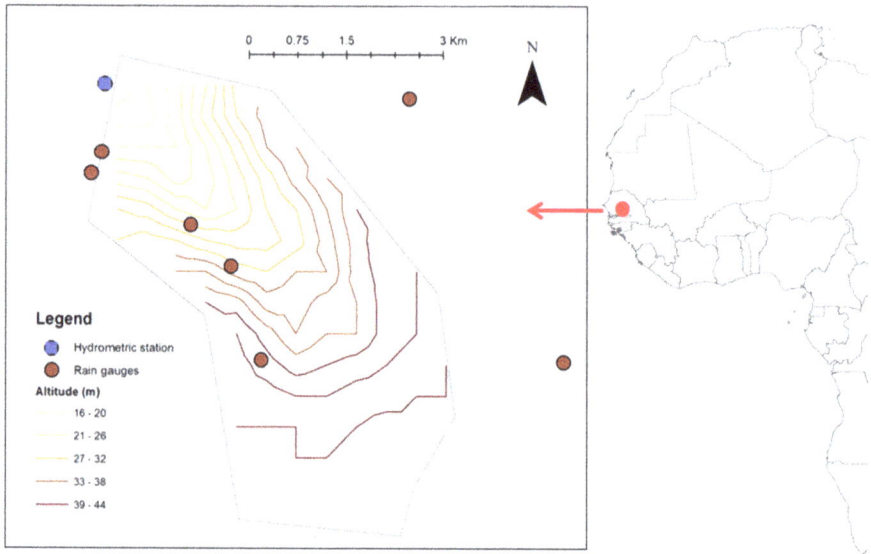

Figure 1. Map of the Ndiba catchment.

2.2. Rainfall and Runoff Data

Ndiba rainfalls were continuously registered from seven rain gauges (Precis-Mecanique with tipping buckets, surface 400 cm^2, graphical recording with daily rotation) in and around the catchment (Figure 1). The mean rainfall for the catchment during the period 1983–1990 was 612.5 mm, and years 1983 and 1984 were particularly dry, as the annual rainfall was less than 500 mm. These years were characterized by a severe drought that affected West Africa since the 1970's. The rainy season occurred during summer months, the maximal rainfall generally occurred in August and its mean value is 183 mm.

Gauging information was available at the hydrometric station of Ndiba (altitude = 13 m). The water levels were continuously recorded by a mechanical OTT 10 stream gauge, between 1983 and 1990. Sixty-four gauging have been performed between 1985 and 1988, up to the water level 160 cm, corresponding to a discharge of 58 m^3·s^{-1}. Those gauging were correctly located around a unique and reliable rating curve. Rainfall and runoff data for Ndiba catchment were extracted from the SIEREM database (http://www.hydrosciences.fr/sierem/) for the period 1983–1990, with a 5-min time interval. Twenty-eight rainfall-runoff events were selected for the study, and were delimited by periods of 48 h when rain intensity did not exceed 1 mm·h^{-1}; in addition, the events were definitely selected if the maximal rainfall amount was more than 10 mm and the peak flow more than 1 m·s^{-1}. Table 1 shows their main characteristics.

Table 1. Flood event characteristics.

Event	Starting Date	Ending Date	Maximum Discharge (m^3/s)	Rainfall Depth (mm)	Maximum Rainfall Intensity (mm/5 min)	Runoff Coefficient (%)	API (mm)	NDVI
1	13/07/1983	14/07/1983	35.1	77.6	139	11.0	146.0	0.164
2	19/07/1983	20/07/1983	1.28	28.8	75	1.2	166.4	0.381
3	24/08/1983	25/08/1983	1.35	50.5	79	0.8	219.5	1.081
4	2/06/1984	3/06/1984	56.6	32	32	37.0	48.0	−0.648

Table 1. *Cont.*

Event	Starting Date	Ending Date	Maximum Discharge (m³/s)	Rainfall Depth (mm)	Maximum Rainfall Intensity (mm/5 min)	Runoff Coefficient (%)	API (mm)	NDVI
5	8/06/1984	9/06/1984	1.06	24.7	22	1.6	69.4	−0.496
6	14/07/1984	15/07/1984	2.06	30.4	85	3.8	189.8	1.219
7	19/07/1985	20/07/1985	15.7	32.1	138	13.0	141.4	0.311
8	18/08/1985	19/08/1985	1.64	56.5	96	1.6	290.3	1.303
9	1/09/1985	2/09/1985	11	63.2	107	7.1	388.4	1.558
10	2/08/1986	3/08/1986	44.2	96.3	112	11.0	179.1	−0.244
11	3/08/1986	4/08/1986	7.77	41.9	34	6.4	230.5	−0.283
12	12/09/1986	13/09/1986	4.07	50.9	116	4.8	446.7	1.684
13	16/06/1987	17/06/1987	3.87	32.2	67	3.9	91.0	−0.457
14	1/07/1987	2/07/1987	1.3	35	48	1.3	152.9	0.767
15	8/08/1987	9/08/1987	2.29	53.3	86	2.2	361.1	0.460
16	13/07/1988	14/07/1988	56.7	78	100	16.0	148.2	0.115
17	28/07/1988	29/07/1988	14.6	90.5	83	5.6	219.2	0.805
18	1/08/1988	2/08/1988	23.9	61.4	100	13.0	245.7	0.176
19	2/08/1988	3/08/1988	12.6	52.4	46	8.8	288.2	0.038
20	8/08/1988	9/08/1988	2.23	29.2	46	2.7	311.9	0.655
21	15/06/1989	16/06/1989	1.33	20.9	32	1.4	23.6	−0.055
22	17/06/1989	18/06/1989	55.5	38.7	77	58.0	70.1	−0.070
23	20/06/1989	21/06/1989	1.04	12.4	40	1.7	81.3	−0.354
24	17/07/1990	18/07/1990	19.8	47.1	146	9.9	116.7	−0.365
25	20/07/1990	21/07/1990	1.03	38.9	105	0.5	163.2	−0.020
26	8/08/1990	9/08/1990	20.5	64.2	140	8.9	197.8	1.229
27	13/08/1990	14/08/1990	1.71	26.7	75	2.8	214.2	1.334
28	17/08/1990	18/08/1990	4.73	39.7	118	3.9	258.6	0.613

The durations of the events ranged from 0.5 to 14.25 h. The rainfall intensity was generally important at the beginning of the precipitations and decreased along the event. The peak flows ranged from 1.06 to 56.7 $m^3 \cdot s^{-1}$ at the outlet of Ndiba catchment. The runoff coefficients ranged from 0.01 to 0.58. Three events had a maximal discharge that was higher than 50 $m^3 \cdot s^{-1}$: these events were the first floods occurring at the beginning of the rainy season, and had the highest runoff coefficient (cf. Table 1).

2.3. Initial Soil Moisture Predictor

Event-based models predictions often depend on the initial moisture condition of the catchment [13], which should be assessed by measurements, or, in our case, reliable predictors. The base flow at the beginning of an event can be used as a predictor of initial moisture conditions in humid climates, but not in most of Sahelian small catchments because there is no flow before the floods. Other predictors are commonly used to define soil moisture conditions, such as the Antecedent Precipitation Index (API), which is given by [21]:

$$API_j = k \cdot API_{j-1} + P_j \tag{1}$$

where P_j is the rainfall occurring the day j and API_{j-1} the index of the previous day. It is multiplied by k, which is a factor that has to be calibrated. The API is computed from daily measurements that were available at the rain gauges.

2.4. Normalized Difference Vegetation Index

For cultivated catchments, the growth of the vegetation often induces a non-stationarity of the hydrological behavior of the catchment, and consequently, of the model parameters [22,23]. Such variability requires to be related to a reliable descriptor of the growth of the vegetation. The Normalized Differential Vegetation Index (NDVI) was used here as a proxy to monitor the seasonal evolution of vegetation, as shown by [11]. The NDVI daily time series have been retrieved

from the NASA's land long term data record (LTDR) project (available online at: https://ltdr.modaps.
eosdis.nasa.gov), which aims to produce a global set of data at a 0.05° spatial resolution, which were
collected from AVHRR and MODIS instruments for climate studies [24]. The algorithms used to derive
the Normalized Differential Vegetation Index (NDVI) from daily surface reflectance are described
in [25,26]. The whole database processing is detailed in [24].

The temporal evolution for the years 1983 to 1988 of NDVI over the Ndiba catchment indicates
a very close annual cycle with the maximum vegetation development during the summer (Figure 2).

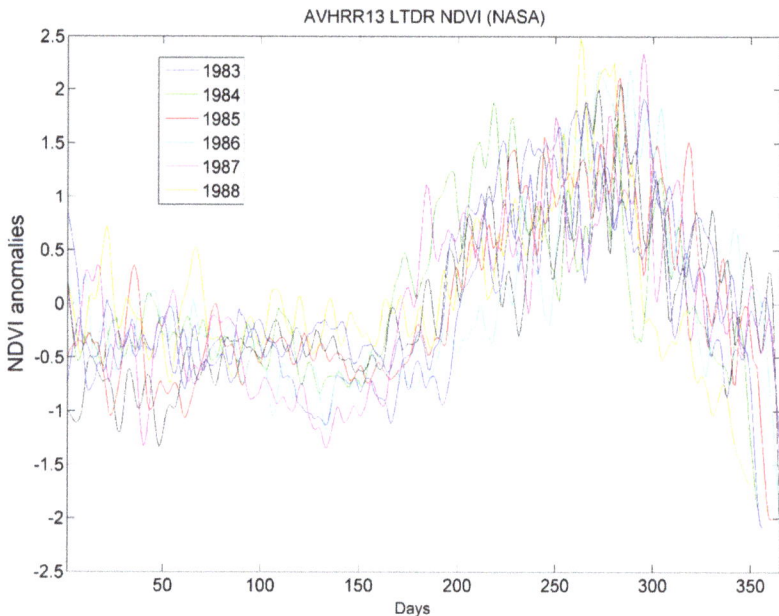

Figure 2. Seasonal cycle of long term data record (LTDR)-normalized difference vegetation index
(NDVI) over the Ndiba catchment.

3. Rainfall-Runoff Modelling

The models were performed within the Atelier Hydrologique Spatialisé (ATHYS) modelling
platform, which was developed at the Hydrosciences Montpellier laboratory [27,28]. ATHYS brings
together a large set of distributed models within a consistent and easy-to-use environment, including
processing of hydrometeorological and geographical data. ATHYS is an open software, which can be
downloaded for Windows or Linux (www.athys-soft.org).

The distributed models operated over a grid mesh of regular squared cells. The basic information
was brought by a Digital Elevation Model (DEM), which supplied elevation, slope, upstream area,
and flowpath for each cell. Here, we used the Advanced Spaceborne Thermal Emission and Reflection
Parameter (ASTER) DEM, projected to a UTM horizontal resolution of 30 m. Then, the rainfall was
interpolated over the grid cells via the inverse distance interpolation method, at every time step,
here 5 min. The runoff models (here, SCS and Green-Ampt) operated thus with 30 m cell size and
5 min time step. Each cell produced a cell-hydrograph at the outlet of the catchment, depending on the
runoff model and the routing model (here, the distributed lag and route model). The complete flood
hydrograph at the outlet of the catchment was finally obtained as the sum of the cell-hydrographs.

3.1. SCS Runoff Model

The United States Soil Conservation Service developed an empirical model to estimate runoff losses [29]. To date, the SCS method is one of the most popular runoff models, and it was the object of many improvements as well in the formulation of the model as in the interpretation of its parameters (see [30] for a review).

Although it was first designed to relate the cumulated runoff and rainfall at the event scale, it is possible to integrate time into this model to predict infiltration rates [31,32]. The SCS model that is considered here gives the instantaneous runoff at any time of the event [32]:

$$P_e(t) = P_b(t)\left(\frac{P(t) - 0.2S}{P(t) + 0.8S}\right)\left(2 - \frac{P(t) - 0.2S}{P(t) + 0.8S}\right) \tag{2}$$

where $P_e(t)$ is the effective rainfall at time t, $P_b(t)$ the precipitation at time t, $P(t)$ the cumulative rainfall since the beginning of the event. S is the maximal water storage capacity at the beginning of the event. The runoff coefficient is expressed by the quantity that multiplies $P_b(t)$, in the second member of the previous equation; it increases with the cumulative rainfall and tends to 1 when the cumulative rainfall tends to infinity.

To consider that the runoff coefficient decreases during periods when no rain occurs, a linear decrease of the cumulated rainfall was considered through a discharge coefficient ds $[T^{-1}]$ [13,33]:

$$\frac{dP(t)}{d(t)} = P_b(t) - ds \cdot P(t) \tag{3}$$

The parameters of the model are S and ds. The S parameter directly controls the main flood peak and volume, while ds contributes to adjust the different flood peaks and volumes resulting from successive rainfalls within a given event. Note that the ds parameter emulates some kind of variable infiltration rate, which depends on the cumulated rainfall $P(t)$. The S and ds parameters were considered to be constant over the catchment, so that the distribution only concerned the rainfall interpolated for each cell.

3.2. Green-Ampt Model

Green-Ampt model [34] is based on physical measurable parameters and describes a hortonian process of water infiltration. The infiltration capacity $f(t)$ is expressed by the following equation:

$$f(t) = Ks \cdot \left(\frac{(\theta_s - \theta_i) \cdot \Psi}{F(t)} + 1\right) \tag{4}$$

where $F(t)$ is the cumulated infiltration [L], θ_s the saturated soil moisture $[L^3 \cdot L^{-3}]$, θ_i the initial soil moisture $[L^3 \cdot L^{-3}]$, K_s the hydraulic conductivity at saturation $[L \cdot T^{-1}]$ and Ψ the suction [L].

As for the SCS runoff model, the GA parameters were considered to be constant over the catchment, although the land use and the soil types were probably not homogeneous over the catchment. The parameters were thus considered as averaged values over the catchment, in order to be coherent with sparse data about the soil properties.

3.3. Routing Model

SCS and Green-Ampt models were coupled to a lag and route model at the cell scale, to produce a cell-hydrograph at the outlet of the catchment, calculated by:

$$Q_m(t) = A \cdot \int_{t_0}^{t - T_m} \frac{P_e(\tau)}{K_m} \cdot exp\left(-\frac{t - T_m - \tau}{K_m}\right) d\tau \tag{5}$$

where A was the cell-area [L²], P_e the effective rainfall [L], T_m the routing time [T]), and K_m the lag time [T] from the cell m to the outlet.

This model, available in ATHYS, was already used by [33]. It simply enables the Unit Hydrograph theory, by using the impulse response function:

$$h(t) = 0 \quad \text{if } t < T_m$$

$$h(t) = \frac{1}{K_m} \cdot exp\left(\frac{t - T_m}{K_m}\right) \quad \text{if } t > T_m$$

The routing time T_m was calculated from the length l_k and the flux velocity V_k of each k-cell between the m-cell and the outlet of the catchment. The lag time K_m was assumed to be linearly dependent of the routing time T_m: $K_m = K_0 \cdot T_m$. In the following, the velocity V_k was considered as the same for all of the cells: $V_k = V_0$, and K_0 as a constant of the catchment.

The routing time T_m and lag time K_m induced an actual distribution of the travel times of the runoff over the catchment: the farther from the outlet the cell m is, the larger the routing time and the lag time are. As mentioned above, the complete flood hydrograph at the outlet of the catchment was obtained by the sum of all the cell-hydrographs.

The complete models were denoted after either SCS-LR or SCS, and GA-LR or Green-Ampt.

3.4. Model Calibration

SCS-LR and GA-LR models parameters were either preset from field measurements or empirical/physical consideration, or calibrated using the BLUE (Best Linear Unbiased Estimator) method. In this latter case, the objective consists of minimizing the cost function, J(x) in order to obtain the optimal values of the variables that are stored in a control vector, x [35]:

$$J(\mathbf{x}) = \frac{1}{2}(\mathbf{x} - \mathbf{x}^b)^T \mathbf{B}^{-1}(\mathbf{x} - \mathbf{x}^b) + \frac{1}{2}(\mathbf{y}^o - H(\mathbf{x}))^T \mathbf{R}^{-1}(\mathbf{y}^o - H(\mathbf{x})) \tag{6}$$

The control vector **x**, size n, contains the set of n variables to be optimized, i.e., the model parameters; $\mathbf{x}^b$ is the background vector of size n, which contains the a priori values of the control vector variables. $\mathbf{y}^o$ is a vector of size p, which contains the p observations to be considered. H is the observation operator, i.e., the rainfall-runoff model, and supplies the predicted runoff. **B** (size n × n) and **R** (size p × p) are the error covariance matrices, respectively, associated to the background, $\mathbf{x}^b$, and the observations, $\mathbf{y}^o$. The cost function J(**x**) thus quantifies both the distance between the control vector, **x**, and the background, $\mathbf{x}^b$, and the distance between the runoff predicted by the model and the observed runoff, $\mathbf{y}^o$, weighted, respectively, by the error covariance matrices **B** and **R**. The smaller the covariance in **B** is, the closer to the background the control vector is; the smaller the covariance in **R** is, the closer to the observation the output of the model is.

As a gradient method, the BLUE method allows for a quick computation of the optimized parameters. The parameters were in our case considered as independent (null covariance) and a large variance was considered in the **B** matrix, in order that the parameters would be low related with the background. This made that the cost function J(**x**) was actually equivalent to any form of quadratic error between the predicted and observed runoff.

For sake of simplicity, the efficiency of the model was also expressed by the Nash-Sutcliffe coefficient NS:

$$NS = 1 - \frac{\sum_i (X_i - Y_i)}{\sum_i (Y_i - \overline{Y})} \tag{7}$$

where X_i and Y_i are the observed and simulated discharges for i-time steps, and $\overline{Y}$ the mean value of the observed discharges during the event. The closer NS is from 1, the better the correlation between the observed and simulated discharges.

4. Results

4.1. SCS-Model Calibration

4.1.1. Estimation of the SCS-Model Parameters

As said above, the SCS-LR parameters were either preset from field measurements or empirical/methodological consideration, or calibrated by using the BLUE method. The *ds* parameter was set in order to remove the small floods due to secondary rainfalls, which were not seen in the observed data. A convenient value $ds = 8$ d^{-1} was kept as a constant for all of the events. The parameters V_0 and K_0 were found to be strongly dependent, so the K_0 parameter was empirically set as $K_0 = 0.75$ for all of the events.

Then, a simultaneous calibration of S and V_0 parameters was realized by using the BLUE method for each event. For the 28 selected events, *NS* values ranged from 0.30 to 0.99 (median = 0.87), showing for most events a good agreement between observed and calibrated discharges. *S* parameter values ranged from 15.2 to 175.2 mm (median = 92 mm) and V_0 from 0.85 m·s^{-1} to 5.8 m·s^{-1} (median = 1.68 m·s^{-1}), Table 2.

Table 2. Models results.

Event	SCS-LR			GA-LR		
	S (mm)	V_0 (m·s^{-1})	*NS*	K_s (mm·h^{-1})	V_0 (m·s^{-1})	*NS*
1	134.1	1.37	0.59	54.2	1.21	0.54
2	104.9	1.98	0.82	38.1	1.71	0.71
3	161.9	4.06	0.83	47.1	3.02	0.45
4	16.8	5.54	0.67	1.8	3.96	0.95
5	56.7	0.86	0.65	6.9	0.42	0.56
6	92.0	1.14	0.92	35.1	1.05	0.94
7	81.4	1.34	0.97	39.5	1.05	0.97
8	136.3	1.19	0.82	30.0	0.87	−2.19
9	159.5	1.70	0.95	64.3	1.58	0.93
10	122.4	2.53	0.96	31.0	2.21	0.9
11	70.2	2.16	0.95	15.7	1.68	0.94
12	136.6	0.94	0.87	78.1	1.03	0.7
13	89.4	1.32	0.7	28.8	1.16	0.48
14	116.8	1.22	0.68	28.9	1.32	0.71
15	145.4	0.90	0.83	53.9	0.77	0.7
16	153.7	14.28	0.77	43.1	4.96	0.88
17	172.9	2.29	0.98	32.5	1.36	0.93
18	159.4	1.88	0.93	48.7	1.62	0.91
19	92.0	1.73	0.84	22.9	1.14	0.89
20	89.9	0.95	0.91	28.8	0.93	0.86
21	62.8	3.70	0.88	6.9	3.43	0.85
22	3.7	1.62	0.61	3.3	1.61	0.9
23	34.8	1.88	0.57	6.6	2.01	0.6
24	73.9	1.44	0.86	35.0	1.17	0.87
25	89.1	2.37	0.97	25.4	0.97	0.83
26	103.6	2.20	0.98	56.5	2.10	0.98
27	74.3	1.37	0.94	28.9	1.20	0.91
28	86.2	1.65	0.94	35.2	1.52	0.93
Median	92.0	1.68	0.87	31.8	1.34	0.88
Average	100.7	2.34	0.84	33.1	1.68	0.70

No correlation could be found between the velocity V_0 and other event feature, such as peak flow and initial water content. The variation of the velocity might be mainly due to uncertainties in time of the recording mechanical gauges. Such uncertainty can reach some tenth of minutes, and due to the short response time of the catchment, can impact the V_0 value a lot.

4.1.2. Relation between S and Antecedent Moisture Indicator

The calibrated values of S for SCS-model for all of the flood events were compared with the previous soil moisture indicator described before (API). For this model, the relation S-API was significant, although weak ($R^2 = 0.38$, $P_{Fisher} < 1\%$), but exhibited an opposite correlation to what could be expected (Figure 3). The maximal capacity of retention was indeed supposed to decrease with increased wetness conditions. For example, in a small mountainous Mediterranean catchment, [13] found a negative correlation between S and API, with a Pearson correlation coefficient r = −0.68). But, in the Ndiba catchment, the wetter the soil initially was, the lesser the runoff was. It was probably due because the classical role of the water content before the flood should be here a secondary factor, and that other phenomena would play a major role in the hydrological response of the catchment. It was clear that the tillage of the soil indeed modified a lot the runoff: for example, the highest runoff, consequently, a small S, was obtained for the first floods of the rainy season, consequently, a low API, when no tillage has been made. But, as we had no data about the tillage practices during the rainy season, it was not possible to precisely quantify the role of the tillage on the runoff.

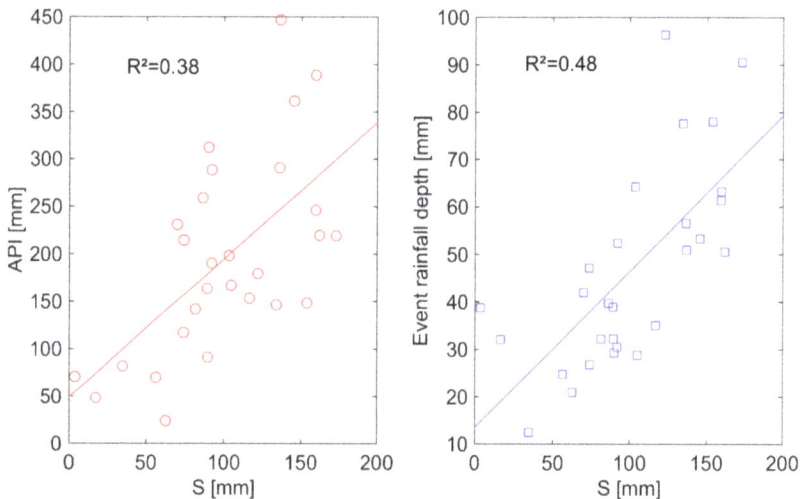

Figure 3. Correlation between S, Antecedent Precipitation Index (API), and rainfall depth.

4.1.3. Relation between S and NDVI

As pointed out by [11], the Sahelian agricultural catchments are prone to highly non stationary conditions due to the tillage of the soils and the growth of the plants. The calibrated values of S were then related to the NDVI index, which was associated to the development of the vegetation. A significant, but weak ($R^2 = 0.29$, $P_{Fisher} < 1\%$), correlation was found again, which denoted the role of the vegetation in the increase of the maximal water storage capacity of the soil when the plants grow. Note that API and NDVI were dependent, since a linear relationship could be drawn between both indexes, with $R^2 = 0.47$, when considering the values that are corresponding to our 28 events data base. Thus, the positive correlation between the calibrated S values and API would be an artefact due to the development of the vegetation.

4.1.4. Relation between S and Rainfall Depth

The calibrated S parameter of SCS-model was also correlated with the rainfall depth. As shown in Figure 3, S tended to increase with rainfall depth giving almost a linear relation between S and rainfall depth ($R^2 = 0.48$, $P_{Fisher} < 1\%$). The relation between rainfall depth and the S parameter was not expected whilst using the SCS-model; but, some studies (among other [36,37]) showed a relation between the *CN* curve number (consequently *S*) and the rainfall depth, for high infiltration sandy soils: *CN* decreases (consequently *S* increases) with rainfall depth. A possible explanation was given by [38], who claimed that higher rainfall depths increase the flooded part of the soil, and that water rises up to the plants bottom and infiltrates more than in the crusted soil.

4.2. Green-Ampt Model Calibration

An alternative model set-up was considered, with the Green-Ampt model for infiltration. As for the SCS model, the Green-Ampt parameters were either preset or calibrated by the BLUE method. A mean constant value of the soil moisture at saturation state was applied, $\theta_s = 0.30$ cm^3 cm^{-3}, according the field measurements [18]. The initial soil moisture value at the beginning of an event was considered empirically as a linear function of the API index: $\theta_i = \text{API}/2000$; the initial water content thus ranged between 0.02 and 0.22 cm^3·cm^{-3}. A suction value of 75 mm was applied, which corresponds to a mean value found in literature for the sandy loam soils [39]. The K_0 parameter was set as $K_0 = 0.75$, the same value than for SCS-LR.

Then, simultaneous calibration of hydraulic conductivity K_s and velocity V_0 was performed by using the BLUE method. The median *NS* coefficient was 0.88. The median calibrated K_s value was 31.8 mm·h^{-1}, which looked in agreement with the values measured in the catchment [19]. However, the calibrated K_s values varied widely between 1 and more than 60 mm·h^{-1} at the event scale. The median calibrated V_0 was 1.34 m·s^{-1}.

4.3. Seasonal Evolution of the K_s Parameter Linked with NDVI Anomalies

The calibrated values of the K_s parameter were correlated with the NDVI anomalies ($R^2 = 0.44$, $P_{Fisher} < 1\%$, Figure 4). This result indicates that the variation that was observed through the year in hydraulic conductivity was partly driven by the evolution of vegetation growth during the rainy season, as pointed out by [11]. This hydrologic behaviour can thus be reproduced by the Green Ampt model with varying infiltration rates during the wet season, according to the NDVI anomalies. Note that the Green-Ampt model scored better than the SCS model, when considering the relationship with NDVI and the main variable parameter of each model: *S* for SCS and K_s for Green-Ampt.

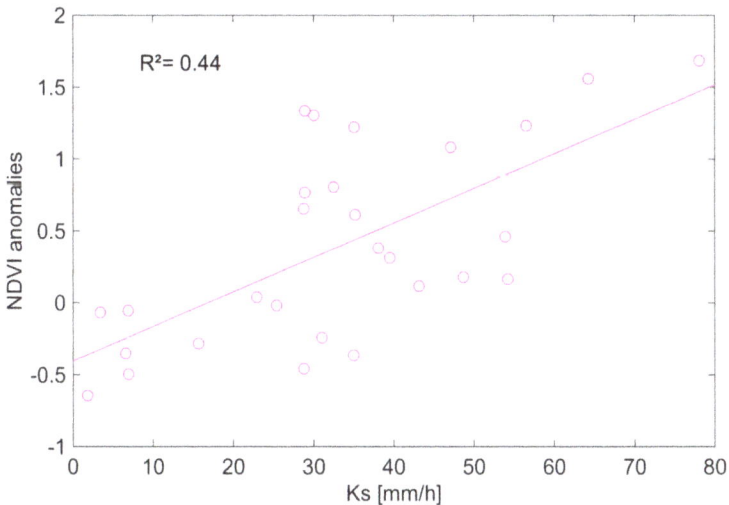

Figure 4. Correlation between hydraulic conductivities (K_s) and NDVI.

4.4. Predictive Scores of the Models

The actual goodness of the models was expressed by the predictive scores, i.e., the NS values that were obtained with the predicted values of S (for SCS) or K_s (for Green-Ampt). The NS values of the 28 events were sorted by descendant order for each model, using the different relationships that were considered as significant (Figure 5): S-NDVI and S-P for SCS, K_s-NDVI for GA. The best predictive scores were achieved for SCS when using the S-P relationship (median $NS = 0.33$), whereas Green-Ampt when using the K_s-NDVI relationship offered the worst predictive score (media $NS = -0.21$); SCS scored better when using NDVI (median $NS = 0.08$), although the R^2 was higher for K_s-NDVI than for S-NDVI. This result is due to the fact that GA is very sensitive to K_s, whereas the sensitivity of SCS with S is less. So, SCS globally scored better than Green-Ampt, in terms of predictive scores. However, using a predictor of S, like the rainfall amount, is not appropriate for applications, such as flood forecasting, because the rainfall amount is not known at the moment of the forecast. In this case, it would be preferable to use SCS with the NDVI predictor for S.

The predictive scores of the model, however, remained rather low. It could be because the calibrations of the models were not optimal: maybe it would be worth to consider that some other parameters had to vary from an event to another (e.g., ds or Ψ); or, because the models were not enough appropriate for this kind of catchment; but we think that the low predictive scores were mostly due because we had not information about the dates and the kind of tillage. Ndiaye et al. [19] showed indeed that tillage had a significant effect on runoff, depending on the type (along or across the slope), and that the effect of the tillage disappeared after nearly 170 mm of cumulated rainfall since the date of tillage. Although vegetation growth should be clearly conditioned by tillage, the NDVI index was probably not efficient enough to restore accurately the tillage practices during the rainy season. We suggest first searching in this direction for improving the models.

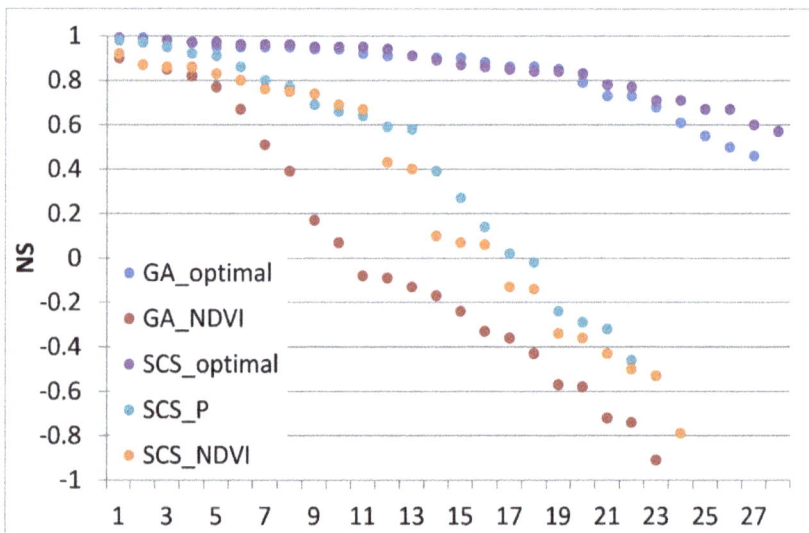

Figure 5. Optimal and predictive scores of the models. GA_optimal corresponds to the calibrated values of K_s; Green-Ampt (GA)_NDVI to the K_s predicted by the relationship K_s-NDVI; Soil Conservation Service (SCS)_optimal to the calibrated values of S; SCS_NDVI to the S predicted by the relationship S-NDVI; SCS_P to the S predicted by the cumulated rainfall of the event. For a given model, the same values of V_0 were applied, and corresponded to the calibrated values of V_0 for SCS or Green-Ampt.

5. Conclusions

The goal of this study was to evaluate the efficiency of event-based rainfall-runoff model, such as SCS-LR or GA-LR for flood prediction in a small catchment that is located in south Senegal. Twenty-eight flood events were chosen to compare the efficiency of both event-based models.

The application of the SCS-model was found satisfactory to reproduce the observed discharges after calibration of its parameters. However, the initial condition of the model could not be predicted as usual, in relation with the water content of the soil at the beginning of the event. The API index, derived from the antecedent rainfalls, was positively correlated with the S parameter, whereas the SCS model usually considers that the correlation should be negative. The S values were also correlated to the NDVI during the event. That is in agreement with the role of the growth of the plants, which induces highly non-stationary conditions for the hydrological response of the catchment, as noted by various authors, and the initial soil moisture conditions should be a secondary factor. The S values were also related to the cumulated rainfall of the event, which finally appeared as the best predictor for S.

The Green-Ampt model results indicated its adequacy to reproduce the flood events in the catchment. As for the SCS-CN model, the correlation of the model parameters with antecedent precipitation was positive, whereas it was expected to be negative. However, a significant correlation between the hydraulic conductivities (K_s) for each event and the normalized difference vegetation index was observed, highlighting the influence of vegetation cover on soil infiltration properties. The Green-Ampt parameters were in agreement with the measured, e.g., hydraulic conductivity at saturation, or estimated values. But, Green-Ampt finally scored less than SCS, because a high sensitivity to the K_s parameter.

At this point, however, the predictive scores of the models are still low when considering the NS values that were obtained with the best predicted values of S or K_s. The main reason seems to be that the tillage practices are not actually taken into account in the models, nor in the predictors (e.g., NDVI) of the parameters of the models. This seems to be the main point to study for the improvement of the predictive scores of the models. In addition, added value could be brought by considering the whole nested set of catchments in Thysse Kaymor.

To conclude, it is worth noting the interest of such existing large data sets dating from the 80's, even before, Thysse Kaymor being an example among other. As these data are still underemployed for modeling benchmark and hydrological processes understanding, they could fill a gap concerning flood predicting in West Africa (e.g., T-years return period flood), and bring more confidence in designing small hydraulic works in small (ponding areas, culverts) as well as large (dams, bridges) catchments.

Acknowledgments: We thank the editors M. Piña and M. Li, and three anonymous reviewers for their valuable help in improving the quality of the paper. We are also very grateful to Nathalie Rouché and Jean-Emmanuel Paturel for providing the rainfall-runoff data used in this study.

Author Contributions: Christophe Bouvier and Yves Tramblay conceived and designed the modelling protocol and contributed to the materials; Christophe Bouvier, Lamia Bouchenaki and Yves Tramblay analyzed the data; Christophe Bouvier and Lamia Bouchenaki wrote the paper.

Conflicts of Interest: The authors declare no conflict of interest.

References

1. Di Baldassarre, G.; Montanari, A.; Lins, H.F.; Koutsoyiannis, D.; Brandimarte, L.; Blöschl, G. Flood fatalities in Africa: From diagnosis to mitigation. *Geophys. Res. Lett.* **2010**, *37*, L22402. [CrossRef]
2. Tschakert, P.; Sagoe, R.; Ofori-Darko, G.; Codjoe, S.N. Floods in the Sahel: An analysis of anomalies, memory, and anticipatory learning. *Clim. Chang.* **2010**, *103*, 471–502. [CrossRef]
3. Amoussou, E.; Tramblay, Y.; Totin, H.; Mahé, G.; Camberlin, P. Dynamics and modelling of floods in the river basin of Mono in Nangbeto, Togo/Benin. *Hydrol. Sci. J.* **2014**, *59*, 2060–2071. [CrossRef]
4. Komi, K.; Neal, J.; Trigg, M.A.; Diekkrüger, B. Modelling of flood hazard extent in data sparse areas: A case study of the Oti River basin, West Africa. *J. Hydrol. Reg. Stud.* **2017**, *10*, 122–132. [CrossRef]

5.	Cappelaere, B.; Vieux, B.E.; Peugeot, C.; Maia, A.; Séguis, L. Hydrologic process simulation of a semiarid, endoreic catchment in Sahelian West Niger: 2. Model calibration and uncertainty characterization. *J. Hydrol.* **2003**, *279*, 244–261. [CrossRef]

6.	Le Lay, M.; Saulnier, G.M.; Galle, S.; Séguis, L.; Metadier, M.; Peugeot, C. Model representation of the Sudanian hydrological processes: Application on the Donga catchment (Benin). *J. Hydrol.* **2008**, *363*, 32–41. [CrossRef]

7.	FAO. *Crues et Apports. Manuel Pour L'estimation des Crues Décennales et des Apports Annuels Pour les Petits Bassins Versants non Jaugés de L'afrique Sahélienne et Tropicale Sèche*; Bulletin FAO D'irrigation et de Drainage: Rome, Italy, 1996; Volume 54, p. 265. ISBN 92-5-203874-4. (In French)

8.	Albergel, J.; Chevalier, P.; Lortic, B. D'Oursi à Gagara: Transposition d'un modèle de ruissellement dans le Sahel (Burkina-Faso). *Hydrol. Cont.* **1987**, *2*, 77–86. (In French)

9.	Bouvier, C.; Desbordes, M. Un modèle de ruissellement pour les villes de l'Afrique de l'Ouest. *Hydrol. Cont.* **1990**, *5*, 77–86. (In French)

10.	Lamachere, J.M.; Puech, C. Cartographie des états de surface par télédétection et prédétermination des crues des petits bassins versants en zones sahélienne et tropicale sèche. In *L'hydrologie Tropicale: Géoscience et Outil Pour le Développement*; International Association of Hydrological Sciences: Wallingford, UK, 1996; p. 238. (In French)

11.	Séguis, L.; Bader, J.C. Modélisation du ruissellement en relation avec l'évolution saisonnière de la végétation (mil, arachide, jachère) au centre Sénégal. *Revue Sci. L'eau* **1997**, *4*, 419–438. (In French) [CrossRef]

12.	Bader, J.C. Modèle analogique de ruissellement à stockage de surface: Test sur parcelles et extrapolation sur versant homogène. *Hydrol. Sci. J.* **1994**, *39*, 569–592. (In French) [CrossRef]

13.	Tramblay, Y.; Bouvier, C.; Martin, C.; Didon-Lescot, J.F.; Todorovik, D.; Domergue, J.M. Assessment of initial soil moisture conditions for event-based rainfall–runoff modelling. *J. Hydrol.* **2010**, *387*, 176–187. [CrossRef]

14.	Berthet, L.; Andreassian, V.; Perrin, C.; Javelle, P. How crucial is it to account for the antecedent moisture conditions in flood forecasting? Comparison of event-based and continuous approaches on 178 catchments. *Hydrol. Earth Syst. Sci.* **2009**, *13*, 819–831. [CrossRef]

15.	Tramblay, Y.; Amoussou, E.; Dorigo, W.; Mahé, G. Flood risk under future climate in data sparse regions: Linking extreme value models and flood generating processes. *J. Hydrol.* **2014**, *519*, 549–558. [CrossRef]

16.	Valentin, C. *Les Etats de Surface des Bassins Versants de Thysse Kaymor (Sénégal)*; Rapport ORSTOM 35.489; Office de la Recherche Scientifique Et Technique Outre-Mer (ORSTOM): Dakar, Senegal, 1990. (In French)

17.	Diome, F. Rôle de la Structure du sol Dans son Fonctionnement Hydrique. Sa Quantification par la Courbe de Retrait. Ph.D. Thesis, Université Cheikh Anta Diop, Dakar, Senegal, 1996; p. 131. (In French)

18.	Albergel, J.; Bernard, A.; Ruelle, P.; Touma, J. Hydrodynamique des Sols. Bassins Expérimentaux de Thysse Kaymor. Rapport de la Campagne de Mesures Fev–Avr 1988. 1989. Available online: http://horizon.documentation.ird.fr/exl-doc/pleins_textes/doc34-05/27469.pdf (accessed on 30 March 2018).

19.	Ndiaye, B.; Esteves, M.; Vandervaere, J.P.; Lapetite, J.M.; Vauclin, M. Effect of rainfall and tillage direction on the evolution of surface crusts, soil hydraulic properties and runoff generation for a sandy loam soil. *J. Hydrol.* **2005**, *307*, 294–311. [CrossRef]

20.	Rodier, J.A. *Caractéristiques des Crues des Petits Bassins Versants Représentatifs au Sahel*; Cahiers ORSTOM, Série Hydrologie; Office De La Recherche Scientifique Et Technique Outre-Mer (ORSTOM): Dakar, Senegal, 1985; Volume XXI, pp. 3–26. (In French)

21.	Kohler, M.A.; Linsley, R.K. *Predicting Runoff from Storm Rainfall*; Res. Paper 34; U.S. Weather Bureau: Washington, DC, USA, 1951.

22.	Zhang, L.; Wang, J.; Bai, Z.; Lv, C. Effects of vegetation on runoff and soil erosion on reclaimed land in an opencast coal-mine dump in a loess area. *Catena* **2015**, *128*, 44–53. [CrossRef]

23.	Hunink, J.E.; Eekhout, J.P.C.; de Vente, J.; Contreras, S.; Droogers, P.; Baille, A. Hydrological Modelling Using Satellite-Based Crop Coefficients: A Comparison of Methods at the Basin Scale. *Remote Sens.* **2017**, *9*, 174. [CrossRef]

24.	Pedelty, J.; Devadiga, S.; Masuoka, E.; Brown, M.; Pinzon, J.; Tucker, C.; Vermote, E.; Prince, S.; Nagol, J.; Justice, C.; et al. Generating a Long-term Land Data Record from the AVHRR and MODIS Instruments. In Proceedings of the IEEE International Geoscience and Remote Sensing Symposium (IGARSS 2007), Barcelona, Spain, 23–28 July 2007; Institute of Electrical and Electronics Engineers: New York, NY, USA, 2007; pp. 1021–1025.

25. Vermote, E.F.; Kaufman, Y.J. Absolute calibration of AVHRR visible and near-infrared channels using ocean and cloud views. *Int. J. Remote Sens.* **1995**, *16*, 2317–2340. [CrossRef]

26. Saleous, N.Z.; Vermote, E.F.; Justice, C.O.; Townshend, J.R.G.; Tucket, C.J.; Goward, S.N. Improvements in the global biospheric record from the Advanced Very High Resolution Radiometer (AVHRR). *Int. J. Remote Sens.* **2000**, *21*, 1251–1277. [CrossRef]

27. Bouvier, C.; Delclaux, F. ATHYS: A hydrological environment for spatial modelling and coupling with a GIS. In Proceedings of the HydroGIS 96, Vienna, Austria, 16–19 April 1996; AIHS Publication: Wallingford, UK, 1996; pp. 19–28.

28. Bouvier, C.; Crespy, A.; L'Aour-Dufour, A.; Crès, F.-N.; Delclaux, F.; Marchandise, A. Distributed Hydrological Modelling—The ATHYS platform. In *Environmental Hydraulics Series 5, Modelling Software*; Tanguy, J.-M., Ed.; Wiley: New York, NY, USA, 2010; pp. 83–100.

29. USDA, Soil Conservation Service. *National Engineering Handbook*; Supplement A, Section 4, Hydrology; Soil Conservation Service: Washington, DC, USA, 1956.

30. Ponce, V.; Hawkins, R. Runoff Curve Number: Has It Reached Maturity? *J. Hydrol. Eng.* **1996**, *1*, 11–19. [CrossRef]

31. Aron, G.; Miller, A.C.; Lakatos, D.F. Infiltration formula based on SCS Curve Number. *J. Irrig. Drain. Div.* **1977**, *103*, 419–428.

32. Gaume, E.; Livet, M.; Desbordes, M.; Villeneuve, J.P. Hydrological analysis of the river Aude, France, flash flood on 12 and 13 November 1999. *J. Hydrol.* **2004**, *286*, 135–154. [CrossRef]

33. Coustau, M.; Bouvier, C.; Borrell-Estupina, V.; Jourde, H. Flood modelling with a distributed event-based parsimonious rainfall-runoff model: Case of the karstic Lez river catchment Nat. *Hazards Earth Syst. Sci.* **2012**, *12*, 1119–1133. [CrossRef]

34. Green, W.H.; Ampt, G.A. Studies on Soil Physics. *J. Agric. Sci.* **1911**, *4*. [CrossRef]

35. Coustau, M.; Ricci, S.; Borrell-Estupina, V.; Bouvier, C.; Thual, O. Benefits and limitations of data assimilation for discharge forecasting using an event-based rainfall–runoff model. *Nat. Hazards Earth Syst. Sci.* **2013**, *13*, 583–596. [CrossRef]

36. Hawkins, R.H. Asymptotic Determination of Runoff Curve Numbers from Data. *J. Irrig. Drain. Eng.* **1993**, *119*, 334–345. [CrossRef]

37. Rezaei-Sadr, H. Influence of coarse soils with high hydraulic conductivity on the applicability of the SCS-CN method. *Hydrol. Sci. J.* **2017**, *62*, 843–848. [CrossRef]

38. Planchon, O.; Janeau, J.L. Le Fonctionnement Hydrodynamique à L'échelle du Versant. In *Equipe HYPERBAV Structure et Fonctionnement Hydropédologique d'un Petit Bassin Versant de Savane Humide*; Etudes et Theses; Office de la Recherche Scientifique et Technique Outre-Mer (ORSTOM): Paris, France, 1990; pp. 165–183.

39. Chow, V.T.; Maidment, D.R.; Mays, L.W. *Applied Hydrology*; McGraw Hill: New York, NY, USA, 1988; p. 572.

geosciences

MDPI

Article

Correlation Analysis of Seasonal Temperature and Precipitation in a Region of Southern Italy

Ennio Ferrari [1],*, Roberto Coscarelli [2] and Beniamino Sirangelo [3]

[1]   Department of Computer Engineering, Modeling, Electronics, and Systems Science (DIMES), University of Calabria, 87036 Rende (CS), Italy
[2]   Research Institute for Geo-hydrological Protection (CNR-IRPI), National Research Council of Italy, 87036 Rende (CS), Italy; r.coscarelli@irpi.cnr.it
[3]   Department of Environmental and Chemical Engineering (DIATIC), University of Calabria, 87036 Rende (CS), Italy; beniamino.sirangelo@unical.it
*   Correspondence: ennio.ferrari@unical.it; Tel.: +39-0984-496-618

Received: 27 March 2018; Accepted: 28 April 2018; Published: 2 May 2018

Abstract: The investigation of the statistical links between changes in temperature and rainfall, though not widely achieved in the past, is an interesting issue because their physical interdependence is difficult to point out. Aiming at detecting possible trends with a pooled approach, a correlative analysis of temperature and rainfall has been carried out by comparing changes in their standardized anomalies from two different 30-year time periods. The procedure has been applied to the time series of seasonal mean temperature and cumulative rainfall observed in four sites of the Calabria region (Southern Italy), with reference to the series which verify the normality hypothesis. Specifically, the displacements of the ellipses, representing the probability density functions of the bivariate normal distribution assumed for the climatic variables, have been quantified and tested for each season, passing from the first subperiod to the following one. The main results concern a decreasing trend of both the temperature and the rainfall anomalies, predominantly in the winter and autumn seasons.

Keywords: seasonal precipitation; seasonal temperature; correlation analysis

1. Introduction

Rainfall and air temperature are among the most investigated meteorological variables in climatic trend studies, mainly due to the serious implications that their spatial and temporal changes can have on several environmental and socioeconomic aspects [1–3]. With regard to the temporal distribution of rainfall, long-term trends have been detected in several areas of the world [4,5]. In the Mediterranean Basin, an alternation of extreme rainy periods and severe droughts or water shortages has been detected [6]. Furthermore, this area is characterized by significant rainfall variability [7,8], caused by synoptic dynamics of extreme events evolving along this basin [9].

In Italy, which has a central position in the Mediterranean area, investigations of long rainfall series showed a decreasing trend, even if not always significant [10]. More detailed analyses have been carried out at smaller scales with varied behaviour: decreasing rainfall amounts in winter versus precipitation increase in the summer months. In particular, these behaviours were detected in the regions of Southern Italy, such as Campania [11], Basilicata [12], Sicily [13], and Calabria [14].

Regarding temperature, several studies evidenced the increase of the mean values of temperatures both at large [15] and local spatial scales [16]. The magnitude of trends varies according to the study area and the studied period. In addition, the increasing rates of maximum and minimum temperatures present a high variability. In the last decades, the analyses of temperature have been focused on the extreme values [17,18]. Salinger and Griffiths [19] showed that the changes in mean and extreme values are closely interconnected between each other and that low trends in average conditions can generate high variations in the extremes, especially in their frequency. Donat and Alexander [20] showed that both daytime (daily maximum) and night-time (daily minimum) temperatures have become higher over the past 60 years, but at different rates: greater for minimum than for maximum values. In Italy, which can be considered as a climate change hotspot [21], variations in the probability density functions of the minimum and maximum daily temperature anomalies were studied by Simolo et al. [22]. Caloiero et al. [23] analysed the minimum and maximum monthly temperatures of 19 stations in Calabria and detected a positive trend for spring and summer months and a marked negative one in September.

In fact, an accurate joint analysis of precipitation and temperature is more difficult to be carried out because of the possible interdependence between them [24]. Nevertheless, Rajeevan et al. [25] found that temperature and rainfall in India were positively correlated during January and May, but negatively correlated during July. Huang et al. [26] showed a negative correlation between rainfall and temperature in the Yellow River basin of China. Cong and Brady [27] applied the copula models to the rainfall and temperature data of a province of Sweden, and they evidenced negative correlations in the months from April to July and in September. Caloiero et al. [14] analysed the spatial and temporal behaviour of monthly precipitation and temperature in the Calabria region (Southern Italy), comparing the Péguy climographs [28] on three subperiods of the whole observation period (1916–2010).

In this paper, a joint analysis of temperature and rainfall has been carried out, comparing time series recorded in some gauges located in Calabria (Southern Italy) over two distinct 30-year subperiods (1951–1980 and 1981–2010). In particular, the anomalies of the seasonal values of temperature and precipitation, standardized by means of the mean values and the standard deviations of the period 1961–1990, were analysed. The series have been selected based on the normality hypothesis. The isocontour lines of the probability density function for the bivariate Gaussian distribution have been considered as ellipses centred on the vector mean of each subperiods. Finally, some statistical tests were applied for verifying the variations of these ellipses passing from one subperiod to the other, aiming at detecting joint trends of the seasonal temperature and rainfall anomalies.

2. Materials and Methods

2.1. Study Area and Data

Calabria is situated in the southern part of the Italian peninsula, with an area of 15,080 km^2 and a perimeter of about 818 km (Figure 1). This region shows high climatic contrasts, due to the geographic position and mountainous nature. Its climate is characterized by typical subtropical summers, with colder snowy winters and fresher summers in the inland zones, typical of Mediterranean areas. Its elongated shape evidences two coastal sides with dissimilar climatic features. The Ionian coast is exposed to the warm African currents, thus experiencing high temperatures with short and heavy precipitation. In contrast, the Tyrrhenian side is more influenced by Western air currents, which cause milder temperatures and orographic precipitation.

The climatic database used in this study, managed by the former Italian Hydrographic Service, concerns the monthly values of cumulated rainfall and mean temperature of some stations characterizing the different climatic conditions of the region for the period 1951–2010. Particular attention has been given to the problems arising from the low quality and inhomogeneities of the data series. Specifically, the monthly database was a part of the high-quality one presented in a previous study, which detected the inhomogeneities through a multiple application of the Craddock test [29]. In particular, four homogeneous monthly cumulated rainfall and mean temperature series were selected, whose percentages of missing data are presented in Table 1. The main statistical features of the seasonal temperature and rainfall series for the reference period 1961–1990 are shown in Table 2.

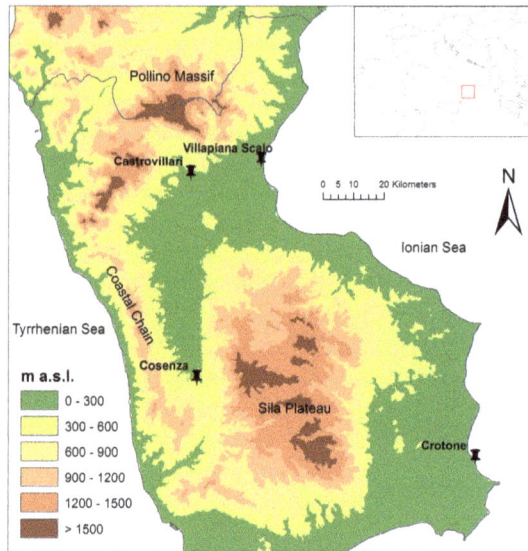

Figure 1. Location of the selected gauges in the Calabria region.

Table 1. Percentages of missing values of monthly mean temperature and cumulated rainfall.

		Missing Data (%)					
		Monthly Temperature			Monthly Rainfall		
Code	Station	1951–1980	1981–2010	Whole Period	1951–1980	1981–2010	Whole Period
930	Villapiana Scalo	0	0	0	5.8	13.6	9.7
1010	Cosenza	0	0.3	0.1	7.5	9.7	8.5
1180	Castrovillari	0	4.7	2.4	5.3	9.4	7.4
1680	Crotone	3.1	5.3	4.2	8.9	11.1	10.0

Table 2. Main statistics of temperature and rainfall for the reference period 1961–1990. ($\overline{T}$: mean daily temperature, SD_T: standard deviation of temperature, $\overline{R}$: cumulated rainfall, SD_R : standard deviation of rainfall).

Season	Station	$\overline{T}$ (°C)	SD_T (°C)	$\overline{R}$ (mm)	SD_R (mm)
Winter	930	10.3	0.8	231.9	81.3
	1010	8.6	1.1	405.4	149.0
	1180	9.2	1.2	371.6	134.7
	1680	11.3	0.8	234.9	89.8
Spring	930	14.9	0.9	125.2	69.0
	1010	14.1	1.4	210.1	74.3
	1180	14.6	1.2	178.0	71.2
	1680	15.4	1.4	117.1	60.3
Summer	930	24.6	1.2	41.4	26.5
	1010	24.0	1.1	64.4	36.4
	1180	24.8	1.3	78.2	57.3
	1680	25.3	1.1	31.9	27.7
Autumn	930	18.5	0.9	203.5	115.1
	1010	17.2	1.2	260.0	99.4
	1180	17.8	1.1	250.1	92.1
	1680	19.4	0.8	250.2	117.1

2.2. Methods

The statistical approach here used to explore the relationships between climatic data series which are not perfectly similar, such as monthly rainfall and temperature, is the correlative analysis applied to the standardized anomalies [30]. This approach also allows for the comparisons of data series of different time periods and lengths [31]. The standardized anomalies of seasonal values of temperature and rainfall were calculated for each site by using their means and standard deviations calculated for the reference period 1961–1990. In this way, the origin of the temperature versus precipitation plots shown in this study corresponds to the mean values of this reference time span, while the anomalies of the two variables extend over the four quadrants of the plot. Specifically, the peculiar climatic conditions of each quadrant are: warm and wet for the upper right quadrant, cold and wet for the upper left one, cold and dry for the lower left one, and warm and dry for the lower right quadrant.

Previous studies on the long-term cumulated rainfall in the Calabria region indicated a shift towards drier conditions around the year 1980 [32,33]. Thus, in order to search for possible temporal trends, the whole 1951–2010 time interval of the data set was fragmented into two 30-year periods: 1951–1980 and 1981–2010. The normality hypothesis was separately tested for both the variables and the 30-year periods by means of the Anderson–Darling test [34]. This is a goodness-of-fit test specially devised to give heavier weights to the distribution tails (where outliers are sometimes located) than the Kolmogorov–Smirnov test.

Only in the cases where the normality hypothesis was not rejected, the correlation analysis has been successively performed. In particular, a bivariate Gaussian distribution was applied to the seasonal variables in each site for the two 30-year periods separately. The probability density function of this distribution can be visualized as isocontour lines in the temperature–rainfall (T–R) plane with prefixed significance levels α_i. These lines are $(1-\alpha_i)\%$ confidence ellipses centered on the mean values of the variables. In this study, the significance level has been fixed as equal to 0.05, thus characterizing each correlation by means of a 95% confidence ellipse, which is oriented according to the sign of its correlation coefficient [31].

Each ellipse can be described through the vector of the means and the variance–covariance matrix, while its eigenvectors provide the directions of its major and minor axes. If the variables show changes between the two subperiods, the results are displacements of the ellipses in the T–R plane, generally formed by rigid transformations (translations and/or rotations) and deformations of the ellipses. Specifically, the translation concerns a modification in the means vector of the ellipses, and the rotation is due to variations in the correlation coefficients, while the deformation is linked to a change in the variance–covariance matrix.

All these cases can be verified through the application of specific statistical tests. In particular, the global statistical significance of the change in the means vector ($\Delta \overline{T}$, $\Delta \overline{R}$) can be assessed through the multivariate Hotelling's test [35], while the statistical significance of the change of each single value of the vector can be verified by means of the univariate t-test of difference in the mean [30].

Concerning the orientation of the ellipses, the statistical significance of the difference in the correlation coefficients passing from a subperiod to the other one can be tested by preliminarily transforming each correlation coefficient between the seasonal temperature and the rainfall anomalies, ρ, into z score through the Fisher Z-transformation [30]:

$$Z = \frac{1}{2} \ln \left(\frac{1+\rho}{1-\rho} \right). \tag{1}$$

In this way, firstly, the Z value obtained for each subperiod can be used to test the significance of its correlation coefficient, given that Z is approximately distributed as a normal law with $\mu = 0$ and $\sigma = (N - 3)^{-0.5}$, where N is the length of the anomalies series. Then, the difference in the correlation coefficients can be tested through the bivariate test statistic:

$$Z_{biv} = \frac{Z_1 - Z_2}{\sqrt{\frac{1}{N_1-3} + \frac{1}{N_2-3}}}, \tag{2}$$

which combines the Z_1 and Z_2 values evaluated for the correlation coefficients ρ_1 and ρ_2 of the two subperiods 1951–1980 and 1981–2010 with lengths N_1 and N_2, respectively. The statistical significance of the difference of the ellipses' orientation can be assessed through the statistic Z_{biv}, which is normally distributed with $\mu = 0$ and $\sigma = 1$. Quantitatively, the change can be expressed as the angle, ϑ (°), between the directions of the main axes of the ellipses in the two subperiods.

The deformations of the ellipses (variations in shape and/or size) corresponding to the two subperiods can be linked to the difference of the variances of the two variables (*F*-test, univariate case). In other terms, the *F*-test can be applied to separately consider the variances for seasonal temperature and rainfall in each subperiod. A measure of the ellipse deformation can be related to the change in the axes' length, which can be expressed as a percentage by:

$$\Delta A = 100(\Delta l_1 + \Delta l_2), \tag{3}$$

where Δl_i ($i = 1, 2$) represents the changes in the major and the minor axes of the ellipse, respectively, which are both related to the eigenvalues of the variance–covariance matrix [31]. This value represents in some way the amount of change in the total variability of the seasonal temperature and rainfall. Finally, displacements of the ellipses were visualized and quantified for both the subperiods, looking for changes in the mean values, the correlation coefficients, and the variances of the seasonal temperature and rainfall anomalies.

3. Results and Discussion

As a first step of the procedure, the normality hypothesis has been separately verified for each seasonal temperature and rainfall series for both the 30-year periods by means of the Anderson–Darling test [34]. The results obtained for each gauge and 30-year period indicate that the normality hypothesis holds for the 1010 and 930 gauges in three out of four seasons, for the 1180 gauge in two out of three seasons, and for the 1680 gauge only in one (Table 3). Regarding the seasons, the normality hypothesis is fully plausible for winter in all the sites and subperiods. In autumn, the hypothesis is acceptable for three out of four gauges, with the exception of the rainfall of gauge 1680, for both the subperiods. In spring, the normality hypothesis is unacceptable in gauge 1680 for both the variables in 1951–1980, and in gauge 1180 for rainfall in 1981–2010. Concerning the summer, all the gauges show at least one series with a behaviour which is not normally distributed. Globally, a lower number of occurrences of non-normal conditions has been detected in 1951–1980 compared to in 1981–2010.

Based on these results, the seasons and time periods for which the normality conditions have been verified were chosen for the correlation analysis. Specifically, this concerns gauges 930 and 1010 for winter, spring, and autumn, gauge 1180 for winter and autumn, and gauge 1680 for winter only. The correlation analysis procedure has been focused on the 95% confidence ellipses drawn for both the two 30-year periods (Figures 2–4).

In the winter season, decreasing values of the means for both the seasonal temperature and rainfall anomalies have been detected passing from 1951–1980 to 1981–2010 in all the selected gauges, except for a weak increase of the seasonal rainfall of gauge 1680 (Table 4). These results are clearly evidenced by the translations of the centroids of the ellipses (Figure 2a–d). The *t*-test, adopted for the verification of these changes in a separate way for $\Delta \overline{T}$ and $\Delta \overline{R}$, provides statistically significant results in six out of eight cases (Table 5). Specifically, the rigid translation of the ellipses proved to be significant for both the winter temperature and rainfall of gauges 930 (with remarkable values of -1.5 and -1.0, respectively) and 1010, while in the other two gauges, this is only verified by one variable ($\Delta \overline{R}$ of gauge 1180 and $\Delta \overline{T}$ of gauge 1680), as shown in Table 4. The statistical significance of the change in the means' vector, jointly assessed by means of the multivariate Hotelling's test, was provided for all the cases, confirming the results obtained though the *t*-test (Table 5). The rotation assumes a high significant value for gauge 930 (109.1°). The deformations are always not significant, with the highest not-significant increase of 61% observed in gauge 1680 (Table 4).

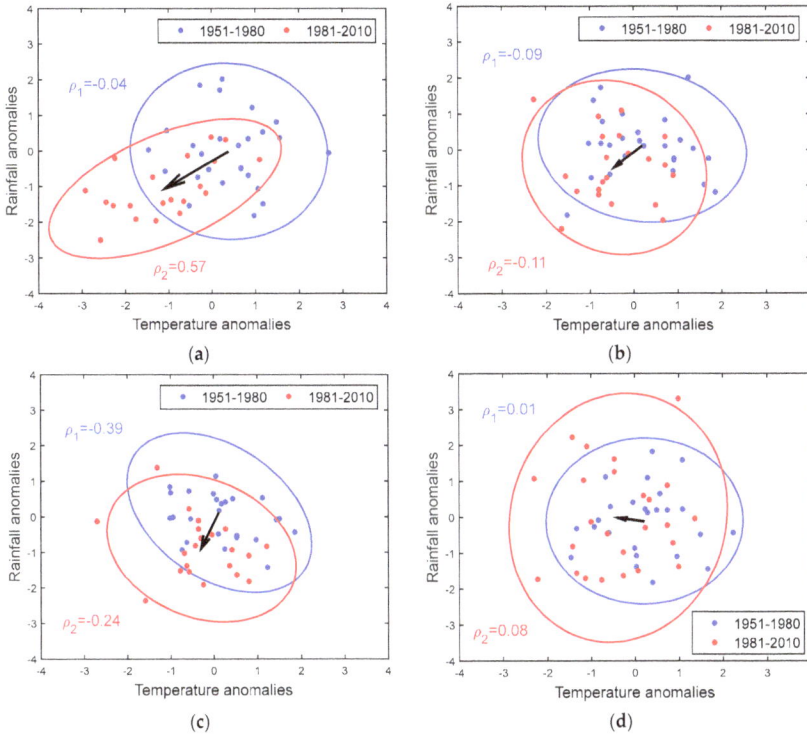

Figure 2. The 95% confidence ellipses for the winter season at gauge (**a**) 930; (**b**) 1010; (**c**) 1180; (**d**) 1680.

In spring, Figure 3a,b shows that both the stations which satisfy the normality hypothesis present decreasing tendencies of temperature and rainfall anomalies, but only gauge 930 (Figure 3a) evidences a significant clear negative tendency (−0.8) of the temperature anomalies (Table 4). The Hotelling's test is verified in spring only for gauge 930, and the *t*-test is verified for the same gauge only for the temperature anomalies. For the rotations, the only significant value has been observed at gauge 930 (69.2°), while not-significant values were detected regarding the deformations.

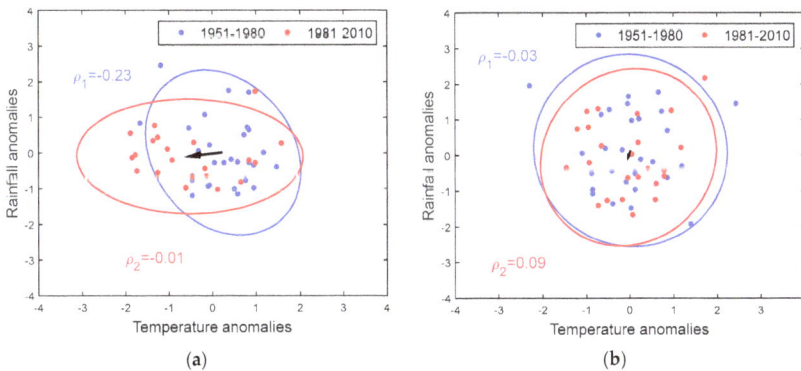

Figure 3. The 95% confidence ellipses for the spring season at gauge (**a**) 930; (**b**) 1010.

In autumn (Figure 4a–c), notable statistically significant decreasing values (Table 4) have been observed. In particular, gauge 930 (Figure 4a) evidences a more pronounced significant negative tendency for the mean temperature anomalies than for the rainfall ones (−0.8 and −0.6 for the temperature and the rainfall anomalies, respectively), as detected through the *t*-test. Gauge 1180 presents the only significant value only for the rainfall anomalies (−0.8). These results have been statistically proved also by means of the multivariate Hotelling's test (Table 5). Not-significant values were observed for the rotations, and the only significant value of deformation has been detected at gauge 930 (−59%).

Table 3. Anderson–Darling statistic for temperature and rainfall series. Critical values for different N have been assumed at the 95% confidence level. (The cases in which the normality hypothesis is rejected are in bold italics.)

Season	Gauges	1951–1980		1981–2010	
		Temperature	Rainfall	Temperature	Rainfall
Winter	930	0.229	0.249	0.189	0.473
	1010	0.474	0.325	0.425	0.196
	1180	0.472	0.707	0.392	0.224
	1680	0.249	0.279	0.279	0.443
Spring	930	0.673	0.490	0.490	0.309
	1010	0.349	0.477	0.257	0.321
	1180	0.303	0.305	0.387	*0.807*
	1680	*0.790*	*0.899*	0.261	0.218
Summer	930	0.216	0.406	*1.047*	*0.791*
	1010	0.264	0.351	0.309	*1.154*
	1180	0.321	*1.229*	0.576	0.448
	1680	0.428	*2.053*	*1.027*	*1.240*
Autumn	930	0.344	0.491	0.338	0.434
	1010	0.242	0.276	0.442	0.223
	1180	0.351	0.196	0.405	0.695
	1680	0.346	*1.015*	0.190	*0.863*

Table 4. Displacements of the 95% contour ellipses corresponding to each station and variable passing from 1951–1980 to 1981–2010. (Statistically significant results at 95% confidence level are in bold italics.)

Season	Site	Translations		Rotations	Deformations
		$\Delta\bar{T}$	$\Delta\bar{R}$	ϑ (°)	ΔA (%)
Winter	930	*−1.5*	*−1.0*	*109.1*	−16
	1010	*−0.6*	*−0.6*	−51.4	0
	1180	−0.4	*−1.0*	9.5	1
	1680	*−0.6*	+0.1	1.8	61
Spring	930	*−0.8*	−0.1	*69.2*	4
	1010	−0.1	−0.2	−14.9	−17
Autumn	930	*−0.8*	*−0.6*	43.5	*−59*
	1010	−0.2	−0.1	13.5	−36
	1180	−0.3	*−0.8*	−42.8	−27

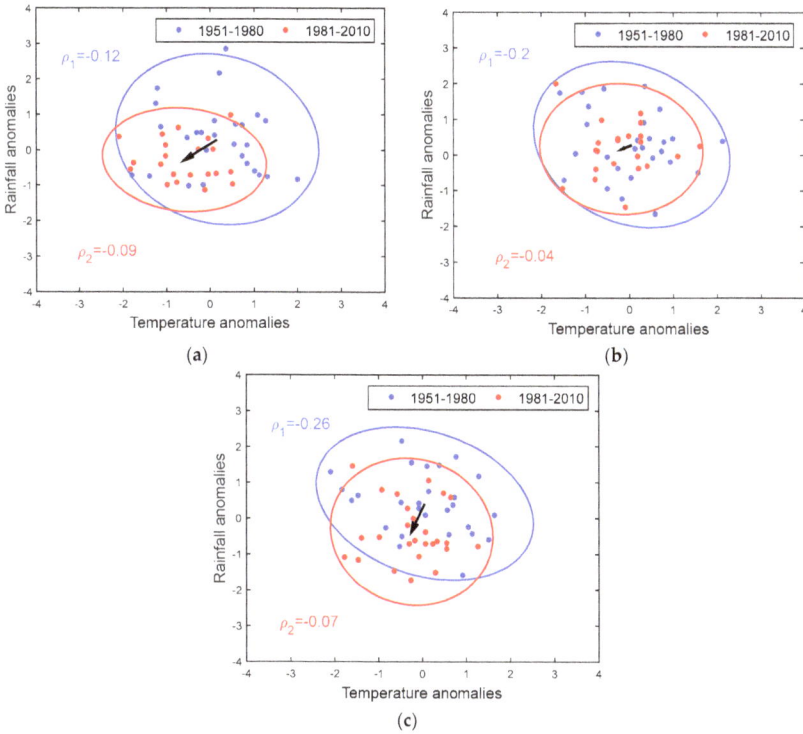

Figure 4. The 95% confidence ellipses for the autumn season at gauge (**a**) 930; (**b**) 1010; (**c**) 1180.

Table 5. Testing results of the change in the means of the seasonal values of temperature (*T*) and rainfall (*R*) observed in 1951–1980 and 1981–2010. (*H*, Hotelling's test; *t*(*T*) and *t*(*R*), two-sample *t*-test. Statistically significant results at the 95% confidence level are in bold italics.)

Season	Site	*H*	*t(T)*	*t(R)*
Winter	930	*30.28*	*−4.84*	*−3.91*
	1010	*11.66*	*−2.39*	*−2.19*
	1180	*23.77*	−1.59	*−3.88*
	1680	*4.91*	*−2.18*	+0.28
Spring	930	*8.70*	*−2.85*	−0.48
	1010	0.46	−0.21	−0.65
Autumn	930	*17.39*	*−3.05*	*−2.51*
	1010	1.37	−0.98	−0.51
	1180	*14.78*	−1.24	*−3.36*

4. Conclusions

The variabilities of the seasonal temperature and rainfall observed in four sites located in Calabria (Southern Italy) have been jointly investigated by means of the analysis of the variations in the 95% confidence ellipses of the bivariate normal distribution, evaluated for two different 30-year periods. The values of the various displacements (translations, rotations, and deformations) detected for the ellipses, estimated passing from the first subperiod to the second one, have been presented. Though this study is limited to a few gauges, the joint variations of the two climatic variables show the same tendencies in most of the considered cases. The results confirm the general negative trend detected for both monthly temperature and rainfall in Southern Italy, detected in previous studies [14,33]. Moreover, the tendency detected for seasonal rainfall confirms the general trend of long-term cumulated precipitation in the Mediterranean area [6,24]. On the contrary, the results here obtained for temperature evidenced that global warming, also revealed in the Mediterranean and the Middle East [24], is not always observed everywhere and in each season of the year, because opposite trends linked to specific local features can be detected, as already evidenced in Calabria by Caloiero et al. [23]. Moreover, it is important to highlight that when the comparison is carried out on results based on datasets with substantial differences, only limited conclusions about trends can be drawn. Nevertheless, even if the obtained results in terms of the joint gradients of rainfall and temperature statistics cannot be used to predict their relationships, because no assessment about the stationarity of the results was carried out, the trends observed, if confirmed in the future, could have potential impacts on several environmental sectors, in particular on agriculture.

Author Contributions: Conceptualization, E.F., R.C. and B.S.; Methodology, Software and Validation, E.F., R.C. and B.S.; Formal Analysis and Investigation, E.F. and R.C.; Writing-Review & Editing, E.F. and R.C.

Conflicts of Interest: The authors declare no conflict of interest.

References

1. Calderini, D.F.; Abeledo, L.G.; Savin, R.; Slafer, G.A. Effect of temperature and carpel size during pre-anthesis on potential grain weight in wheat. *J. Agric. Sci.* **1999**, *132*, 453–459. [CrossRef]
2. Abbate, P.E.; Dardanelli, J.L.; Cantarero, M.G.; Maturano, M.; Melchiori, R.J.M.; Suero, E.E. Climatic and water availability effects on water-use efficiency in wheat. *Crop Sci.* **2004**, *44*, 474–483. [CrossRef]
3. Medori, M.; Michelini, L.; Nogues, I.; Loreto, F.; Calfapietra, C. The impact of root temperature on photosynthesis and isoprene emission in three different plant species. *Sci. World J.* **2012**. [CrossRef] [PubMed]
4. Caloiero, T. Analysis of rainfall trend in New Zealand. *Environ. Earth Sci.* **2015**, *73*, 6297–6310. [CrossRef]
5. Brunetti, M.; Caloiero, T.; Coscarelli, R.; Gullà, G.; Nanni, T.; Simolo, C. Precipitation variability and change in the Calabria region (Italy) from a high resolution daily dataset. *Int. J. Climatol.* **2012**, *32*, 57–73. [CrossRef]
6. Longobardi, A.; Buttafuoco, G.; Caloiero, T.; Coscarelli, R. Spatial and temporal distribution of precipitation in a Mediterranean area (Southern Italy). *Environ. Earth Sci.* **2016**, *75*, 189. [CrossRef]
7. Mehta, A.V.; Yang, S. Precipitation climatology over Mediterranean Basin from ten years of TRMM measurements. *Adv. Geosci.* **2008**, *17*, 87–91. [CrossRef]
8. Reale, M.; Lionello, P. Synoptic climatology of winter intense precipitation events along the Mediterranean coasts. *Nat. Hazards Earth Syst.* **2013**, *13*, 1707–1722. [CrossRef]
9. Lionello, P.; Giorgi, F. Winter precipitation and cyclones in the Mediterranean region: Future climate scenarios in a regional simulation. *Adv. Geosci.* **2007**, *12*, 153–158. [CrossRef]
10. Brunetti, M.; Maugeri, M.; Monti, F.; Nanni, T. Temperature and precipitation variability in Italy in the last two centuries from homogenised instrumental time series. *Int. J. Climatol.* **2006**, *26*, 345–381. [CrossRef]
11. Longobardi, A.; Villani, P. Trend analysis of annual and seasonal rainfall time series in the Mediterranean area. *Int. J. Climatol.* **2010**, *30*, 1538–1546. [CrossRef]
12. Piccarreta, M.; Capolongo, D.; Boenzi, F. Trend analysis of precipitation and drought in Basilicata from 1923 to 2000 within a Southern Italy context. *Int. J. Climatol.* **2004**, *24*, 907–922. [CrossRef]
13. Liuzzo, L.; Bono, E.; Sammartano, V.; Freni, G. Analysis of spatial and temporal rainfall trends in Sicily during the 1921–2012 period. *Theor. Appl. Climatol.* **2016**, *126*, 113–129. [CrossRef]
14. Caloiero, T.; Buttafuoco, G.; Coscarelli, R.; Ferrari, E. Spatial and temporal characterization of climate at regional scale using homogeneous monthly precipitation and air temperature data: An application in Calabria (southern Italy). *Hydrol. Res.* **2015**, *46*, 629–646. [CrossRef]
15. Klok, E.J.; Tank, A. Updated and extended European dataset of daily climate observations. *Int. J. Climatol.* **2009**, *29*, 1182–1191. [CrossRef]
16. Caloiero, T.; Callegari, G.; Cantasano, N.; Coletta, V.; Pellicone, G.; Veltri, A. Bioclimatic Analysis in a Region of Southern Italy (Calabria). *Plant Biosyst.* **2015**. [CrossRef]
17. Liang, K.; Bai, P.; Li, J.J.; Liu, C.M. Variability of temperature extremes in the Yellow River basin during 1961–2011. *Quat. Int.* **2014**, *336*, 52–64. [CrossRef]

18. Caloiero, T. Trend of monthly temperature and daily extreme temperature during 1951–2012 in New Zealand. *Theor. Appl. Climatol.* **2016**. [CrossRef]

19. Salinger, M.J.; Griffiths, G.M. Trends in New Zealand daily temperature and rainfall extremes. *Int. J. Climatol.* **2001**, *21*, 1437–1452. [CrossRef]

20. Donat, M.G.; Alexander, L.V. The shifting probability distribution of global daytime and night-time temperatures. *Geophys. Res. Lett.* **2012**, *39*, L14707. [CrossRef]

21. Giorgi, F. Climate change hot-spots. *Geophys. Res. Lett.* **2006**, *33*, L08707. [CrossRef]

22. Simolo, C.; Brunetti, M.; Maugeri, M.; Nanni, T.; Speranza, A. Understanding climate change–induced variations in daily temperature distributions over Italy. *J. Geophys. Res.* **2010**, *115*, D22110. [CrossRef]

23. Caloiero, T.; Coscarelli, R.; Ferrari, E.; Sirangelo, B. Trend analysis of monthly mean values and extreme indices of daily temperature in a region of southern Italy. *Int. J. Climatol.* **2017**, *37*, 284–297. [CrossRef]

24. Tanarhte, M.; Hadjinicolaou, P.; Lelieveld, J. Intercomparison of temperature and precipitation data sets based on observations in the Mediterranean and the Middle East. *J. Geophys. Res.* **2012**, *117*. [CrossRef]

25. Rajeevan, M.; Pai, D.S.; Thapliyal, V. Spatial and temporal relationships between global land surface air temperature anomalies and Indian summer monsoon rainfall. *Meteorol. Atmos. Phys.* **1998**, *66*, 157–171. [CrossRef]

26. Huang, Y.; Cai, J.; Yin, H.; Cai, M. Correlation of precipitation to temperature variation in the Huanghe River (Yellow River) basin during 1957–2006. *J. Hydrol.* **2009**, *372*, 1–8. [CrossRef]

27. Cong, R.G.; Brady, M. The Interdependence between Rainfall and Temperature: Copula Analyses. *Sci. World J.* **2012**. [CrossRef] [PubMed]

28. Péguy, C.P. Une tentative de délimitation et de schématisation des climats intertropicaux. *Rev. Geogr.* **1961**, *36*, 1–6. [CrossRef]

29. Craddock, J.M. Methods of comparing annual rainfall records for climatic purposes. *Weather* **1979**, *34*, 332–346. [CrossRef]

30. Wilks, D.S. *Statistical Methods in the Atmospheric Sciences*, 2nd ed.; International Geophysics Series; Academic Press: Cambridge, MA, USA, 1995.

31. Rodrigo, F.S. On the covariability of seasonal temperature and precipitation in Spain, 1956–2005. *Int. J. Climatol.* **2014**, *35*, 3362–3370. [CrossRef]

32. Sirangelo, B.; Ferrari, E. Probabilistic analysis of the variation of water resources availability due to rainfall change in the Crati basin (Italy). In *Global Change: Facing Risks and Threats to Water Resources*; Servat, E., Demuth, S., Dezetter, A., Daniell, T., Ferrari, E., Ijjaali, M., Jabrane, R., van Lanen, H., Huang, Y., Eds.; IAHS Publication, N° 340; IAHS Press: Wallingford, UK, 2010; pp. 142–149.

33. Ferrari, E.; Terranova, O. Non-parametric detection of trends and change point years in monthly and annual rainfalls. In Proceedings of the 1st Italian-Russian Workshop on New Trend in Hydrology, CNR-IRPI, Rende, Italy, 24–26 September 2002; pp. 177–188.

34. Kottegoda, N.T.; Rosso, R. *Applied Statistics for Civil and Environmental Engineers*, 2nd ed.; Wiley & Sons: New York, NY, USA, 2008.

35. Hotelling, H. The generalization of Student's ratio. *Ann. Math. Stat.* **1931**, *2*, 360–378. [CrossRef]

geosciences **MDPI**

Article

From Deterministic to Probabilistic Forecasts: The 'Shift-Target' Approach in the Milan Urban Area (Northern Italy)

Gabriele Lombardi [1,*], **Alessandro Ceppi** [1], **Giovanni Ravazzani** [1], **Silvio Davolio** [2] **and Marco Mancini** [1]

[1] Department of Civil and Environmental Engineering, DICA, Politecnico di Milano, 20133 Milano, Italy; alessandro.ceppi@polimi.it (A.C.); giovanni.ravazzani@polimi.it (G.R.); marco.mancini@polimi.it (M.M.)

[2] Institute of Atmospheric Sciences and Climate, National Research Council of Italy, CNR-ISAC, 40129 Bologna, Italy; s.davolio@isac.cnr.it

* Correspondence: gabriele.lombardi@polimi.it; Tel.: +39-02-2399-6416

Received: 6 April 2018; Accepted: 10 May 2018; Published: 15 May 2018

Abstract: The number of natural catastrophes that affect people worldwide is increasing; among these, the hydro-meteorological events represent the worst scenario due to the thousands of deaths and huge damages to private and state ownership they can cause. To prevent this, besides various structural measures, many non-structural solutions, such as the implementation of flood warning systems, have been proposed in recent years. In this study, we suggest a low computational cost method to produce a probabilistic flood prediction system using a single forecast precipitation scenario perturbed via a spatial shift. In fact, it is well-known that accurate forecasts of heavy precipitation, especially associated with deep moist convection, are challenging due to uncertainties arising from the numerical weather prediction (NWP), and high sensitivity to misrepresentation of the initial atmospheric state. Inaccuracies in precipitation forecasts are partially due to spatial misplacing. To produce hydro-meteorological simulations and forecasts, we use a flood forecasting system which comprises the physically-based rainfall-runoff hydrological model FEST-WB developed by the Politecnico di Milano, and the MOLOCH meteorological model provided by the Institute of Atmospheric Sciences and Climate (CNR-ISAC). The areas of study are the hydrological basins of the rivers Seveso, Olona, and Lambro located in the northern part of Milan city (northern Italy) where this system works every day in real-time. In this paper, we show the performance of reforecasts carried out between the years 2012 and 2015: in particular, we explore the 'Shift-Target' (ST) approach in order to obtain 40 ensemble members, which we assume equally likely, derived from the available deterministic precipitation forecast. Performances are shown through statistical indexes based on exceeding the threshold for different gauge stations over the three hydrological basins. Results highlight how the Shift-Target approach complements the deterministic MOLOCH-based flood forecast for warning purposes.

Keywords: floods; urban river basins; Shift-Target approach; hydrological simulations; probabilistic forecasts

---

1. Introduction

From a civil protection point of view, hydro-meteorological forecasts can be seen as a powerful tool of non-structural measures to produce early flood warnings and better counteract potential river flood impacts, whose number is increasing worldwide [1]. Nevertheless, in order to be credible by local authorities and, above all, by citizens, a prediction system must be verified [2], and the verification analysis should be conducted with a large sample of consistent forecasts and observations. In this context,

Demargne et al. [3] proposed the following key questions to guide forecast verification analysis: How suitable are the forecasts for a given application? Are they sufficiently unbiased for the decisions to be made? Are they sufficiently skillful compared to a reference forecast system to justify the method in use?

The ultimate criterion of a good forecast is the decision adopted from it and, from our point of view, it should communicate the information that an end-user needs. Already proposed by Murphy [4] in 1993, good forecasting is not only a matter of "getting it right", but also to make the receivers understand it, and, above all, to be able to draw conclusions from it [5]. Adopting this framework, in this analysis, we are not interested in predicting river discharge with an accurate flood peak in magnitude as well as timing, but in predicting the probability of exceeding any threshold before the event, in order to provide early flood warnings to local authorities.

Nowadays, it is well known in the scientific community around the world that ensemble or probabilistic forecasts contain more information than single-valued forecasts [6,7], a key topic of the EFAS (European Flood Awareness System) and HEPEX (https://hepex.irstea.fr/) projects "to demonstrate the added value of hydrological ensemble predictions (HEPS) for emergency management and water resources sectors to make decisions that have important consequences for economy, public health and safety [8]".

Notwithstanding this, local authorities and citizens continue to disseminate and prefer deterministic hydro-meteorological forecasts, in particular in Italy where this study is set, without including any notion of probability (or chance) of a phenomenon occurring (such as flooding) when the forecast information is communicated. This is possibly an attempt of the authorities to avoid public confusion from multiple, conflicting warnings, while citizens habitually trust a single forecast only, and they are not educated enough to deal with probabilistic prediction.

The use of ensemble prediction systems allows researchers to properly quantify and communicate forecast uncertainties, but from our experience, we are aware that the communication of uncertainties to end-users is difficult and critical [9,10]. For instance, think about a forecast of a 50% probability: users often consider it to indicate that the forecaster is simply "sitting on the fence" [11]. However, if the observed frequency of the event is low, then a 50% probability is a strong signal. Just think of a 50% probability for an airplane to crash before a flight: no passenger would fly on that airplane! Therefore, developments to formalize forecast uncertainty began exploring human expertise and forecasters' capacity to translate forecast uncertainty into statistical confidence intervals [12].

In addition to initial conditions (e.g., missing data, anthropogenic interferences) and hydrological model uncertainty (e.g., calibration of parameters, conceptualization of the model, etc.), another key issue of forecast output uncertainty [13] is the capacity to correctly identify future precipitation both in space and time, which is especially critical when integrating precipitation on small watersheds for hydrological forecasts with a high impact on QPE (Quantitative Precipitation Estimates). Unfortunately, accurate forecasts of deep moist convection and associated extreme rainfall are arduous to be precisely predicted in terms of amount, timing, and target over small hydrological basins due to uncertainties arising from numerical weather prediction (NWP) models, including physical parameterizations and numerical schemes, and to the rapid growth of errors already affecting the initial atmospheric state. Therefore, a probabilistic forecasting approach that can cope and deal with these uncertainties is required [14,15].

Since only a deterministic precipitation forecast is available to produce hydrological forecasts, in this analysis, we tested a pragmatic approach proposed by Thies et al. [16] to account for the precipitation forecast uncertainty: a low computational cost method was set up to produce probabilistic hydrological forecasts based on spatially shifting a single-valued precipitation forecast scenario using different spatial domain shifts.

In particular, we explore an alternative way of the Thies' approach: from a deterministic precipitation forecast issued by the MOLOCH meteorological model (described in Section 2.2), we obtain 40 'ensemble members', equivalent to 40 spatial shifts of the predicted rainfall field in eight directions (North, South, West, East, North-West, North-East, South-West, South-East) at each step of 10 km from 0 to 50 km (which approximately is the entire basin dimensions), maintaining the temperature domain so that it is unchanged. This strategy, called the 'Shift-Target' (ST) approach, provides 40 discharge forecasts which we assumed to be equally likely in terms of occurrence probability in space and it investigates how the spatial uncertainty may impact the flood forecasts and the potential exceedance of flood warning thresholds.

In order to run hydro-meteorological predictions, we use a flood forecasting system which couples the physically based rainfall-runoff hydrological model FEST-WB with the MOLOCH meteorological model as described in Section 2. The area of interest comprises the three hydrological basins of the rivers Seveso, Olona, and Lambro, located in the northern part of Milan, northern Italy: an urban area which has been subjected to a high flood hazard in the past.

This implemented system works every day in real-time and it can be freely consulted at this web site: sol.mmidro.it (MMI, Milano, Italy). This adopted open source policy allows the public to see and exploit the results of our investment in science, and monitoring real-time products can inspire new research that improves techniques; even crowdfunding has been launched between 2017 and 2018.

For a meaningful verification analysis of the system performance, hindcasting (or retrospective forecasting) has been carried out for the period between 2012 and 2015 and the results are based on verification metrics (including contingency scores relative to various exceedance threshold values) for different gauge stations within the three basins.

The paper has a double scope: first, it aims to demonstrate the value of probabilistic hydrological forecasts obtained through the ST procedure in comparison with the single-valued MOLOCH-based hydrological forecast; second, to assess if the proposed shift method can be useful for civil protection services.

The paper is structured as follows: Chapter 2 describes the materials and methods: Section 2.1 shows the area of study which comprises the Milan urban basins; Sections 2.2 and 2.3 present the MOLOCH meteorological model and the FEST-WB hydrological model, respectively; and Sections 2.4 and 2.5 describe the coupling strategy and the verification scores, respectively. Chapter 3 shows the performance of the Shift-Target approach based on MOLOCH shift forecasts and Chapter 4 documents this paper's conclusions.

2. Materials and Methods

2.1. Area of Study

Milan is one of the most densely populated city in Italy with 1,316,000 inhabitants living in 182 km^2. Several rivers and creeks drain to Milan (Figure 1). The main rivers are the Lambro (area of 500 km^2), the Seveso (area of 207 km^2), and the Olona (area of 208 km^2), plus a number of minor tributaries, for a total drainage surface of about 1300 km^2.

Several flood events hit Milan in the past so that, starting from the 1970s, a series of structural flood mitigation measures, such as the Ponte Gurone dam over the Olona basin, the North-West filling channel over the Seveso, and the Pusiano dam over the Lambro, have been adopted with the aim of reducing the flood risk in the urban areas in the last decades. However, despite the complex flood protection system, the city was still impacted by floods in recent years: the 18 September 2010 with 80 M€ damages along the Seveso and Lambro rivers, the 8 July 2014 with 55 M€ damages along the Seveso river, the 15 November 2014 with 6 M€ damages along the Seveso and Lambro rivers, and the 15 July 2009 with 30 M€ along the Olona river.

According to Nemec's [17] recommendation, "to keep the people away from the water, and not the water away from the people", the implementation of a hydro-meteorological prediction system may provide additional support as a non-structural measure for early warning. In fact, since these basins have a response time of a few hours, warnings with a sufficient lead time will enable civil protection authorities and the public to exercise caution and take preventive measures to mitigate the impacts of flooding [18].

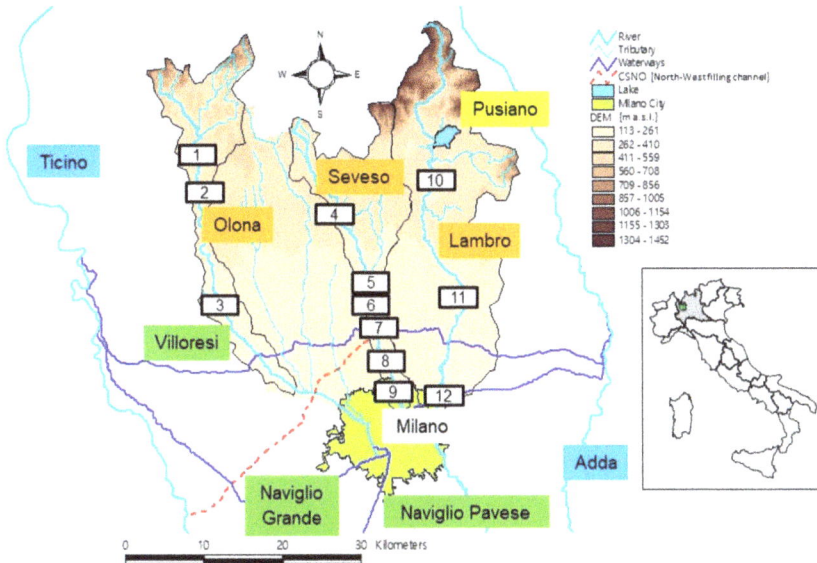

Figure 1. River basins draining to Milano urban area. White rectangles denote sections for which the realtime forecasting system provide warnings: (1) Lozza, (2) Castiglione Olona, (3) Castellanza, (4) Cantù, (5) Cesano Maderno, (6) Bovisio Masciago, (7) Paderno Dugnano, (8) Cormano, (9) Milano Niguarda, (10) Lambrugo, (11) Peregallo, (12) Milano via Feltre. Reprinted from [14], with permission from Elsevier.

At present, hydro-meteorological forecasts, published online, are implemented over the twelve gauge sections shown in Figure 1. However, for the 2012–2015 reforecasting period, not all the observed data were available; hence the verification analysis (shown in Section 2.5) is carried out for only half of the gauge stations, those with at least 900 days of available data.

2.2. The MOLOCH Meteorological Model

MOLOCH [19] is a non-hydrostatic, fully compressible, convection-resolving model, developed at the CNR-ISAC (National Research Council of Italy, Institute of Atmospheric Sciences and Climate). It integrates the set of atmospheric equations using a latitude–longitude rotated grid and a hybrid terrain-following vertical coordinate, depending on air density, which relaxes to horizontal surfaces at a higher elevation from the ground. Details on numerical schemes and model physics, as well as the results of the application to severe weather events and floods, can be found in [20–22]. Time integration is based on a time-split scheme with an implicit treatment of the vertical propagation of sound waves and a forward-backward scheme for the horizontal propagation of gravity and sound waves. Advection is computed using a second order implementation of the Godunov method [23], which is particularly suited to integrate in time the conservation of a scalar quantity [24]. The atmospheric radiation is based on a combined application of the Ritter and Geleyn scheme [25] and the ECMWF scheme [26]. The turbulence scheme is based on an eddy kinetic energy − mixing length (E − l), 1.5-order closure theory [27], where the turbulent kinetic energy equation (including advection) is evaluated. A soil model with seven layers takes into account orography, the geographical distribution of soil types, soil physical parameters, and vegetation coverage, as well as soil physical processes. The microphysical scheme is based on the parameterization proposed by [28] with successive upgrades, and it describes the conversions and interactions of cloud water, cloud ice, and hydrometeors (rain, snow and graupel). MOLOCH is implemented over Italy with a daily operational chain (http://www.isac.cnr.it/dinamica/projects/forecasts) that also comprises the hydrostatic model BOLAM [20], and provides operational forecasts for the following 45 h. The initial and boundary conditions for the BOLAM model are derived from the analyses (00 UTC) and forecasts of the Global Forecast System (GFS, NOAA/NCEP, USA) global model, while MOLOCH is nested (1-way) in BOLAM, initialized with a 3-h BOLAM forecast in order to avoid downscaling based on pure interpolation from the global model. In the period 2012–2017, MOLOCH has undergone continuous development. In particular, its implementation has changed. In fact, its horizontal resolution increased from 2.2 km to 1.25 in October 2016. At present, MOLOCH employs 60 atmospheric levels with output fields available at an hourly frequency.

2.3. The FEST-WB Hydrological Model

For transforming rainfall into runoff, we used the physically-based, spatially distributed FEST-WB (Flash–flood Event–based Spatially distributed rainfall–runoff Transformation, including Water Balance) model, developed by the Politecnico di Milano on top of MOSAICO library [29,30]. The FEST-WB accounts for the main processes of the hydrological cycle: snow melting and accumulation, infiltration, evapotranspiration, surface runoff, flow routing, and subsurface flow. The river basin is discretized with a mesh of regular square cells (200 × 200 m in this study), where water fluxes are calculated at an hourly time step. For further details on the development and application of the FEST-WB, the reader can refer to [31–34].

2.4. The Coupling Strategy

The proposed forecasting cascade system couples (1-way) the MOLOCH meteorological model with the FEST-WB model using the same strategy adopted in [13,33]. Temperature and precipitation outputs are forced into the hydrological model in order to forecast the main hydrological variables (discharge, evapotranspiration, soil moisture, etc.). In this study, we only focus on forecast runoff in the selected gauge sections mentioned in Section 2 for the entire MOLOCH lead time (45 h), adding 12 h for discharge routing at the end of the hydrological forecasting period to get the entire recession limb of the hydrograph. This choice is due to the chance that the precipitation and the observed peak discharge occur before the end of the forecast horizon but the runoff peaks are forecasted later.

Furthermore, since the first few hours of NWP forecasts are not generally reliable, due to the spin-up time of NWP models, we skipped the first 3 h of forecast, hence 54-h flow hindcasts are produced every day between 10 February 2012 and 31 December 2015, which represent up to ~1400 forecasts in total.

Similar to the method proposed by [35], here we aim to account for the spatial uncertainty of the precipitation forecast provided by the meteorological model for the hydrological basins (especially for smaller ones) by applying the ST method to the single-valued MOLOCH forecast. Forecasting the precipitation cells tens of kilometers away from their correct location could lead to significant errors in the hydrological response of the catchments; especially in these watersheds, which have a prolonged North-South shape.

In Figure 2, we show an example of 40 discharge ensemble forecasts produced by the ST method for the Cantù gauging section displayed with the 'peak-box' plot proposed by [36].

Figure 2 shows an example for the 8 July 2014 event, which was one of the most severe episodes in the last 20 years in this area. In this case, the 40-km West-shifted forecast (labelled 'W04' in Figure 2) exceeded the highest critical warning threshold, whereas the forecast based on the original MOLOCH precipitation did not exceed any of the warning thresholds. This means that if the "unperturbed" Moloch forecast of the precipitation system was affected by a location error of about 40 km westward, then the intense precipitation would have fallen over our watershed, producing a forecast discharge peak of 31.8 m^3/s.

To summarize the warning information given by all flow ensemble forecasts, the 'Union Jack' plot (Figure 3) displays the 40 maximum discharges values over the 54-h horizon associated with their spatial shift in all eight directions with a background cell color according to exceeded discharge thresholds.

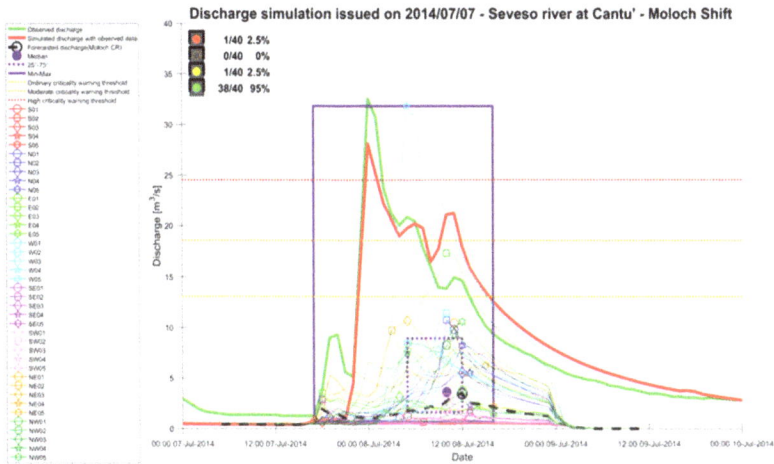

Figure 2. Forty discharge forecasts obtained by forty precipitation domain shifts, issued on 7 July 2014 over the Seveso basin closed at the Cantù gauge section. Observed discharge is shown in the green line and simulated discharge by FEST-WB forced with observed data in red, while different colors identify the hydrological forecasts driven by the Shift-Target (ST) approach.

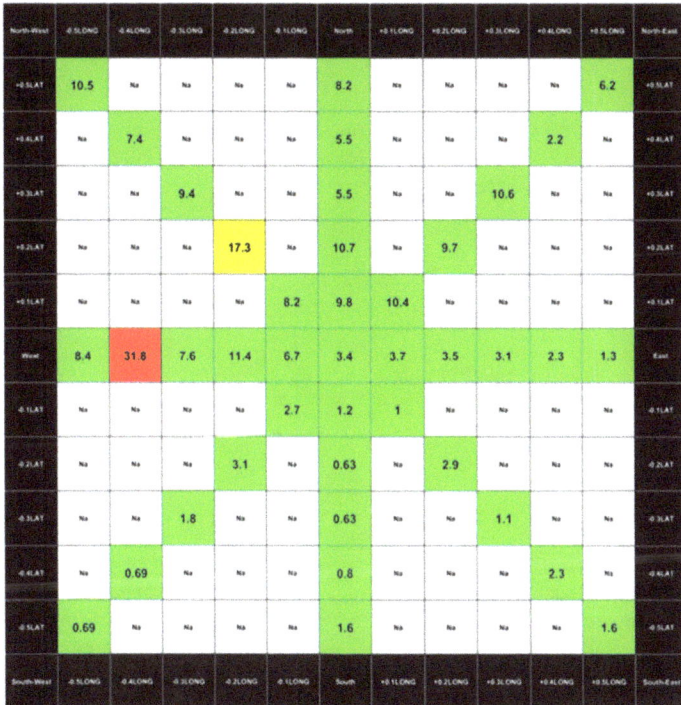

Figure 3. Union Jack plot forecast for the Cantù section issued on 7 July 2014: the 40 values are the maximum discharge forecasts over the entire simulation horizon according to the latitude and longitude shift from 0.1 to 0.5 degree, which is approximately 10 to 50 km.

Representing the 40 forecasted peak discharges on the 'Union Jack' with colour-coded impacts meets the requirements of the civil protection services to quickly assess the worst-case scenario based on the potential spatial error of the MOLOCH precipitation forecast.

When operating in real-time, it is possible to follow the evolution of the storm through cell tracking tools, using weather radar images. Forecasters could then evaluate which of the 40 ensemble members may be more realistic based on the storm evolution over the last hours, leading to a better understanding of the 'most likely' flooding scenario. Therefore, the Shift-Target approach can provide useful information, letting us know about an a-priori possible flood scenario.

It is worth noting that we are not investigating which is the most probable spatial shift or to calculate a spatial weight of these shifts. Concerning this issue, a specific research activity is actually under development, but this is not the aim of our paper. Here, we simply assume that all the 40 combinations are equally likely. Even an error in the peak time is not taken into consideration, since for local Italian civil protection bodies, the most important information is the exceedance of warning thresholds during the 24-h of the following day, in order to implement flood risk protection measures. Hence, in this framework, we would like to evaluate whether this low computational method, which generates a probabilistic precipitation forecast, performs better than the deterministic MOLOCH, and if its performance can generate an added value for civil protection purposes.

2.5. Verification Scores

Common statistical indexes used in scientific literature (www.cawcr.gov.au/projects/verification) [37] are calculated setting up a 4×4 contingency table which compares forecasted and observed events exceeding or not exceeding the three warning thresholds (Table 1); these thresholds are concurrently used by the contingency table and their values are provided for each gauge station by the Regional Civil Protection of Lombardy. Yellow, orange, and red thresholds are related to the discharge with 2-, 5-, and 10-year return periods, respectively.

Table 1. Performance of flows derived from the single-valued Moloch forecast and from the spatially shifted ensemble forecasts. A refers to Hits, B to Misses, C to False Alarm, and D to Correct Negatives, while the subscripts 1, 2, and 3 are levels of wrong prediction.

		Observed			
		No	Yes	Yes	Yes
Forecasted	No	D	B_1	B_2	B_3
	Yes	C_1	A	B_1	B_2
	Yes	C_2	C_1	A	B_1
	Yes	C_3	C_2	C_1	A

In this analysis, we calculate the Accuracy, the Bias Score, the Percent False Alarm (PFA), the Percent Missed Alarm (PMA), and the Correct Negatives Ratio (CNR). Since these three latter scores consider the non-occurred events, which are the majority in the dataset, we also calculate the FAR (False Alarm Ratio), the POD (Probability of Detection), and the Probability of Missed Alarm (POMA), especially to assess the performance in those critical situations when a warning threshold has been exceeded, excluding the other corrected non-events. In fact, for low frequency events such as severe weather warnings, there is a high frequency of "not forecast/not occurred" events. This gives high performance values that are misleading with regard to the forecasting of the low frequency event. For a 4×4 contingency table, statistical indexes are slightly different from the classical 2×2 as in [38,39]. Hence, in Table 2, we report all the equation formulae used to calculate the performance indexes in Section 2.5.

Table 2. Equation formulae used to calculate the common statistical indexes.

	Formula	Best Score	Range
N	$\sum A + \sum B + \sum C + D$	/	/
Accuracy	$(\sum A + D)/N$	100%	0/100%
Bias Score$_{2\times2}$	$(A + C)/(A + B)$	1	$0/+\infty$
Bias Score$_{n\times n}$	Yes_F/Yes_O	1	$0/+\infty$
PFA	$\sum C/N$	0%	0/100%
PMA	$\sum B/N$	0%	0/100%
FAR$_{2\times2}$	$C/(A + C)$	0%	0/100%
FAR$_{nxn}$	$\sum C/Yes_F$	0%	0/100%
POMA$_{2\times2}$	$B/(A + B)$	0%	0/100%
POMA$_{n\times n}$	$\sum B/Yes_O$	0%	0/100%
POD$_{2\times2}$	$A/(A + B)$	100%	0/100%
POD$_{n\times n}$	$\sum A/Yes_O$	100%	0/100%
CNR$_{2\times2}$	$D/(B + D)$	100%	0/100%
CNR$_{n\times n}$	D/No_F	100%	0/100%

Percent False Alarm (PFA), Percent Missed Alarm (PMA), False Alarm Ratio (FAR), Probability of Missed Alarm (POMA), Probability of Detection (POD), Correct Negatives Ratio (CNR).

In particular, No_F is the sum of all terms in the first row of Table 1; Yes_F is the sum of the second, third, and fourth row; and Yes_O is the sum of all terms in the second, third, and fourth column.

Furthermore, another issue relates to verifying the magnitude of our mistakes: i.e., how far are false or missed alarms from observations? If a red alert was issued, was my forecast orange, yellow, or green? This kind of error has a different impact for civil protection authorities. Hence, we introduce new statistical indexes (Table 3) weighted on the distance between observations and false/missed alarms for the three thresholds (yellow, orange, and red). In other words, the closer the wrong prediction is to the Hits, the lesser its error is weighted. The errors are counted as a unit fraction equal to k/s, where k is the step distance from the Hits and s is the number of thresholds (here, it is equal to 3). The worst cases are three steps distance from the Hits, hence dividing it by the number of thresholds, we obtain the unit.

Table 3. Weighted equation formulae used to calculate the common statistical indexes.

	Formula	Best Score	Range
PFA$_{weighted}$	$\sum^k(\sum C_k \times k/s)/N$	0%	0/100%
PMA$_{weighted}$	$\sum^k(\sum B_k \times k/s)/N$	0%	0/100%
FAR$_{nxn\ weighted}$	$\sum^k(\sum C_k \times k/s)/(\sum C + \sum A + \sum B_{YesF})$	0%	0/100%
POMA$_{nxn\ weighted}$	$\sum^k(\sum B_k \times k/s)/(\sum B + \sum A + \sum C_{YesO})$	0%	0/100%

C_{YesO} refers to those cells containing C in "Yes" observed columns in Table 1, while B_{YesF} indicates the cells containing B in "Yes" forecasted rows.

3. Results and Discussion

In this section, we discuss the main results obtained through the comparison between the performances of the single-valued MOLOCH forecasts and the ST ensemble forecasts. First of all, how do these two approaches exactly predict a non-event (green code: no alert)? In Table 4, we report the observed frequency related to the MOLOCH green code prediction, CNR, and the one related to a 95–100% predicted shift probability of the same green code. A good reliability is found: the frequency is always higher than 92% for both the MOLOCH and ST forecasts at every gauge station: this is proof that the two procedures are able to correctly predict non-events with a slight improvement in the ST.

Table 4. Correct Negatives Ratio (CNR) for the MOLOCH deterministic forecast compared with the observed frequency related to the non-event probability higher than 95% for the ST forecasts.

Basin	Section	Moloch (%)	Shift (%)
Seveso	Cantù	96.12	96.89
Olona	Castellanza	97.29	98.27
	Castiglione	99.16	99.46
	Lambrugo	95.76	96.89
Lambro	Milano F.	92.98	93.66
	Peregallo	93.99	94.86

To evaluate the performance of the ST in comparison with the deterministic MOLOCH, we calculate contingency tables for MOLOCH vs. observations, and ST vs. observations (Figures 4 and 5, respectively) in every gauge section. To build these contingency tables for the ST, a categorical forecast has to be assigned from the probabilities of the four alert codes. Our strategy is to choose, as representative of ensembles, the worst critical warning level from red to green, issued with a probability level equal or higher than a given percentage. However, in order to assess a suitable significant percentage to be applied to our instances, we experimented many alternatives. We started with the 33% threshold exceedance probability (at least 14 ensembles out of 40) derived from the Map D-Phase project outcomes [40,41], then we tried with 20% (8 out of 40) and 10% (4 out of 40). Results obtained for the 33% and 20%, for the sake of brevity not shown here, reveal that the unperturbed MOLOCH forecast is slightly preferable: most of the verification scores are similar but better, compared to those of the ST. This can be interpreted as a high skill reached by the deterministic model: i.e., it is not necessary to shift the precipitation domain since its accuracy is itself satisfactory. Maybe, this is not true with other deterministic weather models and this could be an issue for further investigations.

Nevertheless, given the nature of this "non-conventional" ensemble approach, we focus on the contingency scores based on the 10% probability level, since it is not so significant to have a high percentage of exceeding thresholds as to identify that four members out of 40 (equal to 10%) can at least cause dangerous flood scenarios. In Figures 4 and 5, all the contingency tables for the MOLOCH and ST model, respectively, are reported; one for every gauge station analyzed.

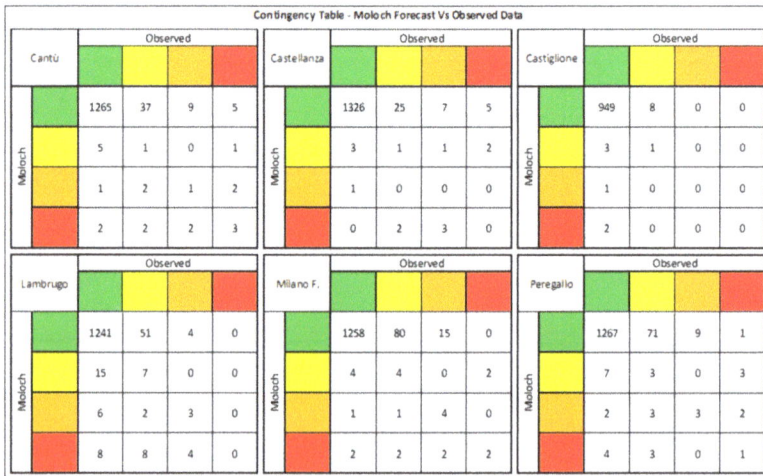

Figure 4. 4 × 4 contingency tables for the gauge section of Cantù, Castellanza, Castiglione, Lambrugo, Milano via Feltre, and Peregallo evaluating the "unperturbated" Moloch forecast vs. the observed data.

Contingency Table - Shift Forecast (33%, discriminating percentage) Vs Observed Data

Cantù		Observed			
Shift		1268	40	9	7
		3	0	1	0
		0	0	1	2
		1	2	1	2

Castellanza		Observed			
Shift		1328	24	6	6
		1	2	1	1
		0	0	0	0
		1	2	4	0

Castiglione		Observed			
Shift		951	9	0	0
		1	0	0	0
		0	0	0	0
		1	0	0	0

Lambrugo		Observed			
Shift		1254	56	6	0
		8	2	0	0
		1	3	1	0
		6	7	4	0

Milano F.		Observed			
Shift		1262	84	16	1
		1	2	2	1
		0	1	1	0
		0	0	2	2

Peregallo		Observed			
Shift		1273	72	9	4
		5	2	1	1
		0	1	2	1
		2	5	0	1

Figure 5. 4 × 4 Contingency tables for the section of Cantù, Castellanza, Castiglione, Lambrugo, Milano via Feltre, and Peregallo evaluating the ST forecast vs. observed data.

Starting from all these data, we have calculated many verification scores reported in Table 5 for MOLOCH and Table 6 for ST. First of all, the Bias Score shows a tendency of the MOLOCH model to underforecast, while the ST underforecasts or overforecasts, depending on the investigated section; in general, this latter marks better values.

In Figures 4 and 5, the high number of Correct Negatives (D in Table 1) that lead to a high Accuracy score in every section for both approaches is evident. In addition, the Percent False Alarm and Percent Missed Alarm scores, which have a percentage lower than 8%, are very favorable since these scores consider both the events and non-events. Unfortunately, the small number of False Alarm and Missed Alarm values is not relevant compared to the one of Correct Negatives, and therefore, these scores are not consistent when we refer to only those cases that have exceeded the thresholds.

Hence, to compare the two approaches in-depth, we take into account FAR, POMA, and POD, since they do not consider the Correct Negatives and they fully highlight the performance differences. The ST has a POD higher in four out of the six gauge sections, equal at Lambrugo, and lower at Milano via Feltre: therefore, this shows the MOLOCH's main tendency to underforecast, as already shown by the Bias Score.

With regard to the FAR, MOLOCH prevails on ST in all sections. The opposite comment can be made for POMA, which is lower for the ST than for MOLOCH. Therefore, the MOLOCH is more suitable to reduce False Alarm (FA), while the ST minimizes Missed Alarm (MA) better, which is more important for civil protection purposes, because flood damages for a missed alarm have a higher economic cost in comparison with counteractions to activate a false alarm; hence this can be considered a plus for the ST approach.

Table 5. Statistical indexes calculated from the MOLOCH "unperturbed" contingency tables.

MOLOCH	Cantù	Castellanza	Castiglione	Lambrugo	Milano F.	Peregallo
Accuracy	94.92%	96.44%	98.55%	92.74%	92.08%	92.39%
Bias Score	0.34	0.28	0.78	0.67	0.21	0.31
PFA	1.05%	0.65%	0.62%	3.19%	0.87%	1.38%
PMA	4.04%	2.91%	0.83%	4.08%	7.04%	6.24%
FAR	63.64%	69.23%	85.71%	81.13%	50.00%	61.29%
POMA	83.08%	86.96%	88.89%	69.62%	86.61%	86.87%
POD	7.69%	2.17%	11.11%	12.66%	8.93%	7.07%

Table 6. Statistical indexes calculated from the Shift-Target contingency tables.

Shift-Target	Cantù	Castellanza	Castiglione	Lambrugo	Milano F.	Peregallo
Accuracy	94.47%	96.00%	96.89%	90.59%	91.50%	91.88%
Bias Score	0.58	0.67	2.78	1.19	0.34	0.56
PFA	2.17%	1.82%	2.39%	6.00%	1.96%	2.83%
PMA	3.36%	2.18%	0.73%	3.41%	6.54%	5.29%
FAR	76.32%	80.65%	92.00%	86.17%	71.05%	70.91%
POMA	69.23%	65.22%	77.78%	58.23%	80.36%	73.74%
POD	12.31%	8.70%	22.22%	12.66%	6.25%	12.12%

By definition, in a 4×4 contingency table, the obtained skill scores are less satisfactory than a traditional 2×2, since the same events, which should be considered as Hits in the 2×2, can also be Hits, but even Missed or False Alarms in a 4×4 instead, due to the discretization of warning levels. Hence, in Tables 7 and 8, we calculate the same indexes shown in Tables 5 and 6 regarding False Alarms and Missed Alarms, but weighted in order to distinguish the level in the error prediction.

Table 7. Weighted statistical indexes calculated from the MOLOCH "unperturbed" contingency tables.

MOLOCH	Cantù (%)	Castellanza (%)	Castiglione (%)	Lambrugo (%)	Milano F. (%)	Peregallo (%)
PFA_W	0.52	0.29	0.38	1.80	0.46	0.77
PMA_W	1.84	1.43	0.28	1.46	2.76	2.42
FAR_W	31.82	30.77	52.38	45.91	26.39	34.41
$POMA_W$	37.95	42.75	29.63	24.89	33.93	33.67

Table 8. Weighted statistical indexes calculated from the Shift contingency tables.

Shift-Target	Cantù (%)	Castellanza (%)	Castiglione (%)	Lambrugo (%)	Milano F. (%)	Peregallo (%)
PFA_W	1.37	1.14	1.63	3.41	1.14	1.72
PMA_W	1.47	1.07	0.24	1.16	2.49	2.05
FAR_W	48.25	50.54	62.67	48.94	41.23	43.03
$POMA_W$	30.26	31.88	25.93	19.83	30.65	28.62

Here, these new indexes improve the percentage by about 30–50% for the two methods. In particular, the FAR decreases an average of 32% for MOLOCH and 30% for ST, while the POMA is 50% for MOLOCH and 43% for ST. This means that when a warning is wrongly issued, the code error is generally not so far from the observed one. Nevertheless, we are not interested in the absolute scores, but in the comparison between the two approaches.

4. Conclusions

Hydro-meteorological systems are nowadays set up with multi-models or multi-analysis approaches gathering deterministic and ensemble forecasts, with the latter being widespread in the scientific community, in order to provide probabilistic information. However, even when using different weather models, a large uncertainty still remains, especially for small river catchments, concerning the location of forecast precipitation. Hence, the proposed study shows the implementation of a different approach using the deterministic high-resolution MOLOCH meteorological model coupled with the FEST-WB hydrological model to obtain probabilistic forecasts. It consists of shifting the precipitation field at in eight directions from 10 to 50 km a step of 10 km, so that the results are 40 discharge forecasts over each analyzed gauge section. The performance of the Shift-Target approach is compared with the "unperturbed" MOLOCH forecast over a period of four years from 10 February 2012 till 31 December 2015. The results show how the ST does not worse the quality of the forecast in comparison with the one by MOLOCH, and in some cases, it is even better.

The potentiality of the ST can be seen as an a-priori rainfall generator that can be used in real-time, above all, during convective events when precise thunderstorm cell forecasting is difficult over small river basins, but probable flood scenarios, obtained by spatial shift forecasts, can already be forecasted. Notwithstanding this, the ST approach is conditioned by the unperturbed MOLOCH forecast in terms of QPF (Quantitative Precipitation Forecasts): if the MOLOCH totally misses the precipitation intensity, e.g., underestimating over the entire area, no shift will improve the forecast. Nevertheless, this approach, obtained with a low computation method, has demonstrated that it is able to provide useful information with respect to the deterministic MOLOCH run, in case of a misplacement of precipitation field.

Future developments will concentrate on enlarging our dataset to investigate more flood episodes and to verify with a more robust stochastic approach, which is the probability distribution of spatial, timing, and intensity of precipitation error.

Author Contributions: G.L., A.C., G.R., and M.M. conceived and designed the experiments; G.L. performed the experiments; G.L. and A.C. analyzed the data; S.D. provided meteorological forecasts; G.L., A.C., G.R., and S.D. wrote the paper.

Funding: This research received no external funding.

Acknowledgments: This work represents a contribution to the HyMeX international program and it has been developed under a research scientific agreement between Politecnico di Milano and ISAC-CNR.

Conflicts of Interest: The authors declare no conflict of interest.

References

1. Munich Reinsurance Company (Munich Re). *Topics Geo Natural Catastrophes 2016: Analyses, Assessments, Positions*; Munich Re: München, Germany, 2017; p. 80.
2. Demargne, J.; Brown, J.D.; Liu, Y.; Seo, D.-J.; Wu, L.; Toth, Z.; Zhu, Y. Diagnostic verification of hydrometeorological and hydrologic ensembles. *Atmos. Sci. Lett.* **2010**, *11*, 114–122. [CrossRef]
3. Demargne, J.; Mullusky, M.; Werner, K.; Adams, T.; Lindsey, S.; Schwein, N.; Marosi, W.; Welles, E. Application of Forecast Verification Science to Operational River Forecasting in the U.S. National Weather Service. *Bull. Am. Meteorol. Soc.* **2009**, *90*, 779–784. [CrossRef]
4. Murphy, A. What is a good forecast? An essay on the nature of goodness in weather forecasting. *Weather Forecast.* **1993**, *8*, 281–293. [CrossRef]
5. Persson, A. Abstract: The ultimate criterion of a "good" forecast is always the decisions made from it. In Proceedings of the 9th HEPEX Webinar: Five Points to Remember in Risk Forecasting, 24 October 2013.
6. Alfieri, L.; Pappenberger, F.; Wetterhall, F.; Haiden, T.; Richardson, D.; Salamon, P. Evaluation of ensemble streamflow predictions in Europe. *J. Hydrol.* **2014**, *517*, 913–922. [CrossRef]
7. Ramos, M.H.; van Andel, S.J.; Pappenberger, F. Do probabilistic forecasts lead to better decisions? *Hydrol. Earth Syst. Sci.* **2013**, *17*, 2219–2232. [CrossRef]

8. Harrigan, S.; Arnal, L. HEPEX Logo Competition: Reveal Your Inner Artist! Available online: https://hepex.irstea. fr/hepex-logo-competition-reveal-your-inner-artist/ (accessed on 28 March 2018).

9. Ramos, M.H.; Mathevet, T.; Thielen, T.; Pappenberger, F. Communicating uncertainty in hydro-meteorological forecasts: Mission impossible? *Meteorol. Appl.* **2010**, *17*, 223–235. [CrossRef]

10. Zalachori, I.; Ramos, M.H.; Garçon, R.; Mathevet, T.; Gailhard, J. Statistical processing of forecasts for hydrological ensemble prediction: A comparative study of different bias correction strategies. *Adv. Sci. Res.* **2012**, *8*, 135–141. [CrossRef]

11. WMO (World Meteorological Organization). *Guidelines on Communicating Forecast Uncertainty*; Technical Document PWS-18 WMO/TD 1422; WMO: Geneva, Switzerland, 2008; p. 25.

12. Houdant, B. Contribution a Lamelioration de la Prevision Hydrometeorologique operationnelle. Pour L'usage des Probabilites dans la Communication Entre Acteurs [Contribution to the Improvement of Operational Hydrometeorological Forecasting. For the Use of Probability in the Communication between Actors]. Ph.D. Thesis, ENGREF, EDF, Grenoble, France, 2004; p. 209. (In French)

13. Jaun, S.; Ahrens, B. Evaluation of a probabilistic hydrometeorological forecast system. *Hydrol. Earth Syst. Sci.* **2009**, *13*, 1031–1043. [CrossRef]

14. Ravazzani, G.; Amengual, A.; Ceppi, A.; Homar, V.; Romero, R.; Lombardi, G.; Mancini, M. Potentialities of ensemble strategies for flood forecasting over the Milano urban area. *J. Hydrol.* **2016**, *539*, 237–253. [CrossRef]

15. Davolio, S.; Miglietta, M.M.; Diomede, T.; Marsigli, C.; Montani, A. A flood episode in northern Italy: Multi-model and single-model mesoscale meteorological ensembles for hydrological predictions. *Hydrol. Earth Syst. Sci.* **2013**, *17*, 2107–2120. [CrossRef]

16. Thies, S.E.; Hense, A.; Damrath, U. Probabilistic precipitation forecasts from a deterministic model: A pragmatic approach. *Meteorol. Appl.* **2005**, *12*, 257–268. [CrossRef]

17. Nemec, J. Global Runoff Data Sets and Use of Geographic Information Systems. In Proceedings of the ISLSCP Conference, Rome, Italy, 2–6 December 1985. ESA SP-248.

18. Yang, T.-H.; Yang, S.-C.; Ho, J.-Y.; Lin, G.-F.; Hwang, G.-D.; Lee, C.-S. Flash flood warnings using the ensemble precipitation forecasting technique: A case study on forecasting floods in Taiwan caused by typhoons. *J. Hydrol.* **2015**, *520*, 367–378. [CrossRef]

19. Malguzzi, P.; Grossi, G.; Buzzi, A.; Ranzi, R.; Buizza, R. The 1966 "century" flood in Italy: A meteorological and hydrological revisitation. *J. Geophys. Res.* **2006**, *111*, D24106. [CrossRef]

20. Buzzi, A.; Davolio, S.; Malguzzi, P.; Drofa, O.; Mastrangelo, D. Heavy rainfall episodes over Liguria of autumn 2011: Numerical forecasting experiments. *Nat. Hazard Earth Syst. Sci.* **2014**, *14*, 1325–1340. [CrossRef]

21. Davolio, S.; Silvestro, F.; Gastaldo, T. Impact of rainfall assimilation on high-resolution hydro-meteorological forecasts over Liguria (Italy). *J. Hydrometeorol.* **2017**, *18*, 2659–2680. [CrossRef]

22. Davolio, S.; Silvestro, F.; Malguzzi, P. Effects of increasing horizontal resolution in a convection permitting model on flood forecasting: The 2011 dramatic events in Liguria (Italy). *J. Hydrometeorol.* **2015**, *16*, 1843–1856. [CrossRef]

23. Godunov, S.K. A difference method for numerical calculation of discontinuous solutions of the equations of hydrodynamics. *Matematicheskii Sbornik* **1959**, *47*, 271–306.

24. Toro, E.F. The weighted average flux method applied to the Euler equations. *Philos. Trans. R. Soc. Lond. A* **1992**, *341*, 499–530. [CrossRef]

25. Ritter, B.; Geleyn, J.F. A comprehensive radiation scheme for numerical weather prediction models with potential applications in climate simulations. *Mon. Weather Rev.* **1992**, *120*, 303–325. [CrossRef]

26. Morcrette, J.-J.; Barker, H.W.; Cole, J.N.S.; Iacono, M.J.; Pincus, R. Impact of a new radiation package, McRad, in the ECMWF Integrated Forecasting System. *Mon. Weather Rev.* **2008**, *136*, 4773–4798. [CrossRef]

27. Zampieri, M.; Malguzzi, P.; Buzzi, A. Sensitivity of quantitative precipitation forecasts to boundary layer parameterization: A flash flood case study in the western Mediterranean. *Nat. Hazards Earth Syst. Sci.* **2005**, *5*, 603–612. [CrossRef]

28. Drofa, O.V.; Malguzzi, P. Parameterization of microphysical processes in a non-hydrostatic prediction model. In Proceedings of the 14th International Conference on Clouds and Precipitation (ICCP), Bologna, Italy, 19–23 July 2004; pp. 1297–3000.

29. Rabuffetti, D.; Ravazzani, G.; Corbari, C.; Mancini, M. Verification of operational Quantitative Discharge Forecast (QDF) for a regional warning system—The AMPHORE case studies in the upper Po River. *Nat. Hazard Earth Syst.* **2008**, *8*, 161–173. [CrossRef]

30. Ravazzani, G. MOSAICO, a library for raster based hydrological applications. *Comput. Geosci.* **2013**, *51*, 1–6. [CrossRef]

31. Feki, M.; Ravazzani, G.; Ceppi, A.; Mancini, M. Influence of soil hydraulic variability on soil moisture simulations and irrigation scheduling in a maize field. *Agric. Water Manag.* **2018**, *202*, 183–194. [CrossRef]

32. Ceppi, A.; Ravazzani, G.; Corbari, C.; Salerno, R.; Meucci, S.; Mancini, M. Real time drought forecasting system for irrigation management. *Hydrol. Earth Syst. Sci.* **2014**, *18*, 3353–3366. [CrossRef]

33. Ceppi, A.; Ravazzani, G.; Salandin, A.; Rabuffetti, D.; Montani, A.; Borgonovo, E.; Mancini, M. Effects of temperature on flood forecasting: Analysis of an operative case study in Alpine basins. *Nat. Hazards Earth Syst. Sci.* **2013**, *13*, 1051–1062. [CrossRef]

34. Ravazzani, G.; Barbero, S.; Salandin, A.; Senatore, A.; Mancini, M. An integrated hydrological model for assessing climate change impacts on water resources of the Upper Po river basin. *Water Resour. Manag.* **2015**, *29*, 1193–1215. [CrossRef]

35. Diomede, T.; Marsigli, C.; Nerozzi, F.; Papetti, P.; Paccagnella, T. Coupling high-resolution precipitation forecasts and discharge predictions to evaluate the impact of spatial uncertainties in numerical weather prediction model outputs. *J. Meteorol. Atmos. Phys.* **2008**, *102*, 37–62. [CrossRef]

36. Zappa, M.; Fundel, F.; Jaun, S. A 'peak-box' approach for supporting interpretation and verification of operational ensemble peak-flow forecast. *Hydrol. Process.* **2013**, *27*, 117–131. [CrossRef]

37. WWRP/WGNE Joint Working Group on Verification cited 2013: Forecast Verification—Issue, Methods and FAQ. Available online: http://www.cawcr.gov.au/projects/verification (accessed on 26 January 2015).

38. Jolliffe, I.T.; Stephenson, D.B. (Eds.) *Forecast. Verification: A Practitioner's Guide*; John Wiley & Sons: Hoboken, NJ, USA, 2012; p. 240.

39. Wilks, D.S. *Statistical Methods in the Atmospheric Sciences*, 2nd ed.; Academic Press: Cambridge, MA, USA, 2006.

40. Zappa, M.; Rotach, M.W.; Arpagaus, M.; Dorninger, M.; Hegg, C.; Montani, A.; Ranzi, R.; Ament, F.; Germann, U.; Grossi, G.; et al. *MAP D-PHASE: Real-Time Demonstration of Hydrological Ensemble Prediction Systems*; Special Issue Article; HEPEX Workshop: Stresa, Italy, June 2007.

41. Rotach, M.W.; Ambrosetti, P.; Ament, F.; Appenzeller, C.; Arpagaus, M.; Bauer, H.-S.; Zappa, M. Supplement to MAP D-PHASE: Real-Time Demonstration of Weather Forecast Quality in the Alpine region: Additional Applications of the D-PHASE Datasets. *Bull. Am. Meteorol. Soc.* **2009**, *90*, s28–s32. [CrossRef]

geosciences

MDPI

Article

Fatalities Caused by Hydrometeorological Disasters in Texas

Srikanto H. Paul *, Hatim O. Sharif and Abigail M. Crawford

Department of Civil and Environmental Engineering, University of Texas at San Antonio, San Antonio, TX 78249, USA; hatim.sharif@utsa.edu (H.O.S.); amc1591@gmail.com (A.M.C.)
* Correspondence: srikanto.paul@utsa.edu; Tel.: +1-210-332-3241

Received: 20 April 2018; Accepted: 15 May 2018; Published: 18 May 2018

Abstract: Texas ranks first in the U.S in number of fatalities due to natural disasters. Based on data culled from the National Oceanic and Atmospheric Administration (NOAA) from 1959 to 2016, the number of hydrometeorological fatalities in Texas have increased over the 58-year study period, but the per capita fatalities have significantly decreased. Spatial review found that non-coastal flooding is the predominant hydrometeorological disaster in a majority of the Texas counties located in "Flash Flood Alley" and accounts for 43% of all hydrometeorological fatalities in the state. Flooding fatalities occur most frequently on "Transportation Routes" followed by heat fatalities in "Permanent Residences". Seasonal and monthly stratification identifies Spring and Summer as the deadliest seasons, with the month of May registering the highest number of total fatalities dominated by flooding and tornado fatalities. Demographic trends of hydrometeorological disaster fatalities indicated that approximately twice as many male fatalities occurred from 1959-2016 than female fatalities, but with decreasing gender disparity over time. Adults are the highest fatality risk group overall, children are most at risk to die in flooding, and the elderly at greatest risk of heat-related death.

Keywords: natural hazards; weather disasters; hydrometeorological fatalities; flooding; tornadoes; extreme temperatures

1. Introduction

Hydrometeorological disasters can result in tremendous damage to infrastructure, significant loss to the economy, and, very often, loss of life. In terms of the human loss, natural disasters resulted in approximately 1.7 million fatalities between 1980 and 2016. More than 49% of these fatalities were due to geophysical events (earthquake, tsunami, volcanic activity), 26% were due to meteorological events (tropical storm, extratropical storm, convective storm, local storm), 14% were due to hydrological events (flood, landslides), and 11% were due to climatological events (extreme temperature, drought, forest fire). Slightly less than 80% of the 16,500 disaster events that caused fatalities were hydrological or meteorological (39%) [1].

Although more research is beginning to shift to multi-hazard analysis [2,3], much of the available natural disaster research focuses on a particular type of disaster (e.g., floods, hurricanes, lightning, earthquakes) or disaster event (e.g., Hurricane Harvey, Northridge Earthquake). The focus on key disaster events is advantageous in that a deeper dive can benefit the preparation and mitigation strategies in the affected areas. Flooding is an exemplary disaster type that is responsible for high fatality rates and has been extensively investigated on a global [4–7] and national scale including the U.S. [8], India [9], Pakistan [10], and Australia [11]. It is also of value to focus at the regional level or the effects of one type of hazard to provide a basis for better allocation of resources to prepare for high risk hydrometeorological disasters with high probability of impact to a specific region.

This study analyzes fatality rates resulting from multiple hydrometeorological disasters that affect the state of Texas at the county level. Texas has a long history of devastation by natural disaster (especially hydrometeorological disasters). The most lethal natural disaster in United States history occurred in Galveston Island, Texas in 1900 in which an estimated 6000–12,000 people died as a result of the "Great Galveston Hurricane". From January 1960 to December 2016, Texas had the highest number of fatalities in the nation in which natural disasters killed an average of 40 people per year [12]. During this period, Texas accounted for 7.4% of all U.S. fatalities (32,289). Flood, heat, and tornado accounted for 60% of all fatalities in Texas during this period. Texas also ranks highest in fatalities per capita (15 fatalities per 100,000 people). During this period two Texas counties ranked in the top ten across all states for the occurrence of disaster events: Harris County (1088 events) and Tarrant County (1009 events). Dallas County, Texas, ranked eighth in the number of fatalities in the U.S. Extensive research has been conducted to investigate the quantitative and qualitative aspect of flooding in the state of Texas [12,13].

Hurricanes are also a critical hydrometeorological disaster that claim many lives in Texas. In 2005, Hurricane Katrina struck the Louisiana coast causing $96 billion in damages and 1833 fatalities. Two-thirds of the fatalities were directly related to more than fifty breaches of the levee and floodwall systems [14]. Most recently in August 2017, Hurricane Harvey made the landfall near Port Aransas on the Gulf Coast as a category-4 storm with wind gusts up to 212 kph (132 mph) and resulted in $200 billion in damages and 103 confirmed deaths in Texas, primarily due to flooding across 11 counties. Thirty-six of the total 68 direct fatalities caused by the hurricane winds and flooding occurred in Harris County (Houston Metropolitan area) [15].

Tornadoes were responsible for 14% of the total number of natural disaster-related fatalities in the U.S. from 1960 to 2015 [16]. Texas leads the nation in the average number of tornadoes between 1991 and 2010 with 155 tornadoes per year followed by Kansas (96), Florida (66) and Oklahoma (62) [17]. Analysis of tornado-induced fatalities and damage in the U.S. between 1880 and 2005 in 2007 identified 1812 tornado-related fatalities caused by 366 fatal tornado events mostly along the northeastern border of the state [18]. The normalized fatality rate per tornado event in Texas is in line with the national normalized averages (2.7 fatalities and 0.54 events).

The impact of disastrous extreme weather to society is a function of both the climatic and local setting. For example, although both the fatality rate and the extent of damage to infrastructure have increasing trends from the 1960's to the present, some studies suggest that population growth and demographic shifts play a greater role in the degree of increase than the increase in intensity and/or frequency of the extreme weather that the Earth has been experiencing in the last several decades [19]. This would suggest that even without any detrimental climate changes, the shifts in U.S. economic development patterns and growth will result in ever increasing losses caused by hydrometeorological disasters. Therefore, it is necessary to recognize spatial and temporal trends of natural disasters to allow for the allocation of resources to the higher risk disasters and their locations.

Supplemental to the intensity of such hazards is the exposure of people in the affected areas. In the last several decades, the U.S. has experienced steady increases in population shifts in rural and coastal development patterns, and economic growth, which have positioned more people in disaster-prone areas [20]. Research on the October 2015 flood event in Columbia, SC, indicated that considerations for public safety were sometimes secondary to profitable land development [21]. Hurricane research that was published in 2018 analyzed the decision biases of persons affected by hurricanes and found that temporal band spatial myopia is a major issue that places a lower priority on long-term decisions (e.g., preparation) than short term routine tasks with the failed intention of addressing the long term need when the disaster event is closer in time [22].

The purpose of this paper is to analyze the fatality rates caused by hydrometeorological disasters in Texas for the period 1959–2016 in an effort to identify counties and metropolitan areas in Texas that have a greater risk for particular hydrometeorological disasters. The hydrometeorological disasters were categorized into "Flooding", "Heat", "Cold Weather", "Tornado", "Lightning", and "Wind Events". Fatalities due to "Tropical Events" (hurricanes and tropical storms) were either classified as flooding or wind events depending on the cause of fatality. The study examines temporal trends, spatial variations, and demographic characteristics of the victims. The paper concludes with a discussion and commentary of considerations that may influence the fatality rate with the goal of providing information and perspectives that would help reduce hydrometeorological disaster fatalities.

2. Materials and Method

2.1. Study Area

Texas is the second largest state in the U.S., both in terms of population and area, with a population of 27,862,596 and a land area of 695,662 km^2. The southeast of Texas shares 591 km (367 miles) of coastline with the Gulf of Mexico and is susceptible to hurricanes and coastal flooding. A major topographical feature that affects the number of hydrometeorological disasters in Texas is the Balcones Escarpment that consists of a series of cliffs dropping from the Edwards Plateau to the Balcones Fault Line. As noted in an article from the Texas Hill Country magazine published in 2016, "This outer rim of the Hill Country is the formation point for many large thunderstorms, which frequently stall along the uplift and then hover over this region" [23]. The "Flash Flood Alley" includes counties having the fastest population growth rates in Texas.

2.2. Data Source

The Texas hydrometeorological disaster fatality information reviewed in this study was culled from the National Oceanic and Atmospheric Administration (NOAA) *Storm Data* reports for the period January 1959 through December 2016 [24]. From 1959–1995, the data were only available via PDF files. Disaster data from 1996–2016 were available via the NOAA searchable database. The data in the *Storm Data* Publication relies on self-reporting from individual states and counties and is dependent upon the verification and validation of the reporting agency. The *Storm Data* had some inconsistencies from year to year and county to county in the classification of the causes of fatalities. For example, deaths by lightning are classified as either electrical deaths or lightning deaths. Similarly, wild fires or prairie fires are listed under either wind events or wildfire events in the database. Heat-related deaths from the homeless or illegal immigrants in rural counties also have a potential to be under-reported since the location of the victims may remain undetected. As an example of a potential under-reporting condition, the *Storm Data* indicates that before 2008 there were no deaths due to heat exposure discovered along the border of Texas and Mexico. This is unlikely given that the U.S. Customs and Border Protection indicate that 7216 people have died from exposure attempting to cross the U.S./Mexico border between 1998 and 2017 [25]. In 2005 alone, more than 500 people died attempting to across the U.S./Mexico border [26]. Fatality information was also reviewed from the Hazards Vulnerability Research Institute (HVRI), U.S Hazard Losses Summary Report (1960–2015), to provide perspective for large scale comparisons of trends between Texas and the national fatality rate [16]. The HVRI data was not used in the numerical analysis of spatial and temporal trends forming the basis of this paper.

2.3. Methodology

Differences in the terminology exists across varying literature sources as it pertains to the effects of natural hazards and disasters on people and the land. This study defines a hazard as a natural event that has the potential to cause harm and a disaster as the effect of the hazard on humanity. Hydrometeorological disasters are defined as "natural processes or phenomena of atmospheric, hydrological or oceanographic nature" [27]. The Texas fatality data used in this study were all caused by hydrometeorological disasters. If the disaster did not result in at least one fatality it was not included. The fatalities also did not have to result from a disaster that was classified through a formal disaster declaration. The definitions of the descriptors and disaster types used in this study for the database files (1996 onward) and the manually aggregated fatality data prior to 1996 are in agreement with the NWS Directive 10-1605 [28].

Only fatalities that were classified as being directly caused by the incident are included in the study. *Storm Data* lists each incident with the date, time, the number of people who died in the incident, the number of people injured, and a brief description of the event. The descriptive narratives provided along with each event were used to get information related to the gender, age, activity, mode of transport, and location of the individual who died. In 1996 and after, the database provided an accompanying chart of the victims. The chart listed the victim's age, gender, and location. If there was a disparity between the description and the accompanying table, the information in the description was used since the descriptions were often retrieved from the police report that was filed with the death.

The data analysis includes temporal and spatial trending using linear trendlines and correlation analysis to verify statistical significance. Moving 10-year averages were also included to in the temporal distribution to support the linear trending. Spatial analysis by county used the ArcGIS (v.10.4) (Esri, Redlands, CA, USA) to generate thematic maps. Fatality rates were normalized by annual population for temporal trends and by the study period median population for the spatial distribution by county. Percentages were generally rounded to the nearest whole number unless otherwise necessary for comparative analysis.

3. Results

3.1. Types of Hydrometeorological Disasters

The *Storm Data* reports 55 disaster event types. For purposes of this study the disaster fatalities reported in Texas from 1959 to 2016 were categorized into one of the following nine hydrometeorological disaster types based on the information provided in the incident report or the NOAA database (Table 1). The definitions are consistent with the general classifications of weather disasters as defined by the National Weather Service.

Table 1. Definitions of Hydrometeorological Disaster Types.

Disaster Type	Characteristics
Flooding	Floods and flash floods due to extreme rain caused by hurricanes *, tropical storms, or other rain storm events
** Tornado	Wind event meeting the minimum classification of wind speed and ground contact
Lightning	Natural high voltage electrical discharge from atmosphere striking person or surface in proximity of person
Heat	Prolonged period of time with extremely high average temperature usually accompanied by drought
Cold Weather	Blizzards, snow storms, ice storms, and prolonged period of time with extremely low average temperatures
Wind	Extreme high winds causing damage but not meeting the minimum criteria of hurricanes or tornados
Other	Hail, water spouts, wildfires, or rain that directly resulted in some major structural damage (e.g., roof collapse)
Rip Current	Coastal specific disaster that includes people killed (drowned) in rip currents. Rip currents have only been tracked as of 1997

* The Saffir–Simpson hurricane wind scale (SSHWS), classifies hurricanes as Western Hemisphere tropical cyclones that exceed the intensities of tropical depressions and tropical storms with sustained winds of at least 74 mph (Category 1). ** The Enhanced Fujita scale (EF) classifies tornadoes based on wind speed and damage (once they have touch ground) from EF-0 (65–85 mph) to EF-5 (>200 mph).

Approximately 80% (205 of the 254) of the counties in Texas reported hydrometeorological fatalities in at least one year during the 58-year study period. Figure 1 shows the primary type of disaster that resulted in fatalities for each Texas county. Seventy-seven of the 205 counties that reported fatalities indicated that flooding was the primary disaster in their county. Thirty-one counties had more than one predominant cause of fatality and so were not categorized as primary disaster.

Counties that reported deaths due to flooding as the predominant hydrometeorological disaster are clustered towards the central region of the state and extend west towards Mexico/New Mexico in the region known as Flash Flood Alley. The disasters that caused the greatest number of fatalities along the Gulf Coast of Texas were wind events and lightning (Figure 1). Heat-related fatality counties were scattered in 11 counties across Texas and 75% of the cold weather fatality counties were in the northwest of Texas (Texas Panhandle) above the 34° N latitude. Fatalities due to tropical events (hurricanes and tropical storms) were all determined to be a result of subsequent flooding and therefore classified as flood-related fatalities. Death due to heat-related events was predominant in Harris County, the most populated county in Texas.

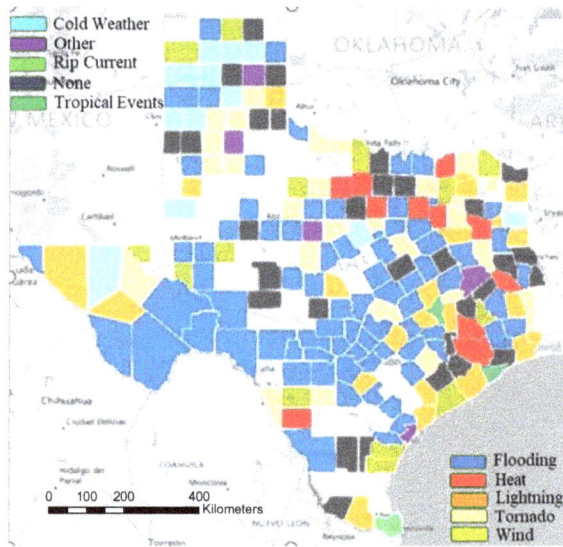

Figure 1. Primary disaster resulting in largest number of deaths per county. Dark grey: no one predominant disaster. Light grey: No hydrometeorological disaster fatalities reported.

A total of 2330 natural disaster-related fatalities occurred in Texas from 1959 to 2016 with 43% due to flooding (991 fatalities) as shown in Table 2. The second most frequent cause of fatalities was extreme heat (16%) followed by tornados (14%) and lightning (10%). The single most fatal natural disaster event during this 58-year period was the tornado of April 1979 that struck Wichita and Wilbarger counties killing 54 people and injuring 1807. This was an EF-4 (Enhanced Fujita scale) tornado with a maximum width of 2.5 km that killed four people along it's northeastern track through the states of Oklahoma (3 deaths) and Indiana (1 death). Seventy-nine percent (79%) of the total tropical storm-related fatalities were caused by hurricanes (108 deaths).

Table 2. Hydrometeorological Disaster Fatalities, source: NOAA (National Oceanic and Atmospheric Administration) *Storm Data* [24].

Disaster Type	Fatalities	% Total
Flooding *	991	42.5
Heat	378	16.2
Tornado	333	14.3
Lightning	222	9.5
Wind	172	7.4
Cold Weather	160	6.9
Other	43	1.9
Rip Current	31	1.3
Total	**2330**	**100**

* Includes fatalities due to hurricanes and tropical storms.

3.2. Temporal Distribution

3.2.1. Annual Distribution of Fatalities

An average of 42 fatalities per year occurred in Texas from 1959 to 2016 with a median of 33 fatalities per year and a total of 2330 hydrometeorological fatalities. The difference between the mean and the median indicates that the annual distribution is positively skewed with long tail in the right direction (towards higher numbers). Although, the raw number of fatalities exhibits a slight increasing trend during the study period, the trend has low linearity ($R^2 = 0.0369$, $p = 0.233$) and the relationship of fatalities over time is statistically weak (Spearman's $p = 0.16$). The lowest number of fatalities (6) occurred in 1963 and the highest number of fatalities (118) occurred in 1998 (Figure 2). Eleven of the 58 years had an annual number of fatalities greater than the 58-year mean plus one standard deviation (40 ± 24).

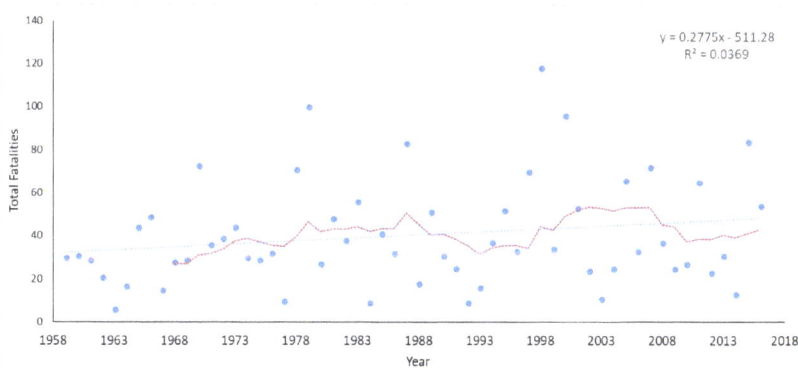

Figure 2. Total Fatalities from natural disasters in Texas from 1959 to 2016 with 10 year rolling average (red dashed line). The solid line represents the linear trend.

The curve of the cumulative annual fatality rate is relatively uniform with two observable spikes in 1978–1979 and 1998 driven by high fatalities resulting from heat and flooding events (Figure 3). Specifically, in 1978 Dallas County had 21 heat-related fatalities and 40 flooding fatalities that occurred in several counties including Bexar, Kerr, Shackelford, Bandera, and Randall counties. From May 1997 to August 1998 a severe heat event hit the southern region of the U.S. from Florida through Texas and into Colorado. Conversely, several flooding events in November 1998 resulted in fatalities in Bexar, Val Verde, Caldwell, Guadalupe, and Real counties.

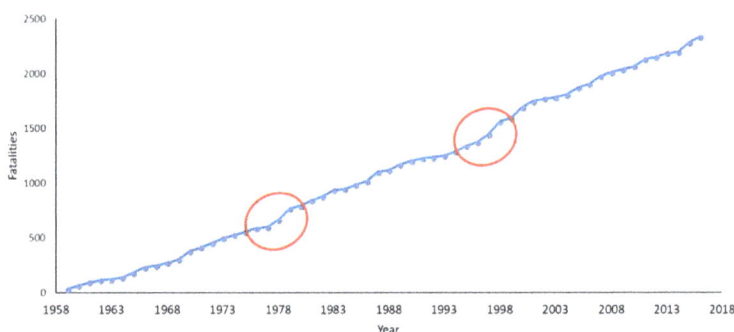

Figure 3. Cumulative number of fatalities (all disasters).

Normalization of the annual fatality rate by population indicates a decreasing trend with slightly better linearity ($R^2 = 0.096$, $p = 0.006$) than the raw trend and a stronger statistically relationship of fatalities over time (Spearman's $\rho = -0.355$) which suggests a gradual decrease in the risk of being killed by hydrometeorological events in Texas over the 58-year study period. The normalized fatality trend can be seen in (Figure 4), which shows the fatality rate due to hydrometeorological disasters per 100,000 Texas residents. Public awareness and educational weather safety campaigns in Texas may have contributed to this reduction of risk [13]. Although the rank order of counties by number of raw fatalities aligns with the highest populated urban centers, these regions are not necessarily the most dangerous. This is evidenced in that several of the counties in immediate proximity to the counties that experienced high fatality have very few fatalities even though the intensity and durations of the disasters were probably very similar between the counties.

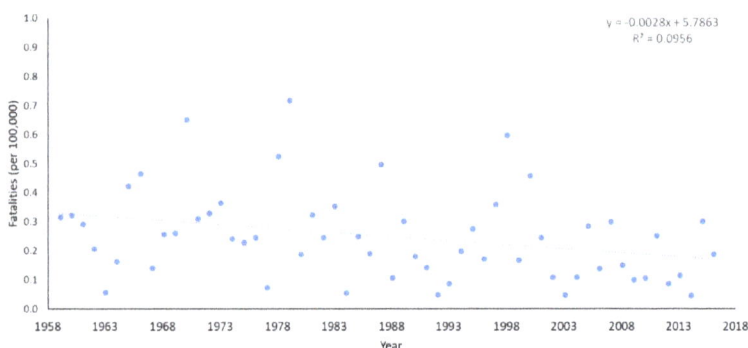

Figure 4. Normalized Fatality Rate from hydrometeorological disasters in Texas from 1959 to 2016. The solid line represents the linear trend.

For example, Loving County has the highest normalized fatality rate (> 4000 per 100,000) in the state due to its small and stagnant population and the fact that the county experienced several multi-fatality events of wind and hailstorm that struck the Red Bluff Lake area killing four persons by drowning when their boat capsized in the lake during a squall. Similarly, on 11 June 1965, the city of Sanderson in Terrell county was devastated by a flash flood. As noted by the Texas State Historical Association, "A wall of water washed down Sanderson Canyon into Sanderson, destroying numerous homes and businesses. Twenty-six people died in the flood. Eleven flood-control dams were constructed to protect Sanderson against another such catastrophe" [29]. The town had a population of 1500 in 1980 and 1128 in 1990. Reduction of fatality risk in these rural counties will require an increase in awareness through weather-related emergency educational programs and resource assistance (financial and physical) to implement safety systems.

The annual fatality rate by disaster type from 1959 to 2016 for six of the eight disaster types is shown in Figure 5. There seems to be a shift in the number of fatalities at the middle of the study period, especially for tornadoes and lightning. Splitting the study period into two equal parts (1959–1987 and 1988–2016) shows the first half having a greater proportion of the total fatalities for all disasters except for heat-related events. Heat-related events show an increasing trend of 0.45 fatalities per year ($R^2 = 0.54$) from 1978 to 2016 with no data available prior to 1978. Eighty-nine percent (89%) of all heat-related fatalities (335 out of 378) occurred after 1994. The increasing trend in heat fatalities is likely a compounded effect of higher than normal average air temperatures, the urban heat island effect, and increasing population in the urban regions. The retention of heat due to the abundance of non-natural building materials results in higher temperatures in the urban center than the surrounding area. As population increases in urban centers and higher temperatures the result will be increased heat fatalities especially for the most physically vulnerable such as the elderly. As noted in Figure 11, heat fatalities occur inside permanent residences about 50% of the time. This statistic is probably a conservative estimate since heat deaths may be under-reported for a variety of reasons. Exposure to extreme heat can cause cardiac or respiratory issues that can be fatal. Therefore, the judgment of the medical professional determines the cause of death as exposure to heat or the underlying medical condition. External factors also may contribute to under-reporting variability of heat fatalities especially along the U.S./Mexico border counties and in the case of chronically ill victims where it is unclear the final cause of death. Immigrant deaths along the border are uncertain due to international policy challenges.

The difference between the early and the latter half of the study period was highest for tornadoes with 71% in the first half of the study, followed by 68% for wind, 63% for cold weather, 62% for Lightning, 57% of tropical storms, and 52% for flooding. Not all years of the 58-year study experienced fatalities from all of the hydrometeorological disasters. Fatalities due to flooding. lightning and wind events were the most consistent occurring in 57, 56, and 47 of the 58 years respectively. Fatalities due to tornadoes, cold weather and heat events occurred in 41, 38, 26 of the 58 years respectively. The year 2011 was the only year that had no reported flood-related fatalities and was also the year that experienced one of the worst droughts in Texas history. Trend analysis over the entire 58-year period indicates statistically significant change (decreasing) for flooding ($R^2 = 0.112$, $p = 0.0104$), wind ($R^2 = 0.190$, $p = 0.0006$), lightning ($R^2 = 0.346$, $p = 1.21 \times 10^{-6}$), and an increasing trend for heat ($R^2 = 0.087$, $p = 0.0243$). Fatality trends due to tornadoes ($R^2 = 0.043$, $p = 0.1190$), and cold ($R^2 = 0.009$, $p = 0.4770$), exhibited slight downward trends but were not statistically significant.

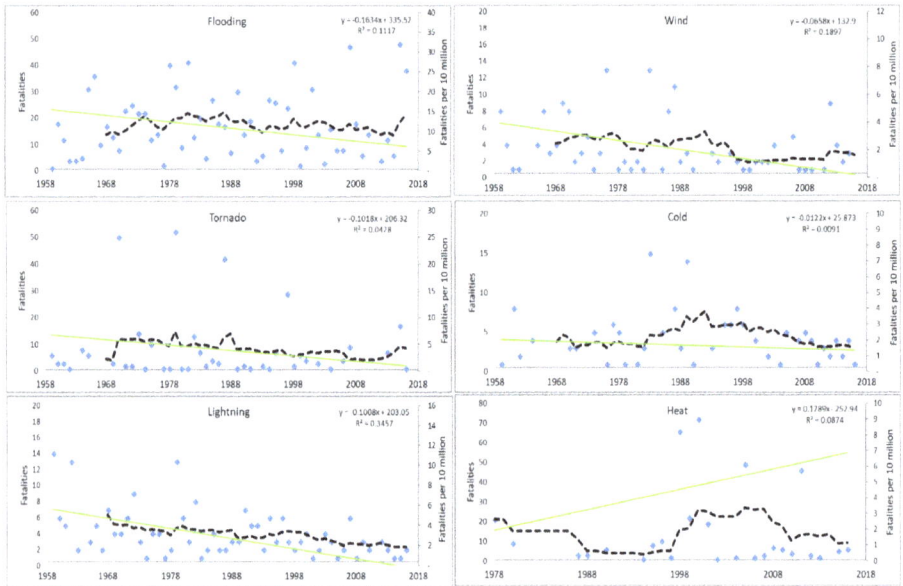

Figure 5. Annual fatality rate in Texas for six disaster types: flooding, wind, tornado, cold, lightning, and heat: 10-year rolling average (dashed line) and normalized (fatalities per 10 million) trend line (green solid). Note: Heat fatalities are reported from 1978 to 2016.

Fatalities due to tropical events (hurricanes and tropical storms) were mostly due to drowning and therefore were integrated into the flooding fatalities unless there was a clear distinction in the fatality record. Only seven tropical event fatalities occurred during the period 2003–2007. The disaster category "Other" includes wildfires and other secondary perils that do not frequently result in death, such as hail, water spouts, or rain that resulted in roof collapse. The "Other" disaster category indicated 71% of the years (41 out of 58 years) had zero fatalities with a steady increase in the fatality rate starting in 2004. Eighty-one percent (81%) of fatalities of this category occurred in 13 years between 2004 and 2016. Rip currents were added to the *Storm Data* in 1998. The first reported fatalities occurred in 2007 with two total fatalities at 31 deaths from 2007 to 2016 with an average of 3.4 per year and a high of 8 deaths in 2011. Five of the 8 deaths were Mexican immigrants visiting the coastal county of Cameron. More years of rip current fatality data is needed to establish any definitive temporal or spatial trends.

3.2.2. Monthly Distribution of Fatalities

The monthly fatality rate due to hydrometeorological disasters illustrates the seasonal variability in the number of fatalities for different types of disasters (Figure 6). A distinct peak is noticeable in May driven by flooding and tornado fatalities, which are responsible for 80% of the fatalities in the month. During the summer months, most fatalities were due to heat events while spring fatalities result primarily from flooding. Flooding fatalities were highest in the months of May, June, and October with 22%, 17%, and 13%, respectively of the total flood-related fatalities. Some disaster-related fatalities are obviously limited to certain seasons such as cold weather fatalities that occur in Winter (85% of all cold weather-related deaths occurred in December, January and February). Seventy one percent (71%) of all fatalities occurred in spring and summer with an even split between the two seasons.

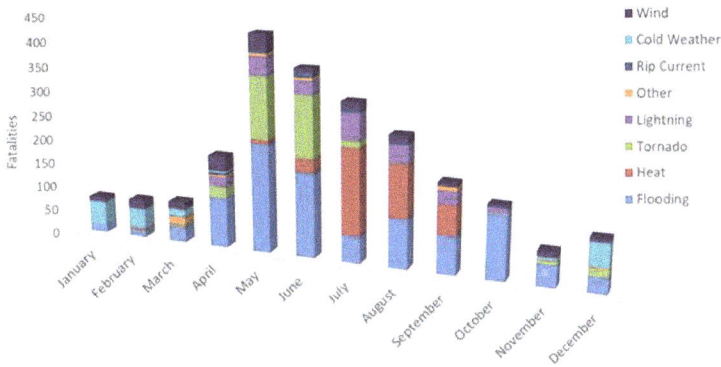

Figure 6. Monthly distribution of hydrometeorological disaster fatalities (all fatalities).

A grouping of the months into four seasons, Winter (December, January, February); Spring (March, April, May); Summer (June, July, August) and Fall (September, October, November), highlights the difference in rolling average fatality rates between the first half of the study period to the second half (Figure 7). The rolling averages show a slight decrease for spring and winter and increase in summer and fall with the largest difference observed in summer due primarily to an increase in heat-related fatalities that occurred between 1998 and 2008. Comparatively, all of the normalized trends for the four seasons have a decreasing trend winter (m = −0.0008, p = 0.687), summer (m = −0.0037, p = 0.600), fall (m = −0.0022, p = 0.563) with only the spring fatality trend (m = −0.0211, p = 0.019) having statistical significance.

3.2.3. Distribution of Fatalities by Time of Day

Time of day was provided for 68% of the hydrometeorological fatalities culled from the *Storm Data*. Each time of death was assigned to one of the four periods in a day: morning (6 a.m.–12 p.m.), afternoon (12 p.m.–6 p.m.), evening (6 p.m.–12 a.m.), and night (12 a.m.–6 a.m.). Of the fatalities with known time of death, 36% occurred in the afternoon, 26% in the evening, 21% in the morning, and 18% at night. Eighty percent (80%) of the fatalities with unknown time of the day were caused by flooding or heat-related events. Fifty percent (50%) of the total fatalities (562) that occurred in the afternoon were due to tornadoes and flooding. Detailed analysis shows that flooding events have a slightly higher chance of causing death at night or in the morning hours. However, tornados are much more likely to fatally strike in the afternoon/evening hours (Figure 8).

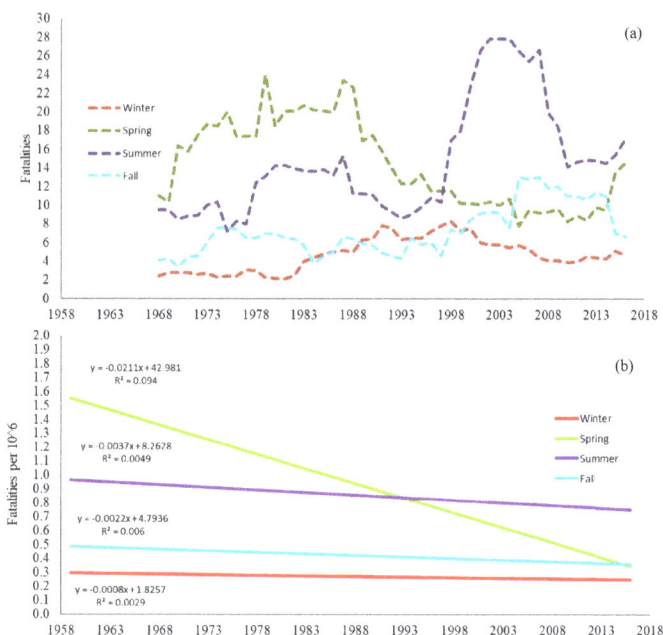

Figure 7. (**a**) Number of total fatalities (rolling 10-year averages) by season; (**b**) Normalized fatality rates by season.

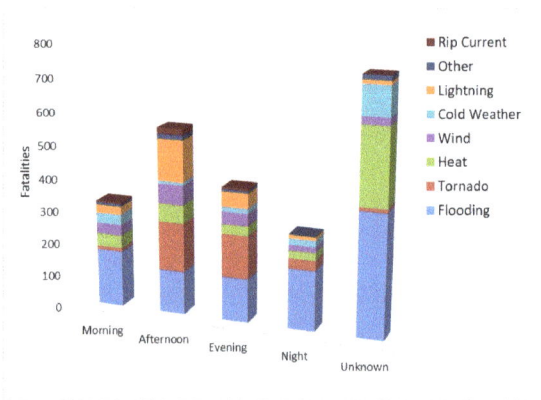

Figure 8. Distribution of hydrometeorological disaster fatalities by time of day.

3.3. Spatial Distribution

Most hydrometeorological fatalities occurred in populated counties (Harris (Houston), Bexar (San Antonio), Dallas and Tarrant (Dallas area), Travis and Williamson (Austin) as well as rural counties in west Texas with low populations high fatality numbers are noted in the Flash Flood Alley counties and some coastal counties (Figure 9).

Table 3 provides the ranking of the top 5 counties with the highest number of hydrometeorological disaster fatalities which combined, account for 32% of the total hydrometeorological disaster fatalities. Slightly more than 3% of the total reported fatalities did not include county information.

Forty-eight percent (48%) of the fatalities in Harris County were caused by heat-related events followed by flooding (33%) and lightning (11%). Heat events also caused the highest percentage of deaths (50%) in Dallas County followed by flooding (29%). Bexar County ranked third with 103 fatalities of which 82% were caused by flooding making Bexar the county the most dangerous in the state for death from flooding (per capita).

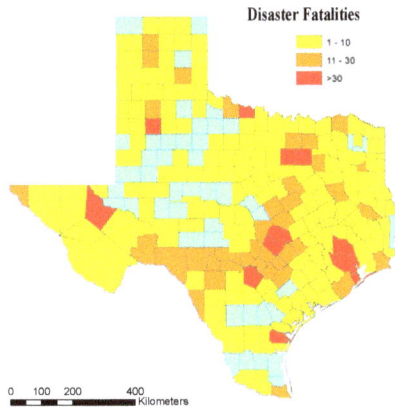

Figure 9. Raw number of hydrometeorological disaster fatalities by county.

Table 3. Top 5 Texas counties (Raw Fatality Rate). source: NOAA *Storm Data* (1959–2016) [24].

Rank	County	Fatalities
1	Harris	259
2	Dallas	228
3	Bexar	103
4	Tarrant	87
5	Travis	76

The top 5 counties within each hydrometeorological disaster type represent a significant percentage of the overall fatality rate within each of the category ranging from a high of 76% of all heat-related fatalities to a low of 28% of all wind event fatalities (Table 4). Dallas County is in the top five for six of the eight disaster types (heat, flooding, lightning, cold, wind, and others not shown in table). Harris is the only county that tops the list for more than one type of disaster and ranks number one for heat, flooding, and lightning fatalities and ranks second in wind fatalities. The top counties with the highest number of fatalities (and most populated counties) identified in Table 3 also dominate the top 5 ranking of counties in Table 4 for flooding, lightning and heat fatalities. Interestingly, although flooding is responsible for 43% of all disaster fatalities in the state, the top five counties only account for 34% indicating that flooding fatalities are extant over a high number of counties.

Table 4. Top 5 Texas counties with highest fatality rate (and % of total) by hydrometeorological disaster. source: NOAA *Storm Data* (1959–2016) [24].

Heat (76%)		Tornado (45%)		Wind (28%)		Flooding (34%)		Lightning (31%)		Cold (30%)	
Harris	124	Wichita	47	Nueces	16	Harris	85	Harris	29	Dallas	16
Dallas	113	Reeves	30	Harris	10	Bexar	84	Jefferson	12	Potter	12
Tarrant	22	Williamson	29	Dallas	8	Dallas	65	Dallas	11	McLennan	7
Montgomery	16	Lubbock	26	Brazoria	7	Travis	59	Tarrant	9	Tarrant	7
Travis	12	Donley	17	Denton	7	Tarrant	43	Bexar	8	Castro	6

3.4. Fatalities by Age and Gender

Age was provided for 57% of the total reported deaths (1333 fatalities) in which "Adults" made up 52%, the "Elderly" 28%, and "Children" 20% of the known age fatalities. In this study, "Children" are defined as newborns up to 17 years, "Adults" from 18 years to 64 years, and "Elderly" as persons above 65 years of age. Adults made up 53% and children made up 27% of all flooding fatalities. This fatality statistic requires more data (age aggregation) and further research into the specifics of the situation before any defensible conclusions can be drawn. On open conjecture it can be suggested that flooding is responsible for death of families which typically includes a higher number of children than adults that either did not evacuate or were killed during the evacuation process on transportation routes. Comparing the number of fatalities and the number of people in each age group provides a measure of relative risk of death within each age group. Using the 1990 census population in each age group: children (4,857,469), adults (10,420,598), and elderly (1,708,443) the relative risk of fatality for all hydrometeorological disasters during the period covered by this paper indicates a similar risk between children (0.006%) and adults (0.007%). But the risk of the elderly dying is more than three times the level at 0.022% due to the much lower elderly population (approximately 10% of total population). The elderly were mostly at risk for heat events, accounting for 52% of all heat-related fatalities (Figure 10).

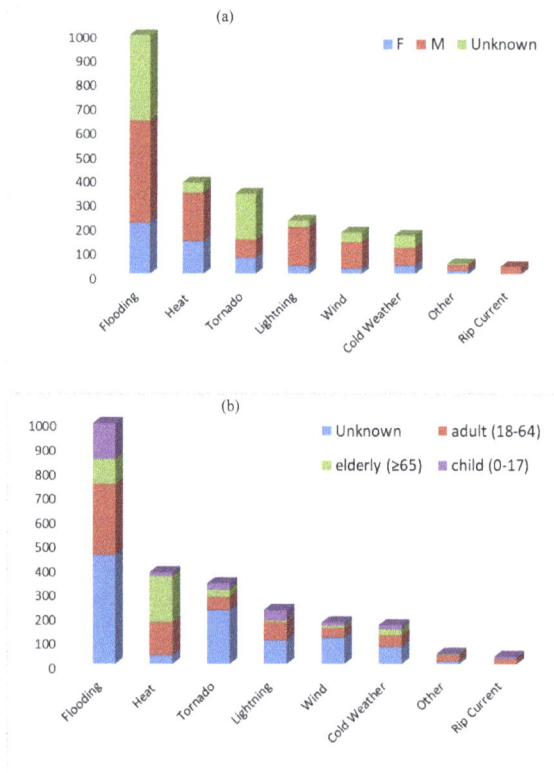

Figure 10. Total natural disaster fatalities considered in this study classified (**a**) by gender and (**b**) by age group

The gender of the victim was provided in 69% of the reported fatalities. Among those, males made up 68% and females 32% representing an approximate ratio of 9:5. This gender disparity has also been observed in other research, for example in flood-related fatalities [12,13,30,31] and in lightning-related fatalities [32–35]. In all cases there was a high male to female ratio of fatalities. For the current study, the ratio of male to female fatalities is approximately 2:1. The greatest disparity was found in wind and lightning fatalities that show a 5:1 ratio of male to female fatalities. Rip currents have only been tracked since 1998 but the data thus far indicates a 30:1 ratio of male to female fatalities.

3.5. Fatalities by Activity Location

The *Storm Data* describes 18 potential activity fatality locations. For purposes of this study each of the disaster events reported in Texas from 1959 to 2016 was categorized within one of the following nine locations identified in based on the information provided in the incident reports (Table 5).

Table 5. Definitions of Hydrometeorological Disaster Fatality Locations.

Location	Definition
In Water	Streams, river, bayous, oceans, floods, etc. and includes activities such as swimming, boating, surfing, and working on oil rigs
By Water	Boat docs, levies, beaches or other types of shoreline appurtenances
Temporary or non-Permanent Shelters	Tents, car ports, trees, and other temporary shelters that do not have a foundation (excluding umbrellas)
Outside	People who were outside but not in or near water, people standing in lawns, in construction sites that did not offer shelter, in ball fields, parks, golf courses, etc. People seeking shelter under umbrellas are also included. People standing/sitting near or on top of personal vehicles that are not along a transportation rout are included in outside (e.g., people walking from their home to their car who died before reaching their vehicle, people sitting on top of trucks in fields)
Transportation Route	Roadways, freeways or toll ways, parking lots, sidewalks or air travel routes. People walking along roads who hid behind a vehicle right before the disaster are categorized under transportation routes. Fatalities in vehicles were not assumed to be along transportation routes and were classified as unknown unless the description indicated a transportation route. Exclusion: people hiking or traveling along non-established routes by foot were not included in this category, and instead were classified as "outside"
Mobile Home	Standard and double-wide mobile homes
Permanent Residence	Domiciles that have a foundation, including but not limited to brick houses, frame houses, and apartment buildings
Public and Permanent Buildings	Schools, restaurants, airports, and other buildings with foundations that are not residences
Other/Unknown	All other locations not described by any of the other location categories listed or if the location was not specified

The activity location in which the fatality occurred was provided in 75% of the total fatalities that were reported in Texas from 1959 to 2016. Figure 11 shows the stratification of fatalities by location of occurrence and disaster types considered in this study. Fatalities with known locations, occurred most often (38%) on transportation routes such as roadways, freeways or toll ways, parking lots, sidewalks or air travel routes. Automobile accidents are not categorized as transportation routes unless they were specified as such in the report. Eighteen percent (18%) of known location fatalities occurred in "Permanent Residence", followed by "Outside" and "In Water" at 16% and 15%, respectively. The high fatality rate in and around certain activity locations observed in Texas is also highlighted in research conducted in Switzerland for the period 1946–2015 [36] in which the researcher noted the greatest number of natural disaster fatalities occurring on transportation routes (33%), followed by in or around buildings and open terrain.

Sixty-five percent (65%) of fatalities on transportation routes were caused by flooding. This percentage is potentially underestimated since 65% of the fatalities in an unknown location were caused by flooding. As noted in other Texas flooding fatality studies [31] driving into flash flooding conditions is a significant occurrence that would make it very difficult to assign a location with no clear transportation route known. Also 25% of tornado fatalities are reported with an unknown location. Forty-eight percent (48%) of heat-related fatalities occurred in permanent residences. Tornados caused 73% of all hydrometeorological disaster fatalities reported in mobile homes. It must not be overlooked that the "Other" category included 11 deaths of children as result of being left in a car unattended and succumbing to heat exposure, a very preventable tragedy.

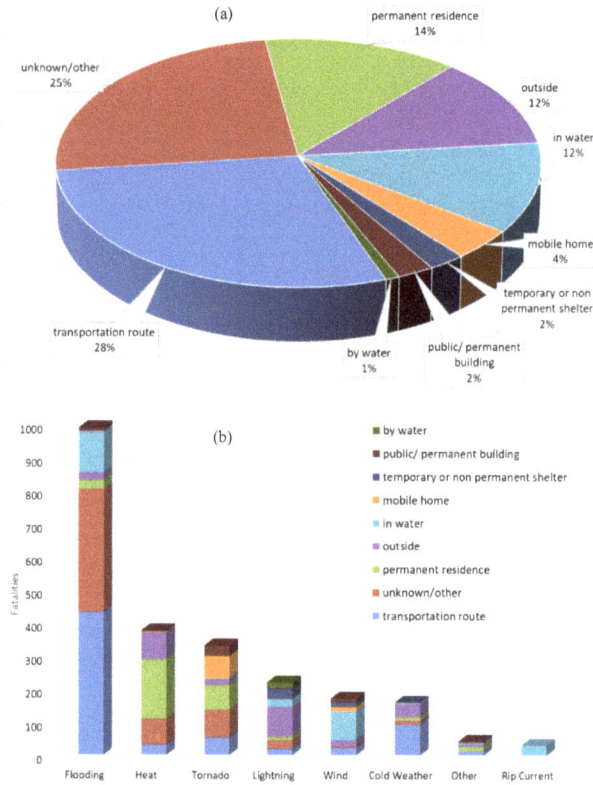

Figure 11. Hydrometeorological disaster fatalities classified (**a**) by reported location of occurrence and (**b**) by disaster type and location

4. Discussion

The predominant types of natural disasters in Texas that result in fatalities are those initiated by weather conditions such as flooding, tornadoes, and extreme temperatures. This study did not analyze the climate conditions or associate global warming to disaster events, but rather its intent was to analyze the spatial and temporal distribution of fatalities by disaster type. Regardless of the reasons for changes in frequency or intensity of the hydrometeorological disaster events, the parametric shifts can challenge the preparedness and resiliency of a region and in many cases impact the number of fatalities incurred. Analysis of these types of natural disaster trends based on historic data can enhance predictability and preparedness planning to reduce the loss of life. Regional mortality and morbidity is also affected by the demographics and behavior of the people in the region of impact, specifically age, gender, and behavior patterns (location) have an observable relationship to the fatality rate due to hydrometeorological disasters.

4.1. Population and Fatality Rates

Texas exhibits regional variability in the hydrometeorological disaster fatality rate that is weighted to regions of high population. This suggests that as more people continue to move into populated urban areas or into regions that are at higher risk for hydrometeorological disasters such as flood plains, tornado alleys, or coastal regions, fatality rate will likely increase with or without an increase in the number of disaster events. Highly populated regions are more susceptible to a higher number of natural disaster fatalities than lower populated regions due to the sheer number of persons per area. As the population of Texas and the number of hydrometeorological disasters continues to increase, the result will likely be a continuing increase trend in the number of hydrometeorological fatalities. The current population growth rate for Texas is 1.8% which is the third in the U.S. According to the Texas Demographic Center [37], the vast majority of population growth since 1850 has occurred in metropolitan areas while the population in non-metropolitan counties has declined. This urban population increase coincides with an increasing trend of the annual fatalities as noted in Section 4.2.1.

The counties with the greatest population density: Harris, Dallas, Bexar, and Travis had the highest actual fatality rate, but each less than 15 fatalities per 100,000 persons over the study period. In contrast, some counties with lower populations had much higher per capita fatalities (higher risk for fatalities) although they were adjacent to the high population counties and experienced similar hydrometeorological disaster frequency and intensity. For example, Bexar county had 8.7 fatalities per 100,000 while surrounding county of Comal had 188 fatalities per 100,000 people. Harris county had 9.5 fatalities per 100,000 and the surrounding counties of Brazoria Chambers had 55 and 40 fatalities per 100,000 respectively. Figure 12 shows that per capita fatality rates are highest in sparsely populated counties in the southwestern portion of Flash Flood Alley and the Texas Panhandle in the northwestern part of the state.

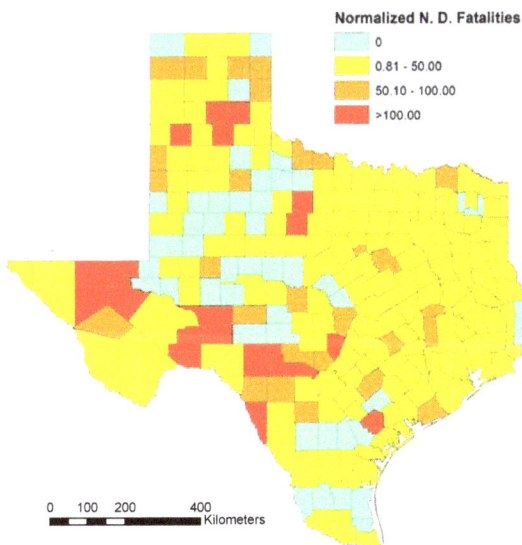

Figure 12. Fatalities Normalized by Population (per 100,000).

4.2. Activity Locations for Fatality Occurrences

Twenty-five percent (25%) of the reported fatalities did not include a specific activity location of occurrence. But even with this uncertainty the available data strongly suggests that transportation routes are a leading fatality location. Approximately 38% of the total number of hydrometeorological disaster fatalities that reported a location (1756) identified transportation route as the activity location in which the fatality occurred. Approximately two-thirds (65%) of the fatalities on transportation routes were due to flooding which implies that driving into flood conditions is a frequent high-risk activity that likely contributed to many of these deaths. This conjecture is similar to conclusions drawn from research conducted in a study of natural hazard fatalities in Switzerland for the period 1946–2015 [36]. Additionally, research conducted at the University of Texas found that 73% of all flood fatalities in Texas during the period 1959 to 2008 were vehicle related and 16.5% were due to people walking into floodwaters from [13]. Further analysis of hydrometeorological fatalities in the U.S. suggests that flood fatalities are likely underestimated. Sixty-five percent (65%) of fatalities with an unknown location were due to floods, suggesting probable vehicle related incidents with no specific transportation route established.

The number of fatalities on transportation routes are directly related to the number of people on transportation routes which is related to the economic development of the affected region and the demographics of the people professionally and personally committed to the transportation routes. Developed economies result in more transportation between place of employment, schools, commercial and recreational destinations. This skew may be offset since regions of greater wealth and communication networks are more likely (have the ability) to evacuate out of harm which reduces the fatality rate when compared to regions of lower wealth and economic development who are unable or unwilling to evacuate. The employment rate, family structure (dual income/single income), cultural norms, and habits and behavior of the affected population significantly impacts the location of individuals at any given time and thereby impacts the number of fatalities experienced by hydrometeorological disaster events.

4.3. Gender and Age

Although gender and age were not reported for a large number of disaster fatality victims, the available information indicates more male fatalities than female fatalities. The gender gap in the hydrometeorological fatality rate exhibited a decreasing trend from the early years of the study to the more recent years suggesting a change in exposure possibly due to shifting roles and responsibilities of men and women in society.

Gender and age are two key demographics that differentiate in lifestyle, behavior and risk tolerance that ultimately affect the fatality rate. Female fatalities due to natural disasters increased significantly from the first to the second half of the study period. The 10-year rolling average more than quadrupled from three deaths in 1969 to 13 deaths in 2016. For the male population, the 10-year rolling average only increased (180%) from 10 deaths to 28 deaths during this same period. If it is generally accepted that contemporary (2017) societal gender roles and responsibilities are not the same as they were in 1959, some reasons for this difference may be in changing situational exposure of the workforce coupled with the level of risk accepted by males versus females. A change in situational exposure is evident in the recent years which is experiencing more role reversal in traditional male/female work roles (e.g., outdoor labor/indoor service).

The impact of gender and risk tolerance is most obvious in the rip current fatality numbers albeit culled from a limited time period (1997–2016) with a 30:1 ratio of male to female fatalities. Deaths from rip currents are more likely a factor of the difference in male / female risk tolerance than in a change of societal roles with more males swimming and pursuing water sports in an area and time of rip current activity. Tornado fatalities are an outlier to the general trends exhibiting a higher female to male fatality ratio, but more data is needed since a potential skew may exist with only 50% of all tornado fatalities reporting gender.

Risk tolerance also changes with age. Within the fatalities that included age (54% of total fatalities), two high risk groups stand out with the largest number of fatalities; 20–29 years. young adult age group and the 70–79 years age group. Changes in priori-ties, family, education, and responsibilities occur for most in this 50-year span and the accepted risk taking from youthful invincibility (such is likely the case for this age group that frequently drive into flash flood conditions), typically progresses to a longer period of more stable (less risky) lifestyle, until the later years when human vulnerabilities of age-related health limitations and immobility issues results in an increase potential for succumbing to hydrometeorological disasters. The limited number of fatality reports that includes age information in this study increases the uncertainty within the trend analysis and is an area that warrants further research. But within the confines of the given data, the elderly appear to be most susceptible to hydrometeorological disaster fatalities based on the percentage of total fatalities and taking into account the relatively low overall percentage of the Texas elderly population. The Texas Demographic Center statistics indicate that Texas population is 27.3% (<18 years, child), 62.4% (18–64 years, adult), and 10.3% (≥65 years, elderly) [37]. The age of the national and state populations overall is increasing and therefore the 2010 census estimates are conservative when comparing the fatality rates among the age groups from this study taking into account the much smaller number of elderly population versus younger age groups.

This study identified that the top two locations for elderly fatalities due to hydrometeorological disaster occurred in permanent residences (45%) and transportation routes (24%) which suggests a high risk to the homebound elderly segment of the population and to elderly when evacuating a disaster. Heat events and flooding were the top two disasters killing the elderly accounting for almost 80% of all elderly fatalities. The number of heat related fatalities is on the rise across all age groups but is particularly evident in the southwest of the U.S. and is usually combined with an extended drought period. Increasing global temperatures may be a factor in this trend with 2016 having the highest average temperature on record as well as the highest monthly temperatures in eight of the 12 months (January–May and July–September) [38]. With heat-related deaths heavily weighted towards the elderly that sometimes do not have family or social networks to acknowledge and report their demise, the fatality rate in this age group may be under-reported

The elderly are not only at risk of heat-related fatalities, but are also very vulnerable to flooding triggered by heavy rains and/or high winds such as exist in hurricane or tropical storm conditions. The fatality rate due to flooding ranked second for the elderly in the study but 45% of all flooding fatalities not reporting age of the victim, there is some uncertainty in this statistic. More certain is that flooding devastates a community on many levels including power interruptions and blockage of transportation routes. Medical attention is a necessity for many elderly, whether it is within a medical facility, nursing home or the need to obtain prescription drugs. All these are typically inhibited during a flooding condition. Although there is debate on whether it is safer to evacuate or shelter the elderly in place during a flood, improvement strategies to reduce fatalities should not be stalled until final consensus.

The study also found that flooding was responsible for 53% of the total hydrometeorological disaster fatalities where the victims were known to be children (<18 years). Young children tend to outnumber adults in a family unit and are dependent on the care and good decisions of parents or guardians. If the family guardian makes a decision (e.g., driving into dangerous flood waters) that result in a fatal outcome it is likely that the number of children will be greater than adult fatalities for the same incident. Data from this study indicates the top activity location for child flood fatalities is on transportation routes (38%) and seems to support the conjecture.

4.4. Evacuation or Shelter in Place

In response to a pending flood or hurricane event, unless there is a mandatory evacuation order given by the city or county jurisdiction, the critical questions to consider is whether it is safer to evacuate by driving or walking to the evacuation location or if there is less risk of harm to shelter in place. For example, although flood fatalities are most likely to occur on transportation routes, all 68 fatalities (except one) during Hurricane Harvey occurred inside homes. Several factors should be considered when deciding to evacuate or shelter in place such as the perceived risk to the specific area of residence, number of children and elderly in the family, health condition and mobility of each family member, condition of the shelter residence, condition and availability of transportation, and evacuation destination distance. The decision to evacuate during a hurricane or flood can be the difference between life and death.

This decision becomes especially critical with regards to the elderly residing in nursing homes and long-term care facilities (LTCF). Research conducted in 2017 by Pierce on disaster preparedness of LTCF [39] found deficiencies in integrated and coordinated disaster planning, staff training, practical consideration before governments order mandatory evacuations, and accurate assessment of the increased medical needs of LTCF residents following a disaster. Previous research on the management of nursing home residents [40] found that, "the decision to completely evacuate, partially evacuate (including transfers of individual residents), or to shelter in place must be based on the integration of real-time data regarding the disaster event, the facility in question, and the clinical profiles of the residents at risk". Similar research by Dosa et al. [41] specific to Hurricane Katrina and Rita based on a survey of LCTF administrative directors noted a much higher mortality rate with evacuation actions than with shelter in place that was attributed to lack of governmental assistance, unsupported technical and physical requirements for transportation, and difficulty in retaining adequate staff.

4.5. Temporal Distribution

The fatality rate of annual raw fatalities increased from 1959 to 2016 with a maximum of 118 fatalities in 1998 due primarily to heat events in the months of July and August and flooding in August and October. Stratification of total fatalities by season indicated that the majority (70%) of all fatalities occurred in spring and summer with floods as the predominant disaster in spring and heat-related deaths in Summer. Monthly variation indicates the highest risk for flooding and tornado fatalities in April and May and the highest risk of heat events in August and September. Within the 68% of the hydrometeorological disaster fatalities that reported the time of day, the data in this study suggests that the afternoon period has the highest risk of fatality from tornado, flooding, or lightning.

Based on the current level of understanding in the relationship between earth sciences and meteorological conditions there is limited scientific predictability of disaster impacts. Predicting hydrometeorological disasters is challenging not only from a scientific basis but also because the fatality rate is not solely a factor of the type of the disaster but is impacted by societal activities of the region in which the disaster event may occur. In general, hurricanes have some level of temporal probability and typically make landfall at night when the storm strengthens due to the latent heat release in the upper and middle atmosphere. Tornadoes also tend to occur in the late afternoon and early evening hours, when the atmospheric conditions are most ripe for supercell thunderstorms and are most common from 4 p.m. to 9 p.m. in the evening [42]. Disaster events such as flooding are dependent on the amount and rate of precipitation and location of adjacent bodies of water (coastal, riverine, or inland). The resulting impact of such a disaster is a function of the activities occurring in the community affected at the time of the flood such as transportation density. Other factors include the level of early warning and evacuation, the time of the day of the flooding to accommodate or hinder rescue and transport efforts.

Similarly, if tornadoes strike during the day (especially a weekday) more people are at work or school and are in buildings where there are adequate public shelter facilities that are typically more disaster resilient than a private residence. Although one quarter of tornado deaths did not include an activity location, within the known study data, only 12% of the fatalities occurred in public/permanent buildings to support this conjecture. The study data also identified that more than 80% of lightning fatalities occur outside, in or around water, and temporary shelters. The extent of fatalities due to lightning is an example of the combined effect of the disaster and the victims location.

Lightning fatalities have decreased significantly on a national and state level in the last several decades as a result of a decrease in exposure (outdoor labor, agricultural work) and the strengthening of OSHA (Occupation-al Safety and Health Agency) safety protocols. Children and adults are the high-risk age groups for lightning fatalities and mitigation efforts to reduce the fatality rate can include increased public awareness in school and at the workplace to move or stay indoors during lightning events. This is especially critical for early morning lightning storms which have the greatest killing potential due to the electric charge build-up overnight [43].

5. Conclusions and Recommendations

This study reviewed the fatality rates due to hydrometeorological disasters in Texas over a 58-year study period (1959–2016) with the objective of providing perspectives and information to enhance public awareness, support investment in infrastructure improvement, and serve as input to state and regional disaster mitigation plans. The ability to reduce the number of hydrometeorological fatalities in Texas should not be underestimated. Resources are available but require political will to drive prioritized allocation to ensure weighted coverage in the highest risk areas. Information gleaned from the review of trends from historic hydro-meteorological disasters analyzed in this study can assist decision-makers in determining the best allocation of resources to provide maximum mitigation potential for high risk disasters and regions.

Based on the *Storm Data* analyzed in this study the normalized fatality rates are decreasing for all hydrometeorological disasters except for heat fatalities. The overall growth in population and urban centers plays a key role in the decreasing normalized fatality rates. But population growth appears to have an increasing effect on heat fatalities. The study results show that heat fatalities have a strong correlation to counties with high population density as well as disproportionately effecting the elderly segment of the population. Dedicated financial support can improve emergency preparedness for the elderly in nursing homes, long-term care facilities and private residences to ensure backup power, channels of communication, and available transportation to address immobility issues for the elderly in the case of mandatory evacuation or the necessity to shelter in place.

In addition to the elderly being susceptible to heat fatalities, they are also vulnerable to flooding. The main reasons appear to be related to mobility issues and interruption in medical care. Senior residences and long-term care facilities must have the ability to safely evacuate all their residents if required or be able to shelter in place with all necessary medical staff, medication and back-up power for prolonged medical assistance. Adequate early warning and funding to build a preparedness plan and inventory are vital components to this cause. Requirements for emergency staffing and assistance must be mandated through policy with preparation and training funded before a disaster strikes. Flooding is also the leading killer of adults and children in Texas especially on transportation routes. Adequate road, bridge, and waterway maintenance and improvement to reduce roadway flooding should be an ongoing approved budget item in lieu of recreational upgrades or other low risk projects.

Conversely to populated urban centers that have a higher number of actual fatalities, regions with a low population density exhibit a higher normalized fatality risk. Although the normalized fatality risk is inversely proportional to the population due to the low number of people, the options for survival can still be improved through better preparation. Low population counties are typically rural and do not receive as much funding for road and water management projects. Engineering building codes also maybe more lax contributing to devastation from tornado or other high wind events especially on the coast during hurricane season. Coastal land development must be managed to avoid permanent or non-permanent housing being established in high risk hurricane and storm surge zones. Rural poverty should not be directly related to the risk of death due to a hydrometeorological disaster. The county and state should focus disaster preparation awareness and ensure basic funding is made available to those low population areas with high fatality rates.

Flooding, heat, and tornado events rank as the top three causes of hydrometeorological disaster fatalities in Texas. Regions that are prone to non-coastal flooding are predominantly in the counties within the regions known as Flash Flood Alley and incur a high number of fatalities on transportation routes. Therefore, risk reduction can be supported by investment in roadway flood control improvement including early warning flash flood signage, establishing alternate routes in case of emergencies and mandatory evacuation, preemptive emergency public transportation protocols, and public awareness through educational programs. Tornadoes occur most often in the northeastern counties of Texas, particularly in the months of April and May, and predominantly affect those in temporary or non-permanent shelter (e.g., mobile homes). Contingency planning for the segment of society that is vulnerable to tornadoes can include more frequent public awareness and information campaigns during these months along with practice drills for what to do and where to go when a tornado touchdown is likely. Ensuring that emergency shelters in proximity to mobile home communities are available, accessible, and publicized during these high-risk months also has the potential to save lives. Similar basic considerations can also reduce the risk of fatalities for cold weather, wind events and other types of natural hazards. It is imperative that research builds on historic data to better understand the synergy between high risk disasters, regions and vulnerable segments of society to reduce the risk of hydrometeorological disaster fatalities in Texas.

Author Contributions: A.M.C. provided manual aggregation of fatality data contained in archived pdf files and electronic files from the NOAA *Storm Data* repository from 1959 to 2016. H.O.S. provided interim review, comments, and professional guidance in all aspects of writing this paper. S.H.P. performed the quantitative data analysis, qualitative interpretation of results and discussion, and wrote this paper

Funding: This research was funded by the Nuclear Regulatory Commission (NRC) Fellowship Grant #NRC-HQ-60-17-G-0036.

Acknowledgments: We are grateful to the University of Texas at San Antonio for faculty and technical support and the Nuclear Regulatory Commission for financial support of this research.

Conflicts of Interest: The authors declare no conflict of interest.

References

1. Munich Re. NatCatSERVICE, Natural Catastrophe Know-how for Risk Management and Research. Natural Catastrophe Online Tool. Available online: http://natcatservice.munichre.com/ (accessed on 17 May 2018).
2. Hahn, D.; Viaud, E.; Corotis, R. Multihazard Mapping of the United States. *ASCE-ASME J. Risk Uncertain. Eng. Syst. A Civ. Eng.* **2016**, *3*, 04016016. [CrossRef]
3. Borden, K.; Cutter, S. Spatial patterns of natural disasters mortality in the United States. *Int. J. Health Geogr.* **2008**, *7*, 1–13. [CrossRef] [PubMed]
4. Chowdhury, A.; Mushtaque, R.; Bhuyia, A.; Choudhury, A.; Sen, R. The Bangladesh cyclone of 1991: Why so many people died. *Disasters* **1993**, *17*, 291–304. [CrossRef] [PubMed]
5. Gerritsen, H. What happened in 1953? The Big Flood in the Netherlands in retrospect. *Philos. Trans. R. Soc.* **2005**, *A363*, 1271–1291. [CrossRef] [PubMed]
6. Jonkman, S.; Maaskant, B.; Boyd, E.; Levitan, M. Loss of life caused by the flooding of New Orleans after hurricane Katrina: Analysis of the relationship between flood characteristics and mortality. *Risk Anal. Int. J.* **2009**, *29*, 676–698. [CrossRef] [PubMed]
7. Kure, S.; Jibiki, Y.; Quimpo, M.; Manalo, U.; Ono, Y.; Mano, A. Evaluation of the Characteristics of Human Loss and Building Damage and Reasons for the Magnification of Damage Due to Typhoon Haiyan, Coastal. *Eng. J.* **2016**, *58*, 1640008. [CrossRef]
8. Ashley, S.; Ashley, W. Flood fatalities in the United States. *J. Appl. Meteorol. Climatol.* **2008**, *47*, 805–818. [CrossRef]
9. Singh, O.; Kumar, M. Flood events, fatalities and damages in India from 1978 to 2006. *Nat. Disasters* **2013**, *69*, 1815–1834. [CrossRef]
10. Paulikas, M.; Rahman, M. A temporal assessment of flooding fatalities in Pakistan (1950–2012). *J. Flood Risk Manag.* **2015**, *8*, 62–70. [CrossRef]

11. FitzGerald, G.; Du, W.; Jamal, A.; Clark, M.; Hou, X. Flood fatalities in contemporary Australia (1997–2008). *Emerg. Med. Australas.* **2010**, *22*, 180–186. [CrossRef] [PubMed]

12. Sharif, H.; Jackson, T.; Hossain, M.; Zane, D. Analysis of Flood Fatalities in Texas. *Nat. Disasters Rev.* **2014**, *16*, 04014016. [CrossRef]

13. Sharif, H.; Jackson, T.; Hossain, M.; Bin-Shafique, S.; Zane, D. Motor Vehicle-related Flood Fatalities in Texas, 1959–2008. *J. Trans. Saf. Secur.* **2010**, *2*, 325–335. [CrossRef]

14. Fox News. Fox Facts: Hurricane Katrina Damage. 2006. Available online: http://www.foxnews.com/story/2006/08/29/fox-facts-hurricane-katrina-damage.html (accessed on 2 April 2018).

15. Wikipedia: Hurricane Harvey. Available online: https://en.wikipedia.org/wiki/Hurricane_Harvey (accessed on 2 April 2018).

16. Spatial Disaster Events and Losses Database for the United States (SHELDUS); Hazards and Vulnerability Research Institute. U.S. Hazard Losses (1960–2015) Summary Report. 2017. Available online: http://hvri.geog.sc.edu/SHELDUS/index.cfm?page=reports (accessed on 2 April 2018).

17. National Oceanic and Atmospheric Administration. NOAA. U.S. Tornado Climatology. 2017. Available online: https://www.ncdc.noaa.gov/climate-information/extreme-events/us-tornado-climatology (accessed on 2 April 2018).

18. Ashley, W.S. Spatial and temporal analysis of tornado fatalities in the United States: 1880–2005. *Weather Forecast.* **2007**, *22*, 1214–1228. [CrossRef]

19. Changnon, A.; Pielke, R.; Changnon, D.; Sylves, R.; Pulwarty, R. Human factors explain the increased losses from weather and climate extremes. *Bull. Am. Meteorol. Soc.* **2000**, *81*, 437–442. [CrossRef]

20. Cutter, S.L.; Finch, C. Temporal and spatial changes in social vulnerability to natural disasters. *Proc. Natl. Acad. Sci. USA* **2008**, *105*, 2301–2306. [CrossRef] [PubMed]

21. Cutter, S.; Emrich, C.; Gall, M.; Reeves, R. Flash Flood Risk and the Paradox of Urban Development. *Nat. Disasters Rev.* **2017**, *19*, 05017005. [CrossRef]

22. Milch, K.; Broad, K.; Orlove, B.; Meyer, R. Decision Science Perspectives on Hurricane Vulnerability: Evidence from the 2010–2012 Atlantic Hurricane Seasons. *Atmosphere* **2018**, *9*, 32. [CrossRef]

23. Sault, S. Why the Hill Country Is A.K.A. 'Flash Flood Alley' 2016. Texas Hill Country. Available online: http://texashillcountry.com/why-the-hill-country-is-a-k-a-flash-flood-alley/ (accessed on 2 April 2018).

24. National Oceanic and Atmospheric Administration (NOAA). National Centers for Environmental Information. Storm Events Database 2017. Available online: https://www.ncdc.noaa.gov/stormevents/ (accessed on 2 April 2018).

25. Border Patrol. Southwest Border Sectors. 2018. Available online: https://www.cbp.gov/sites/default/files/assets/documents/2017-Dec/BP%20Southwest%20Border%20Sector%20Deaths%20FY1998%20-%20FY2017.pdf (accessed on 2 April 2018).

26. Lomonaco, C. U.S. Mexico Border: The Season of Death. *PBS Frontline World.* 2006. Available online: https://www.pbs.org/frontlineworld/blog/2006/06/usmexico_border_1.html (accessed on 2 April 2018).

27. United Nations Education, Scientific, and Cultural Organization (UNESCO), Disaster Risk Reduction. Available online: http://www.unesco.org/new/en/natural_sciences/special-themes/disaster-risk-reduction/natural-hazards/hydro-meteorological-hazards/ (accessed on 2 April 2018).

28. National Weather Service (NWS). Storm Events Database. Directives Systems. 2018. Available online: http://www.nws.noaa.gov/directives/010/010.php (accessed on 2 April 2018).

29. Pitts, Swanya H., Texas State Historical Association (TSHA). Sanderson, Tx. Available online: https://tshaonline.org/handbook/online/articles/hjs07 (accessed on 2 April 2018).

30. Coates, L. Flood Fatalities in Australia, 1788–1996. *Aust. Geogr.* **1999**, *30*, 391–408. [CrossRef]

31. Sharif, H.; Hossain, M.; Jackson, T.; Bin-Shafique, S. Person-Place-Time Analysis of Vehicle Fatalities Caused by Flash Floods in Texas. *Geomat. Nat. Disasters Risk* **2012**, *3*, 311–323. [CrossRef]

32. Singh, O.; Singh, J. Lightning fatalities over India: 1979–2011. *Meteorol. Appl.* **2015**, *22*, 770–778. [CrossRef]

33. Navarrete-Aldana, N.; Cooper, M.A.; Holle, R.L. Lightning fatalities in Colombia from 2000 to 2009. *Nat. Disasters* **2014**, *74*, 1349–1362. [CrossRef]

34. Elsom, D. Deaths and injuries caused by Lightning in the United Kingdom: Analyses of two databases. *Atmos. Res.* **2001**, *56*, 325–334. [CrossRef]

35. Curran, E.; Holle, R.; Lopez, R. Lightning Casualties and Damage in the United States from 1959 to 1994. *J. Clim.* **2001**, *13*, 3448–3464. [CrossRef]

36. Badoux, A.; Andres, N.; Techel, F.; Hegg, C. Natural Disaster Fatalities in Switzerland from 1946 to 2015. *Nat. Disasters Earth Syst. Sci.* **2016**, *16*, 2747–2768. [CrossRef]

37. Texas Demographic Center. Projections of the Population of Texas and Counties in Texas by Age, Sex and Race/Ethnicity for 2010–2050. 2014. Available online: http://txsdc.utsa.edu/Data/TPEPP/Projections/ (accessed on 2 April 2018).

38. Shaftel, H. NASA, Global Climate Change, Vital Signs of the Planet, (2018). Climate Change: How Do We Know? Available online: https://climate.nasa.gov/evidence/ (accessed on 2 April 2018).

39. Pierce, J.; Morley, S.; West, T.; Upton, L.; Banks, L. Improving Long-Term Care Facility Disaster Preparedness and Response: A Literature Review. *Disaster Med. Public Health Preparedness* **2017**, *11*, 140–149. [CrossRef] [PubMed]

40. Dosa, D.; Hyer, K.; Brown, L.; Artenstein, A.; Polivka-West, L.; Mor, V. The controversy inherent in managing frail nursing home residents during complex hurricane emergencies. *J. Am. Med. Dir. Assoc.* **2008**, *9*, 599–604. [CrossRef] [PubMed]

41. Dosa, D. To Evacuate or Not to Evacuate: Lessons Learned from Louisiana Nursing Home Administrators Following Hurricanes Katrina and Rita. *J. Am. Med. Dir. Assoc.* **2007**, *8*, 142–149. [CrossRef] [PubMed]

42. Weather Underground. Prepare for a Tornado. Available online: https://www.wunderground.com/prepare/tornado (accessed on 16 April 2018).

43. Woollaston, V. Lightning is at its most powerful at 8am in the morning but more storms occur in the afternoon, Daily Mail 2015. Available online: http://www.dailymail.co.uk/sciencetech/article-2998781/Lightning-powerful-8am-morning-storms-occur-afternoon.html (accessed on 16 April 2018).

geosciences

MDPI

Article

Assessment of the Performance of Satellite-Based Precipitation Products for Flood Events across Diverse Spatial Scales Using GSSHA Modeling System

Chad Furl, Dawit Ghebreyesus and Hatim O. Sharif *

Department of Civil and Environmental Engineering, University of Texas at San Antonio, 1 UTSA Circle, San Antonio, TX 78249, USA; chad.furl@gmail.com (C.F.); dawit.ghebreyesus@my.utsa.edu (D.G.)
* Correspondence: hatim.sharif@utsa.edu; Tel.: +1-210-458-6478

Received: 21 April 2018; Accepted: 24 May 2018; Published: 28 May 2018

Abstract: Accurate precipitation measurements for high magnitude rainfall events are of great importance in hydrometeorology and climatology research. The focus of the study is to assess the performance of satellite-based precipitation products against a gauge adjusted Next-Generation Radar (NEXRAD) Stage IV product during high magnitude rainfall events. The assessment was categorized across three spatial scales using watershed ranging from ~200–10,000 km^2. The propagation of the errors from rainfall estimates to runoff estimates was analyzed by forcing a hydrologic-model with the satellite-based precipitation products for nine storm events from 2004 to 2015. The National Oceanic and Atmospheric Administration (NOAA) Climate Prediction Center (CPC) Morphing Technique (CMORPH) products showed high correlation to the NEXRAD estimates in all spatial domains, and had an average Nash-Sutcliffe coefficient of 0.81. The Global Precipitation Measurement (GPM) Early product was inconsistent with a very high variance of Nash-Sutcliffe coefficient in all spatial domains (from −0.46 to 0.38), however, the variance decreased as the watershed size increased. Surprisingly, Tropical Rainfall Measuring Mission (TRMM) also showed a very high variance in all the performance statics. In contrast, the un-corrected product of the TRMM showed a relatively better performance. The errors of the precipitation estimates were amplified in the simulated hydrographs. Even though the products provide evenly distributed near-global spatiotemporal estimates, they significantly underestimate strong storm events in all spatial scales.

Keywords: hydrology; NEXRAD; remote sensing; GSSHA; flooding; GPM

1. Introduction

Accurate precipitation measurements for high magnitude events are of key importance to a number of areas in hydrometeorology and climatology research. In addition to research pursuits, these measurements have great value to public well-being by providing the backbone of rainfall-runoff prediction systems aimed at forecasting floods [1,2]. Over the past couple of decades in operational settings, these datasets have primarily been generated with radar and rain gauge networks [3]. Radar networks have the advantage of providing near real-time information over a continuous region at very fine scales, mostly unattainable with ground-based gauge networks. Numerous validation studies showed good performance of radar measurements, especially when combined with gauge networks for bias adjustments/quality control (e.g., Wang, Xie [4], Habib, Larson [5]). However, lack of even global distribution of radar network and problems such as beam blockage in complex terrain introduced significant gaps in radar coverage that pushed researchers to explore robust solution [6].

Satellite precipitation estimates provide a means for timely, near-global precipitation estimates, and much of the recent effort has been put into their validation and verification [7–13]. Several products, including those provided by the recently launched Global Precipitation Measurement (GPM)

mission, now provide the spatiotemporal resolution needed to forecast or conduct post-event analysis of flash floods. Even though the potential of satellite-based products was highly regarded, their poor performances were reported widely across the globe, especially, in their ability to accurately capture high magnitude precipitation events. Nikolopoulos, Anagnostou [14] demonstrated mean areal precipitation is consistently underestimated in their satellite ensemble analysis of a high magnitude precipitation event in Italy. AghaKouchak, Behrangi [15] examined several operational satellite precipitation products across the southern Great Plains with respect to precipitation thresholds and demonstrated the detection skill reduces as the choice of extreme threshold decreases. Mehran and AghaKouchak [16] reported similar findings when comparing three operational satellite products across the conterminous United States. Mei, Anagnostou [17] showed that satellite precipitation estimates are more biased for frontal events than for short-duration events. However, the error statistics of the products showed higher variability for the latter. Moreover, the products showed high inconsistency across different terrain [12] and climatic conditions [11]. These and other studies stress the need for more analysis and evaluation of the accuracy and performance of recent satellite products in capturing the behavior of extreme precipitation events by comparing them against products from ground-based measurement networks (radar or rain gauges).

Satellite-based precipitation products were found to be more accurate in a dry season and in wet tropical and dry zones than in semi-arid and mountainous regions. The uncertainty amongst the products was higher in estimating heavy rainfall storms in a semi-arid area. Moreover, the products, in general, overestimate the number of rainy days and underestimate the heavy rainfall storms [11]. Amongst the highly cited satellite-based products in the literature, the National Oceanic and Atmospheric Administration (NOAA) Climate Prediction Center (CPC) Morphing Technique (CMORPH) and Precipitation Estimation from Remotely Sensed Information using Artificial Neural Networks (PERSIANN)were reported to be spatially inconsistent [10–12,18–20]. The Tropical Rainfall Measuring Mission (TRMM) and its continuation mission GPM were found in many studies to be relatively consistent and more accurate but overestimated the average rainfall events and underestimated the heavy storm events in general [11–13,19,21].

The potential of high-resolution satellite precipitation estimates in hydrological applications is supported by the facts that satellite measurements are not inhibited by local topography and are available at a global scale. Forcing hydrological models with high-resolution satellite-based precipitation products can provide a streamflow forecast for ungauged, complex terrain basins. The manner in which rainfall errors propagate through a hydrologic model has important implications for building operational flow forecasts for such basins. Propagation of errors is influenced by spatial and temporal resolution of the satellite estimate, basin scale, and complexity of the physical interactions represented by the watershed model, among others. Presently, the majority of detailed error propagation studies were forced with radar rainfall data (e.g., Sharif, Ogden [22], Sharif, Ogden [23,24], Vivoni, Entekhabi [25]) with comparatively less work done for satellite-based precipitation (e.g., Nikolopoulos, Anagnostou [14], Gebregiorgis, Tian [26], Maggioni, Vergara [27], Chintalapudi, Sharif [28]). Moreover, most of the studies forced by satellite-based precipitation on propagation error into hydrologic predictions were focused on grid-based evaluation or long-term basin-averaged runoff response (e.g., Su, Gao [29], Wu, Adler [30]).

Spatial scale (with respect to both satellite resolution and basin size) is an important aspect in rainfall-to-runoff error propagation for satellite precipitation, and a more comprehensive understanding of it plays a vital role in mitigation of natural disasters. Nikolopoulos, Anagnostou [14] developed satellite rainfall ensembles for a single flood event and showed error propagation is strongly related to the size and characteristics of the watershed and the satellite product resolution. A rainfall-runoff process reduces the satellite-precipitation error variance in a mild-sloped catchment, and this effect exhibits the basin-scale dependence [31]. However, many other factors also have a significant impact, such as precipitation type, magnitude, and spatiotemporal pattern, and basin characteristics interact with the scale effect [31–33].

The Gridded Surface Subsurface Hydrologic Analysis (GSSHA) model, which is fully distributed and physically-based, was developed by the Department of Defense in order to simulate surface flows in non-Hortonian watersheds and watersheds with diverse characteristics of runoff production [34]. The model employs a mass-conserving solution of partial differential equations to produce the different components of hydrologic processes. The model was able to reproduce stream flows from a very diverse watershed with reasonable accuracy [35]. Moreover, the grid size can be used to optimize the required accuracy with the required computational power [36].

In the present study, the performance of several satellite precipitation products with respect to gauge corrected ground-based radar estimations for nine moderate to high magnitude events across the Guadalupe River system in south Texas was investigated. The analysis was conducted across three nested watersheds (ranging from 200 to 10,000 km^2 in area) to capture and quantify the effect of the scale on the propagation of the error. Satellite-based precipitation data sets were used to force a fully distributed physics-based Gridded Surface Subsurface Hydrologic Analysis (GSSHA) model to examine error propagation through the hydrologic model. Both gauge-corrected and uncorrected satellite products were used, encompassing a variety of latency times, spatial resolutions, and temporal resolutions. Satellite-based precipitation datasets used in the study include various products from GPM, PERSIANN system, CMORPH, and TRMM.

2. Materials and Method

2.1. Watershed

The Guadalupe River originates in south-central Texas and flows southeasterly until emptying into the Guadalupe estuary/Gulf of Mexico. In this study, the testbed is the middle and upper portions of the basin, with the watershed outlet taken near Gonzales, TX past the confluence of the Guadalupe and San Marcos rivers (herein referred to as Guadalupe basin). At the outlet, the basin drains approximately 9000 km^2. Two additional catchments within the watershed were delineated for scale effect analysis: Little Blanco River (178 km^2) and the Blanco River (1130 km^2). The spatial extent of the Guadalupe watershed along with the two nested watersheds is shown in Figure 1. Canyon Lake reservoir is formed by an impoundment along the Guadalupe River and contains significant flood storage, thus, we removed the dam from our watershed model to simulate a naturally flowing river for the analysis of hydrograph error propagation.

The Guadalupe River flows across distinct landscapes with varying hydrological characteristics. The upper portion of the watershed is located in an area known as the Texas Hill Country. This region is comprised of a karstic landscape with steep surfaces, exposed bedrock, and very thin clayey soils. As the river passes through the Balcones Escarpment, it encounters the Edwards Aquifer recharge and artesian zones. In these regions, soils are permeable and there is much groundwater-surface water interaction. The lower portion of the river crosses the Blackland Prairies before entering the Coastal Plain. A generalized soil map of the study area including the recharge and artesian areas of the Edwards Aquifer is displayed in Figure 1. There are a number of studies available describing the surface characteristics of the watershed in detail along with its flood hydrology [36–38].

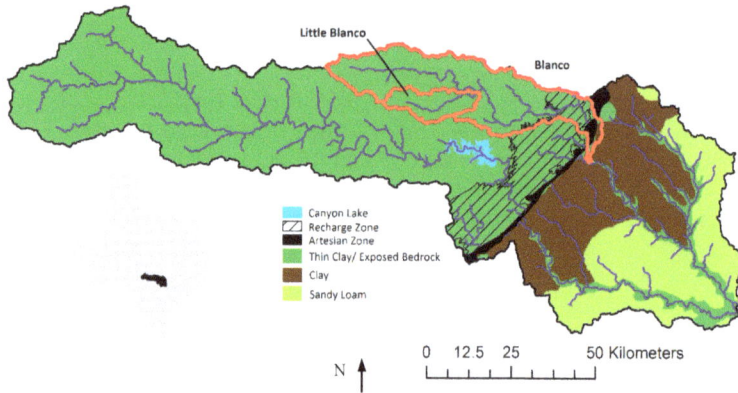

Figure 1. Location and area map of the study watersheds along with a generalized soil map. Each of the three interior watersheds are outlined, and the Edwards Aquifer recharge and artesian zones are displayed.

2.2. Storm Events

The Texas Hill Country is one of the most flash flood-prone areas of the entire United States due to its flood-prone physiography and susceptibility to extreme precipitation [38,39]. Although not considered among the very humid regions of the U.S., proximity to the Gulf of Mexico allows for extremely moist tropical air masses to reach the Balcones Escarpment where they can be subjected to orographic lift [40]. The region holds or has held several precipitation world records on time scales less than 24 h (USGS 2014). The precipitation envelope curve for Texas is comprised mostly from events in this region with others from the coastal plain. Once precipitation falls, the availability of steep slopes, high drainage density, exposed bedrock, and clay-rich soils have the ability to produce extremely high runoff coefficients with short lag times [41].

Here, nine large precipitation events from 2004–2015 across the middle Guadalupe basin were selected to examine satellite precipitation estimate performance and hydrologic model error propagation. All of the storm event accumulations from the Stage IV precipitation record are presented in Figure 2. The hydrometeorology of several of these events has been examined in detail including Furl, Sharif [42] (May 2015 event), Furl, Sharif [40] (September 2010 event), and Sharif, Sparks [36] (November 2004 events).

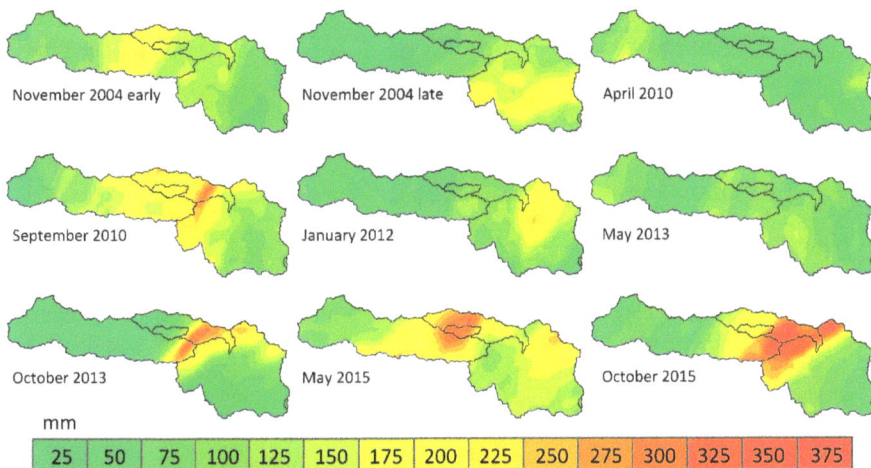

Figure 2. Total accumulations from stage IV data for individual storm events used in the analysis (mm). The month and year in which the storm occurred is displayed along with outlines of the interior watersheds. Numbers along the legend represent the maximum value from each category.

2.3. Precipitation Datasets

In total, ten satellite precipitation products were examined, encompassing a variety of spatiotemporal resolutions. Moreover, the examined products include gauge corrected and uncorrected products to assess the impact of the adjustment. A brief description of the precipitation products is included below.

2.3.1. NEXRAD Stage IV

Each of the satellite precipitation datasets was compared to the National Weather Service (NWS) and the National Centers for Environmental Prediction (NCEP) stage IV Quantitative Precipitation Estimate (herein Stage IV) [13]. The precipitation estimate is a quality controlled multi-sensor product (radar and gauges) produced by NCEP from the NEXRAD Precipitation Processing System [44] and the NWS River Forecast Center precipitation processing [45]. Precipitation bins are 4 km × 4 km and have an hourly temporal resolution. The primary radar operating across the study area is National Weather Service in Austin/San Antonio (KEWX) station approximately 70 km from the watershed outlet.

The authors acknowledge the inherent biases that accompany radar-based precipitation estimates. However, the relatively fine space-time scales of the dataset provide the best means to describe the spatiotemporal heterogeneity of the rainfall across the basin and make satellite comparisons. Moreover, previous studies by the authors demonstrated that Stage IV products were more suitable than observations by typical rain gauge networks as inputs to physically based distributed-parameter models (e.g., [28,36]).

2.3.2. GPM

The GPM core observatory was launched on 27 February 2014 providing a new means of satellite global precipitation measurement. The GPM consists of a core-satellite and numerous others in its constellation. The GPM mission is based on a constellation of microwave radiometers and integrated IR sensors to cover the blind spot of the microwave sensors. The Integrated Multi-Satellite Retrievals for GPM (IMERG) is the precipitation product developed by the GPM network. The core GPM satellite carries a dual-frequency precipitation radar along with multichannel microwave imagers and is used for calibration of the constellation satellites. Additionally, GPM can integrate infrared (IR) measurements from geostationary data to cover areas not seen by constellation satellites. The data produces a near global precipitation product with a spatial resolution of 0.1° and 30-min temporal resolution [46,47].

IMERG output is available in Early, Late, and Final runs, with a latency of approximately 4 h, 18 h, and 4 months, respectively. The Final IMERG run is calibrated by monthly gauge precipitation data following a certain procedure (Huffman, Bolvin [46]). In the present study, version 3 processing algorithms were used, and each of the three IMERG products were examined.

2.3.3. PERSIANN

The PERSIANN system estimates rainfall from infrared image data provided by geostationary satellites. PERSIANN data are calibrated in real time from independent microwave precipitation estimates. The calibration process is based on an adaptive training technique which updates neural network parameters when microwave data are available [48]. The data are available in 0.25°, 30-min resolution approximately 2 days after the gridded IR images are collected. The rainfall product covers tropical and middle latitudes from 50 S to 50 N [48,49].

PERSIANN-Cloud Classification System (PERSIANN-CCS) allows for precipitation estimates at the same temporal resolution and a finer spatial resolution (0.04°). Additionally, the data are available in near real-time. The system allows for the discernment and classification of cloud patch features based on height, areal extent, and variable texture. These classifications are used to further refine the assignment rainfall within each cloud. The product with a latency of two days was used for PERSIANN-CCS in this study.

2.3.4. CMORPH

CMORPH estimates precipitation from microwave-based precipitation images advected in time using infrared images from geosynchronous satellites. The product combines the positive side of the two satellites: estimated precipitation from low orbited satellites using microwave images and transportation of the estimated precipitation in time using the IR from the geosynchronous satellites. Microwave images are much better in estimating precipitation but they are not continuous, and IR from geosynchronous satellites are available and are continuous in time. The precipitation product is available at 30 min intervals with 8 km resolution as well as 3-h, 0.25° resolution. Precipitation estimates are available approximately 18 h past instrument measurement [50]. CMORPH products used in this study include the raw satellite-only precipitation product (CMORPH_RAW), the climate data record (CDR) version (CMORPH CDR) and the published 8 KM resolution product CMORPH 8KM.

2.3.5. TRMM

The Tropical Rainfall Measuring Mission (TRMM) employs a combination of microwave and IR data to estimate precipitation at 0.25° every 3 h. The TRMM product is produced by combining microwave estimates which are used to calibrate IR estimates from geosynchronous satellites. The IR estimates are used to fill gaps left by the microwave sensors. TRMM 3B42 V7 and TRMM-RT 3B42 V7 were used in the study. Gridded monthly rain gauge values are used to adjust the TRMM 3B42 V7 estimates [51]. The TRMM-RT (Real-Time) product is a near real-time dataset with no gauge adjustments. An overview of the availability of the entire dataset is shown in Table 1.

Table 1. Description of satellite precipitation dataset availability.

	Nov. 2004 Early	Nov. 2004 Late	Apr. 2010	Sept. 2010	Jan. 2012	May 2013	Oct. 2013	May 2015	Oct. 2015	Gauge Adjusted
TRMM B42	x	x	x	x	x	x	x	x	x	Y
TRMM-RT B42	x	x	x	x	x	x	x	x	x	N
PERSIANN	x	x	x	x	x	x	x	x	x	N
PERSIANN CCS	x	x	x	x	x	x	x	x	x	N
CMORPH CDR	x	x	x	x	x	x	x	NA	NA	Y
CMORPH 8KM	x	x	x	x	x	x	x	NA	NA	Y
CMORPH RAW	x	x	x	x	x	x	x	x	x	N
GPM IMERG EARLY	NA	NA	NA	NA	NA	NA	NA	x	x	N
GPM IMERG LATE	NA	NA	NA	NA	NA	NA	NA	x	x	N
GPM IMERG FINAL	NA	NA	NA	NA	NA	NA	NA	x	x	Y

2.4. Hydrologic Model

Precipitation datasets were used to force the fully distributed physics-based GSSHA model [34,52]. Hydrological processes simulated included infiltration, landscape retention, overland flow, and stream routing. Evapotranspiration and deep aquifer contributions were assumed to be insignificant relative to the processes since the simulation is event based. Model preprocessing was conducted using ArcGIS and Aquaveo's Watershed Modeling System. Watershed terrain was constructed from USGS 10 m digital elevation models (DEM) filled using the Cleandam algorithm distributed with the GSSHA model. Land use and land cover data were extracted from the National Land Cover Database 2011 (NLCD 2011) dataset. Soils data were prepared from SSURGO datasets along with maps from the Edwards Aquifer Authority defining the Edwards Aquifer recharge zone.

Infiltration calculations were conducted using Green and Ampt with redistribution [53] and pre-calibrated saturated hydraulic conductivity values taken from Rawls, Brakensiek [54]. Grid cells were assigned to one of four land use classes for retention and overland roughness. Stream channels were modeled using irregular cross sections for the main channel and large tributaries. The irregular channel and floodplain geometry were extracted from a triangular irregular network constructed from the DEM allowing for control of floodplain simulation. Upland tributaries were modeled as a uniform trapezoidal profile. Reach specific Manning's n values were assigned based on field observations and prior modeling experience in this region of Texas. Routing was calculated using the diffusive wave equation in 1D for streams and 2D for overland flow. The hydrological model was run on a 150-m grid cell size with a 1-min simulation time step.

Distributed models have the distinct advantage of allowing examination of hydrologic properties at any point in the basin. In this study, three watershed models were constructed: Blanco watershed, Upper Guadalupe watershed, and Middle Guadalupe watershed. Results from the Little Blanco watershed were harvested from the proper interior node of the Blanco River watershed model. The Middle Guadalupe model (i.e., implementation of the hydrologic model over Middle Guadalupe) used streamflow from the outlets of the Blanco and Upper Guadalupe as boundary condition inflows, thereby allowing a very fine gridded distributed model over a 9000 km² basin. The Upper Guadalupe model discharge hydrograph was input into the Middle Guadalupe at the outlet of Canyon Lake, bypassing the reservoir.

The Blanco watershed model was the primary model calibrated. Furl et al. [42] calibrated the model to the November 2004 "early" event used here and achieved r2, Nash-Sutcliffe model efficiency (NSE), and percent bias (PBIAS) values of 0.91, 0.90, and 10.2%, respectively for the calibration run. Similar model parameter values were used for the Upper Guadalupe model. The setup for the Middle Guadalupe followed hydrologic parameters described by Sharif, Sparks [36], which described the November 2004 "late" event. It should be noted that our main objective with model calibration is to provide realistic rainfall-runoff mechanisms such that error propagation analysis can be conducted. Surface properties for the Blanco River watershed are shown in Table 2. The readers are directed to Sharif, Sparks [36] and Furl et al. [42] for detailed descriptions of the watershed models and their comparisons with measured flows.

Table 2. Gridded Surface Subsurface Hydrologic Analysis (GSSHA) infiltration and overland flow parameters for the Blanco watershed model.

Soil Texture/ Land Use	Saturated Hydraulic Conductivity (cm·hr^{-1})	Capillary Head (cm)	Effective Porosity	Manning's Roughed Coefficient	Retention Depth (mm)
Recharge zone	10.0	23.6	0.417	-	-
Clay	1.2	0.06	0.385	-	-
Loam	0.01	1.3	0.434	-	-
Fine loam	0.02	2.18	0.412	-	-
Fine silt	0.01	0.68	0.486	-	-
Fine sand	0.03	23.6	0.417	-	-
Urban	-	-	-	0.18	5.0
Forest	-	-	-	0.25	5.0
Shrub	-	-	-	0.20	5.0
Grasslan/agriculture	-	-	-	0.30	5.0

2.5. Evaluation Criteria

Satellite precipitation results were analyzed by comparing mean areal precipitation hyetographs with those generated from the Stage IV precipitation record. Here, the reference hydrographs were those driven by the reference precipitation product (radar). It will not be appropriate to use observed hydrographs as a reference since we do not have a precipitation product that will perfectly produce the observed hydrographs. A weighting method was used in the averaging routine to account for rainfall bins only partially covering a portion of the basin. For comparisons, satellite hyetographs were scaled to a one-hour time step using a simple linear transformation in order to match the Stage IV record. The comparison period was confined to when the Stage IV record indicated 1 mm of precipitation had fallen across the basin until rainfall ceased. Streamflow hydrograph comparisons were conducted in a similar manner by comparing satellite generated model output with the hydrograph generated by the Stage IV record. The analysis period was determined by visually examining the Stage IV generated hydrographs and capturing from just before the rising limb of the hydrograph until after the falling limb. The comparisons for hyetographs and hydrographs were completed using the percent bias (PBIAS), normalized root-mean-square-error (nRMSE), and Nash-Sutcliffe model efficiency (NSE) statistics. Simple relative error in precipitation, peak flow, and volume of flow was calculated for error propagation analysis. Calculations were completed using the hydroGOF package [55] in R environment as follows:

$$\text{PBIAS} = 100 \times \frac{\sum_{i=1}^{N}(S_i - O_i)}{\sum_{i=1}^{N} O_i} \tag{1}$$

$$\text{nRMSE} = 100 \times \frac{\sqrt{\frac{1}{N}\sum_{i=1}^{N}(S_i - O_i)^2}}{nval} \tag{2}$$

$$where\ nval = \begin{cases} sd(O_i), & norm = \text{"sd"} \\ O_{max} - O_{min}, & norm = \text{"maxmin"} \end{cases}$$

$$\text{NSE} = 1 - \frac{\sum_{i=1}^{N}(S_i - O_i)^2}{\sum_{i=1}^{N}(O_i - \overline{O})^2} \tag{3}$$

$$relative\ error(\delta x) = \frac{\Delta x}{x} = \frac{x_0 - x}{x} = \frac{x_0}{x} - 1 \tag{4}$$

where:

S_i is the simulated rainfall (estimated by the product),
O_i is the estimated rainfall by NEXRAD stage IV,
x_0 is the estimated rainfall by the product/precipitation/simulated peak flow with the product, and
x is the estimated rainfall NEXRAD stage IV/simulated peak flow with NEXRAD stage IV rainfall.

3. Results and Discussion

3.1. Precipitation

Precipitation from the Stage IV record averaged over the entire watershed ranged from approximately 50–150 mm, and storm durations lasted from just a few hours to approximately 72 h. Among the 27 isolated storm event and watershed size combinations (9 storm events × 3 watersheds), the satellite-based precipitation products showed a wide range level of accuracy when compared to the Stage IV estimates. As shown in Figure 3, for the largest spatial domain, satellite precipitation estimates showed the ability to very closely match radar results (November 2004 (late), May 2013), and consistently overestimate (November 2004, early) and also systematically underestimate precipitation (October 2015). The products tended to significantly underestimate in four events and only once overestimated when compared to NEXRAD Stage IV estimates. In addition to the inherent errors in the satellite products due to calibration and the rainfall estimation technique (i.e., microwave or infrared), the relatively coarse resolution of the products may have contributed to the underestimations errors. Underestimation is more pronounced for the large events where satellite underestimates the high intensity periods. Interestingly, the Final GPM product underestimates rainfall more than the earlier products. This can also be attributed to the nature of the events where climatology and gauge adjustments did not capture the localized intensity of the events. In the rest of the four storm events, the NEXRAD product seemed to fit the average of all the satellite-based products (Figure 3). The satellite-based products failed to capture the storm events that occurred in the Fall (September, October) with the exception of the 2004 storm where they tended to overestimate the storm. In contrast, the margin of error was very low in storm events that occurred in May.

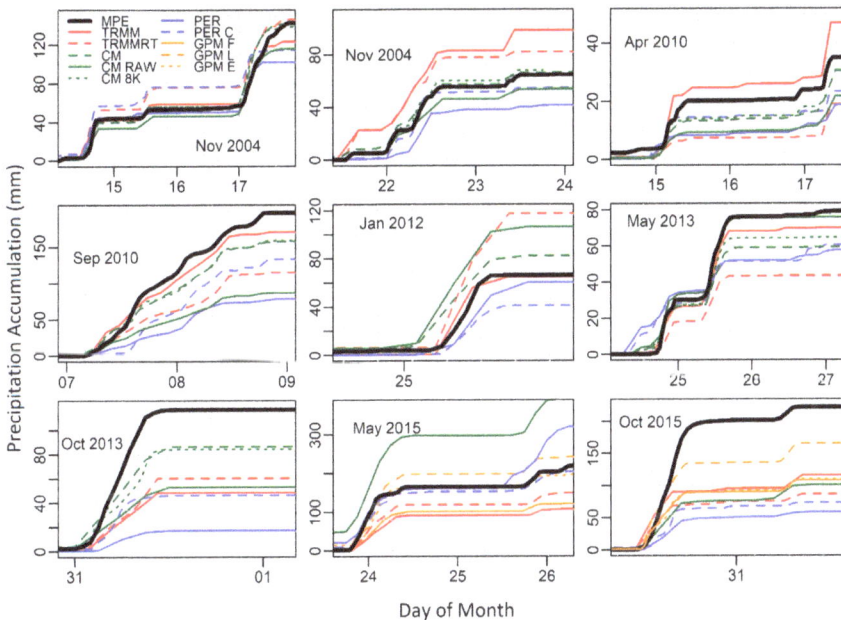

Figure 3. Storm event rainfall accumulations averaged over the Blanco watershed. Legend labels are abbreviated as follows: MPE—Stage IV, TRMM—TRMM 3B42, CM—CMORPH CDR, GPM F—GPM IMERG FINAL, CM 8K—CMORPH 8KM, PER C—PERSIANN CCS, GPM E—GPM IMERG EARLY, GPM L—GPM IMERG LATE, CM RAW—CMORPH RAW, TRMM-RT—TRMM-RT 3B42, and PER—PERSIANN.

In general, the satellite products (adjusted and unadjusted) underestimated the storm events from the stage IV record at all spatial scales with the exception of some storm events. This is not surprising given the small sample size focused on events on the tail side of the distribution. Other researchers have noted similar satellite underestimations for high magnitude events [14–16,56]. However, it should be noted there was no strong correlation between percent bias and total accumulated precipitation for any of the three spatial domains. Moreover, satellite-based products underestimated heavy storm events in larger spatial domains (0.4 to 1.3 million km^2) in several regions of Africa [11].

Generally, the satellite-based precipitation products showed less variability in the case of the Guadalupe basin (Large) relative to the two smaller watersheds (Figure 4). This could be mainly because of the smoothing power of mean value over the large spatial domain (filtering the noise introduced by the products). Both products from CMOPRH (labeled as CM and CM 8K) showed very high correlation with the stage IV product in all spatial domains with very high Nash coefficient. GPM Early was found to be inconsistent with a very high variance of Nash coefficient in all spatial domains, however, the variance was decreased as the watershed size increase. Surprisingly, TRMM showed a very high variance in all the performance statics, especially in the two small watersheds. In contrast, the TRMM-RT product showed relatively better performance. As described above, the performance of GPM Final product was inferior to the earlier ones. The whole distribution of the performance statistics is provided in Figure 4.

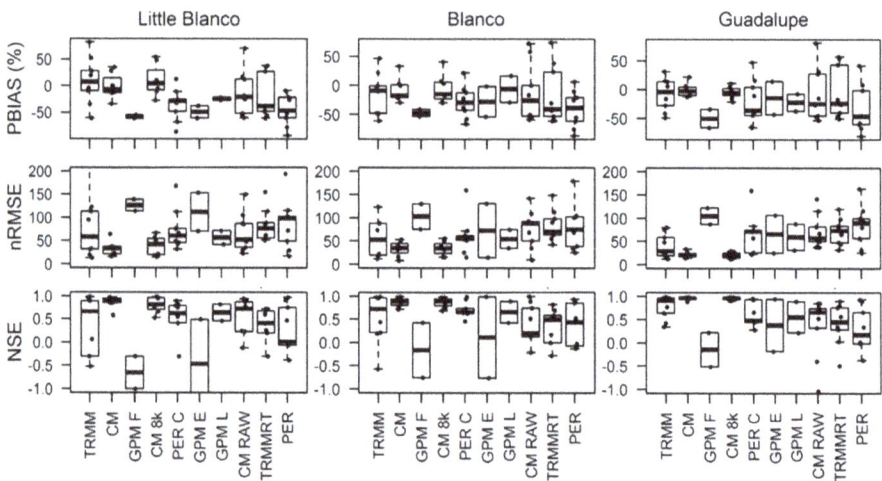

Figure 4. Boxplots for hyetograph performance statistics for all spatial domains. *X*-axis labels are abbreviated as follows: TRMM—TRMM 3B42, CM—CMORPH CDR, GPM F—GPM IMERG FINAL, CM 8K—CMORPH 8KM, PER C—PERSIANN CCS, GPM E—GPM IMERG EARLY, GPM L—GPM IMERG LATE, CM RAW—CMORPH RAW, TRMM-RT—TRMM-RT 3B42, and PER—PERSIANN. Boxplots display the lower and upper quartiles and median. Whiskers extend to the data point nearest ± 1.5 * interquartile range.

In order to provide some comparison between satellites products, performance statistic results were pooled from all spatial domains for each individual satellite product. Table 3 displays the median, average, and range of the statistics after this aggregation. For the 0.25° uncorrected products, performance statistics indicated CMORPH RAW > TRMM-RT > PERSIANN for the nine events examined. Sapiano and Arkin [10] found that correlations were highest with CMORPH in an inter-comparison and validation study on sub-daily satellite precipitation data. For the gauge corrected products at 0.25°, there was very little difference between TRMM and CMORPH when the same events were compared (2015 events unavailable for CMORPH). It is difficult to draw conclusions about the performance of GPM given that only two events were measured. GPM results are compared to the other products for the 2015 events below.

Table 3. Satellite hyetograph performance statistics aggregated across the three spatial domains.

		Gauge Correction					No Gauge Correction				
		TRMM	CMORPH	GPM FINAL	CMORPH 8KM	PERSIANN CCS	GPM EARLY	GPM LATE	CMORPH RAW	TRMM-RT	PERSIANN
	PBIAS	−3.9	−2.9	−51.9	0.2	−28.2	−30.4	−17.6	−12.2	−14.2	−39.7
	nRMSE	58.2	29.5	111.2	31.4	66.9	83.5	57.3	72.3	79.3	84.1
Average	NSE	0.42	0.89	−0.32	0.87	0.38	0.01	0.62	0.32	0.26	0.06
	PBIAS	−4.4	−7.0	−54.1	−4.6	−29.7	−40.3	−24.3	−24.7	−37.2	−46.3
	nRMSE	48.0	26.9	117.8	28.0	59.8	88.7	57.2	67.9	73.2	90.0
Median	NSE	0.77	0.93	−0.40	0.92	0.63	0.16	0.64	0.53	0.45	0.18
	PBIAS	143.1	69.3	32.1	84.3	132.5	74.8	54.0	140.4	135.2	134.0
	nRMSE	210.7	56.5	63.5	54.7	153.2	138.2	56.1	140.9	123.0	176.0
Range	NSE	5.0	0.4	1.4	0.5	2.9	2.4	0.7	2.3	2.4	3.9
Count		27	21	6	21	27	6	6	27	27	27

3.2. Impact of Spatial Resolution

Several papers have noted a scale dependence of error caused by the inability of coarse-resolution products to adequately represent mean areal precipitation in smaller basins because their sampling involves an area much larger than the basin (e.g., Nikolopoulos, Anagnostou [14]). Here, we investigate the scale dependence of rainfall error first by comparing the CMORPH and PERSIANN products with their fine-scale counterparts and then by examining changes in PBIAS as a function of watershed size.

The PERSIANN CCS product has a spatial resolution of 0.04 degrees and has a similar size to stage IV bins across the study area. When compared to PERSIANN, the PERSIANN CCS product consistently performed better in each of the three watersheds for all performance statistics. However, the gap in performance statistics did not grow as watershed size decreased, as may be expected if scale issues were the root cause of the discrepancy. It is difficult to identify the primary causes for the differing performance given PERSIANN CCS uses different processing algorithms.

Unlike the PERSIANN products, there was virtually no difference between CMORPH and the CMORPH 8KM product with regard to performance statistics. This suggests the downscaling techniques employed by the CMORPH 8KM product are not adequate if their intent is to provide a more detailed spatial representation of rainfall.

3.3. GPM Rainfall Events

The two largest rain events from the dataset occurred in May and October of 2015, and both resulted in significant flash flood events along the Blanco River [42]. These events were captured by GPM and offer an initial look at GPM performance for short duration high magnitude storms. Figure 5 shows performance statistic results for each of the real-time products (PERSIANN, PERSIANN-CCS, TRMM-RT, and CMORPH RAW) along with the Early and Late GPM runs. Generally, the GPM products performed better than did the other real-time satellite products. For the May 2015 event, the Early product produced better estimates than the Late run, with the opposite pattern for the October 2015 event. Gauge corrected estimates (Table 3) showed a significant underestimation of the events, which is not surprising given it is adjusted to monthly values. The Early GPM product failed to capture the storm event of October 2015 showed by the negative value of Nash coefficient (Figure 5).

As anticipated, hydrograph results closely mimic the rainfall fields with respect to their ability to overestimate and underestimate the reference Stage IV forcing. The hydrographs driven by Stage IV rainfall along with satellite results for the nine storm events at the Guadalupe basin outlet are shown in Figure 6 (same as Figure 3). The hydrographs from the products were able to capture the bi-modal behavior of the hydrograph but with a high range of accuracy levels for November 2004 and May 2014 events. Some of the errors were quite high, indicating that the rainfall errors were amplified in the resulting runoff hydrograph. In the case of the events that have a one peak hydrograph, most of the products tend to underestimate the hydrograph by a considerable amount. As expected, the hydrographs driven by the GPM Early and Late products had less errors than those driven by the Final products due to the severe underestimation of rainfall by the latter as described above (see Figures 3 and 4). In all the fall events, the pattern of the hydrograph was more or less captured, but with very significant underestimation of the precipitation. For the events that occurred in May, the magnitude and the pattern of the stage IV seem to be the mean value of the hydrographs from the satellite-based products.

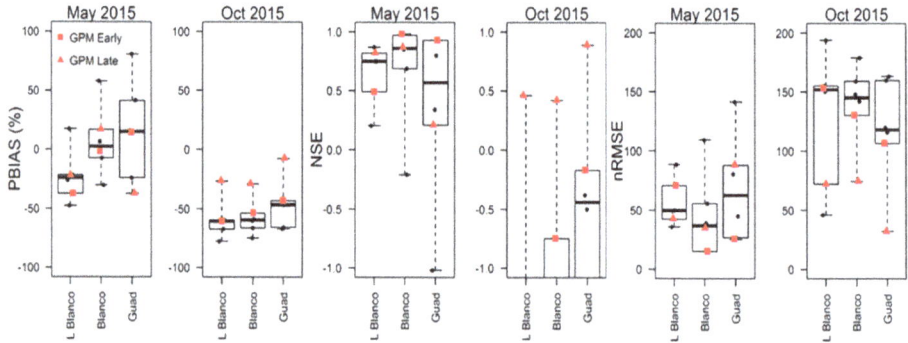

Figure 5. Performance statistic results for the non-gauge adjusted satellite results for the May and October 2015 storms that included the GPM products.

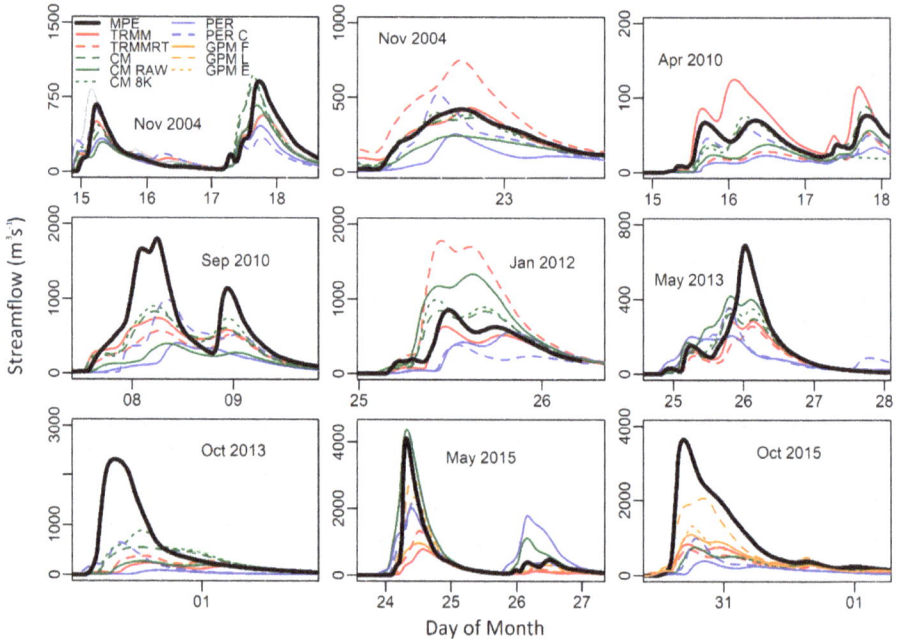

Figure 6. Streamflow hydrographs at the Blanco watershed outlet. Legend labels are abbreviated as follows: MPE—Stage IV, TRMM—TRMM 3B42, CM—CMORPH CDR, GPM F—GPM IMERG FINAL, CM 8K—CMORPH 8KM, PER C—PERSIANN CCS, GPM E—GPM IMERG EARLY, GPM L—GPM IMERG LATE, CM RAW—CMORPH RAW, TRMM-RT—TRMM-RT 3B42, and PER—PERSIANN.

The performance of the precipitation products in the simulated hydrograph followed a similar pattern as described in the precipitation analysis. However, the variability of the products seems to increase as the scale of the watershed increases. Boxplot results showing performance statistics at each of the three basins are displayed in Figure 7. The CMORPH product (labeled as CM) showed higher Nash coefficient in Little Blanco (the smallest basin) but as the size of the watershed increased, the performance was seen to plummet. A similar pattern was observed in most the products when moving from Little Blanco to Guadalupe. All evaluation criteria showed a very wide range and high variability and error magnitude in the case of the Guadalupe Basin (Figure 7). The accumulated effect of all the discrepancies in the products across the watershed caused a significant increase in variability at the outlet. However, the increase in spatial domain of the watershed improved the performance of the GPM Late product across all the criteria.

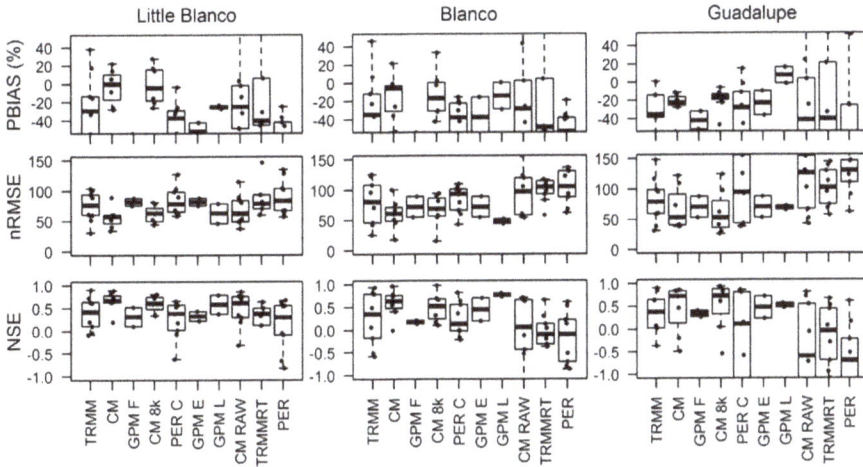

Figure 7. Boxplots for hydrograph performance statistics for all spatial domains. X-axis labels abbreviations and boxplot representations are the same as those described in Figure 4.

3.4. GPM Model Simulations

GPM products showed a higher Nash Coefficient than their counterparts in both events in all spatial scales except in the case of Little Blanco for the May 2015 event. Moreover, the Late GPM product outclassed the Early GPM product in almost all the criteria and in all spatial domains. There is no clear pattern of the impact of scale effect on a single product but, in case of GPM products, the performance of the PBIAS was improved as the spatial domain increased. However, the variability of the performance of the non-gauge adjusted products increased as the spatial domain size increased (Figure 8).

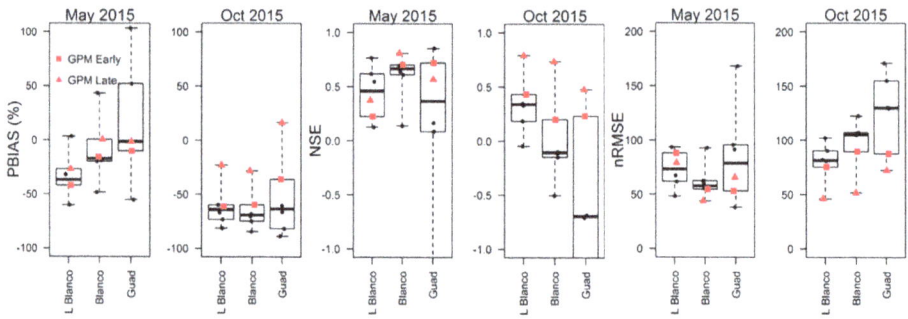

Figure 8. Performance statistic results for the non-gauge adjusted satellite results for the May and October 2015 storms that included the GPM products (for the hydrographs).

3.5. Error Propagation

The error was seen to propagate from the precipitation dataset to the hydrograph at the outlet. The propagation was magnified in all of the criteria shown in Figure 9 except in the case of the streamflow PBIAS. Moreover, the pattern was seen across all the spatial domains in the same manner. The scale effect of the spatial domains does not seem to affect the error propagation, as they were very close in all the evaluation criteria (Figure 9).

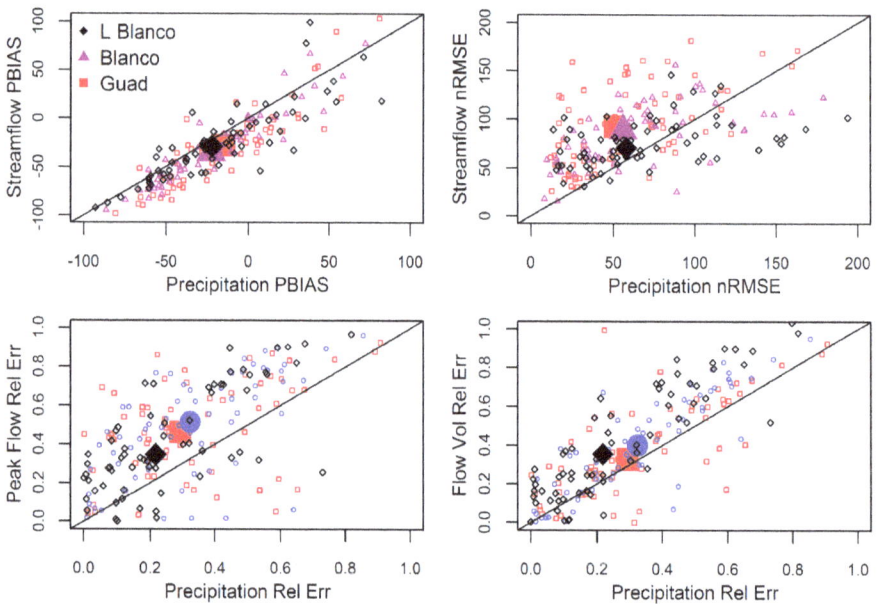

Figure 9. Annual Error propagations descriptions comparing streamflow and precipitation percent bias (PBIAS) and normalized root-mean-square-error (nRMSE) (**top left** and **top right**, respectively) and relative error in precipitation versus peak flow relative error (**bottom left**) and streamflow volume relative error (**bottom right**) N.B. the big markers represent the mean value.

4. Conclusions

Precipitation is the main driver of all the hydrologic models that are used to predict/forecast the relationship between rainfall and runoff. Moreover, rainfall amount and distribution represent the major components of the floodplain analysis and water resource management practices. That is why it is a significant achievement to capture the spatial and temporal distribution of rainfall, since the accuracy of almost all hydrologic processes depends on the accuracy of the precipitation estimates. Rain gauges are only reliable for a very small area because of the intermittent behavior of precipitation. Radars have problems with beam blockage in complex terrain and lack even distribution across the globe. Satellite-based precipitation estimation with high spatiotemporal resolution has a potential to capture the spatiotemporal distribution of precipitation if the products and algorithms are improved to a reasonable accuracy.

The assessment of ten satellite-based precipitation products was carried in relation to the radar stage IV (NCEP product) over Guadalupe river basin with a drainage area of around 9000 km^2. Moreover, the assessment was done in two smaller sub-watersheds of the Guadalupe river basin (Little Blanco River (178 km^2) and the Blanco River (1130 km^2)). This procedure was done to assess the scale impact on the accuracy of the products. Nine significantly large events with a wide spatial coverage were used in the analysis.

Furthermore, to understand the propagation of rainfall error into the predicted runoff, hydrologic model simulations were implemented. GSSHA, a physically-based fully distributed hydrologic model, forced with those ten satellite-based precipitation products, was used to simulate the rainfall-runoff relationship for the basins. The most widely used model evaluation criteria such as Nash-Sutcliffe, PBIAS, nRMSE, and relative error were used in the assessment of both precipitation and hydrographs of the outlet.

The products underestimated the storm events in relation to the radar product Stage IV. This pattern was seen in several other studies over various regions of the world [14–16,56]. Moreover, the satellite-based precipitation products showed a very compact distribution in all the evaluation criteria in the case of the largest basin. Both products of CMORPH showed a very high correlation in all spatial domains and was reflected with an average Nash-Sutcliffe coefficient of 0.81. GPM Early was found to be inconsistent with a very high variance of Nash coefficient in all spatial domains (from −0.46 to 0.38), however, the variance was decreased as the watershed size increased. This is mainly due to the smoothing caused by averaging over a larger area. Among all GPM products, the Final product underestimated rainfall most, indicating that the methodology used to prepare the product (using climatology and rain gauges) probably was not able to capture the areas and/or periods of very intense localized rainfall. Surprisingly, TRMM also showed a very high variance in all the performance statics, especially in the two small watersheds (from −4.0 to 0.99 with an average of 0.16). In contrast, the TRMM RT (non-gauge corrected product of TRMM) product showed relatively better performance of Nash-Sutcliffe with an average of 0.39 and a range from 0.05 to 0.82.

The pattern of the precipitation estimates was also reflected on the simulated hydrograph forced by the precipitation products. The average Nash-Sutcliffe coefficient was reduced from 0.81 in precipitation to 0.58 in the runoff for CMORPH. CMORPH product showed higher Nash coefficient in Little Blanco (the smallest basin) but as the size of the watershed increased, the performance was seen to plummet. A similar pattern was observed in most of the products when moving from Little Blanco to Guadalupe. However, the increase in the spatial domain of the watershed improved the performance of the GPM Late product across all the criteria.

The error was seen to amplify as it propagated from the precipitation dataset to the hydrograph at the outlet. The propagation was magnified in all of the evaluation criteria except in the case of the streamflow PBIAS. Moreover, the pattern was seen across all the spatial domains in the same manner. The scale effect of the spatial domains does not seem to affect the error propagation as it was very close in all of the evaluation criteria.

In summary, the satellite-based precipitation products provide very high spatiotemporal resolution precipitation estimates. However, the estimates lack accuracy, especially at a local scale. The products underestimate heavy storm events significantly, and the errors were amplified in the runoff hydrographs generated.

Author Contributions: H.O.S. and C.F. designed the overall study. C.F. downloaded the remote sensing products and prepared and performed the hydrologic model simulations with input from H.O.S. D.G. helped C.F. with post-analysis of the model outputs and preparation of the first draft. C.F. and H.O.S. reviewed and revised the manuscript. H.O.S. did the final overall proofreading of the manuscript.

Funding: This research was funded by U.S. Army Research Office (Grant W912HZ-14-P-0160).

Acknowledgments: This research funded in part by the U.S. Army Research Office (Grant W912HZ-14-P-0160). This support is cordially acknowledged.

Conflicts of Interest: The authors declare no conflict of interest.

References

1. Dai, A. Precipitation Characteristics in Eighteen Coupled climate models. *J. Clim.* **2006**, *19*, 4605–4630. [CrossRef]
2. New, M.; Todd, M.; Hulme, M.; Jones, P. Precipitation measurements and trends in the twentieth century. *Int. J. Clim.* **2001**, *21*, 1889–1922. [CrossRef]
3. Krajewski, W.; Smith, J. Radar hydrology: Rainfall estimation. *Adv. Water Resour.* **2002**, *25*, 1387–1394. [CrossRef]
4. Wang, X.; Xie, H.; Sharif, H.; Zeitler, J. Validating NEXRAD MPE and Stage III precipitation products for uniform rainfall on the Upper Guadalupe River Basin of the Texas Hill Country. *J. Hydrol.* **2008**, *348*, 73–86. [CrossRef]
5. Habib, E.; Larson, B.F.; Graschel, J. Validation of NEXRAD multisensor precipitation estimates using an experimental dense rain gauge network in south Louisiana. *J. Hydrol.* **2009**, *373*, 463–478. [CrossRef]
6. Maddox, R.A.; Zhang, J.; Gourley, J.J.; Howard, K.W. Weather radar coverage over the contiguous United States. *Weather Forecast.* **2002**, *17*, 927–934. [CrossRef]
7. Nicholson, S.E.; Some, B.; McCollum, J.; Nelkin, E.; Klotter, D.; Berte, Y.; Diallo, B.M.; Gaye, I.; Kpabeba, G.; Ndiaye, O.; et al. Validation of TRMM and other rainfall estimates with a high-density gauge dataset for West Africa. Part II: Validation of TRMM rainfall products. *J. Appl. Meteorol.* **2003**, *42*, 1355–1368. [CrossRef]
8. McCollum, J.R.; Krajewski, W.F.; Ferraro, R.R.; Ba, M.B. Evaluation of biases of satellite rainfall estimation algorithms over the continental United States. *J. Appl. Meteorol.* **2002**, *41*, 1065–1080. [CrossRef]
9. Ebert, E.E.; Janowiak, J.E.; Kidd, C. Comparison of near-real-time precipitation estimates from satellite observations and numerical models. *Bull. Am. Meteorol. Soc.* **2007**, *88*, 47–64. [CrossRef]
10. Sapiano, M.; Arkin, P. An intercomparison and validation of high-resolution satellite precipitation estimates with 3-hourly gauge data. *J. Hydrometeorol.* **2009**, *10*, 149–166. [CrossRef]
11. Thiemig, V.; Rojas, R.; Zambrano-Bigiarini, M.; Levizzani, V.; de Roo, A. Validation of satellite-based precipitation products over sparsely gauged African river basins. *J. Hydrometeorol.* **2012**, *13*, 1760–1783. [CrossRef]
12. Hirpa, F.A.; Gebremichael, M.; Hopson, T. Evaluation of high-resolution satellite precipitation products over very complex terrain in Ethiopia. *J. Appl. Meteorol. Climatol.* **2010**, *49*, 1044–1051. [CrossRef]
13. Mantas, V.; Liu, Z.; Caro, C.; Pereira, A.J.S.C. Validation of TRMM multi-satellite precipitation analysis (TMPA) products in the Peruvian Andes. *Atmos. Res.* **2015**, *163*, 132–145. [CrossRef]
14. Nikolopoulos, E.I.; Anagnostou, E.N.; Hossain, F.; Gebremichael, M.; Borga, M. Understanding the scale relationships of uncertainty propagation of satellite rainfall through a distributed hydrologic model. *J. Hydrometeorol.* **2010**, *11*, 520–532. [CrossRef]
15. AghaKouchak, A.; Behrangi, A.; Sorooshian, S.; Hsu, K.; Amitai, E. Evaluation of satellite-retrieved extreme precipitation rates across the central United States. *J. Geophys. Res. Atmos.* **2011**, *116*. [CrossRef]
16. Mehran, A.; AghaKouchak, A. Capabilities of satellite precipitation datasets to estimate heavy precipitation rates at different temporal accumulations. *Hydrol. Process.* **2014**, *28*, 2262–2270. [CrossRef]

17. Mei, Y.; Anagnostou, E.N.; Nikolopoulos, E.I.; Borga, M. Error analysis of satellite precipitation products in mountainous basins. *J. Hydrometeorol.* **2014**, *15*, 1778–1793. [CrossRef]

18. Shen, Y.; Xiong, A.; Wang, Y.; Xie, P. Performance of high-resolution satellite precipitation products over China. *J. Geophys. Res. Atmos.* **2010**, *115*. [CrossRef]

19. Vernimmen, R.; Hooijer, A.; Mamenun; Aldrian, E.; van Dijk, A.I.J.M. Evaluation and bias correction of satellite rainfall data for drought monitoring in Indonesia. *Hydrol. Earth Syst. Sci.* **2012**, *16*, 133–146. [CrossRef]

20. Wehbe, Y.; Ghebreyesus, D.; Temimi, M.; Milewski, A.; al Mandous, A. Assessment of the consistency among global precipitation products over the United Arab Emirates. *J. Hydrol. Reg. Stud.* **2017**, *12*, 122–135. [CrossRef]

21. Omranian, E.; Sharif, H.O. Evaluation of the Global Precipitation Measurement (GPM) Satellite Rainfall Products Over the Lower Colorado River Basin, Texas. *J. Am. Water Resour. Assoc.* **2018**. [CrossRef]

22. Sharif, H.O.; Ogden, F.L.; Krajewski, W.F.; Xue, M. Numerical simulations of radar rainfall error propagation. *Water Resour. Res.* **2002**, *38*, 15-1–15-14. [CrossRef]

23. Sharif, H.O.; Ogden, F.L.; Krajewski, W.F.; Xue, M. Statistical analysis of radar rainfall error propagation. *J. Hydrometeorol.* **2004**, *5*, 199–212. [CrossRef]

24. Borga, M. Accuracy of radar rainfall estimates for streamflow simulation. *J. Hydrol.* **2002**, *267*, 26–39. [CrossRef]

25. Vivoni, E.R.; Entekhabi, D.; Hoffman, R.N. Error propagation of radar rainfall nowcasting fields through a fully distributed flood forecasting model. *J. Appl. Meteorol. Climatol.* **2007**, *46*, 932–940. [CrossRef]

26. Gebregiorgis, A.S.; Tian, Y.; Peters-Lidard, C.D.; Hossain, F. Tracing hydrologic model simulation error as a function of satellite rainfall estimation bias components and land use and land cover conditions. *Water Resour. Res.* **2012**, *48*. [CrossRef]

27. Maggioni, V.; Vergara, H.J.; Anagnostou, E.N.; Gourley, J.J.; Hong, Y.; Stampoulis, D. Investigating the applicability of error correction ensembles of satellite rainfall products in river flow simulations. *J. Hydrometeorol.* **2013**, *14*, 1194–1211. [CrossRef]

28. Chintalapudi, S.; Sharif, H.O.; Xie, H. Sensitivity of distributed hydrologic simulations to ground and satellite based rainfall products. *Water* **2014**, *6*, 1221–1245. [CrossRef]

29. Su, F.; Gao, H.; Huffman, G.J.; Lettenmaier, D.P. Potential utility of the real-time TMPA-RT precipitation estimates in streamflow prediction. *J. Hydrometeorol.* **2011**, *12*, 444–455. [CrossRef]

30. Wu, H.; Adler, R.F.; Tian, Y.; Huffman, G.J.; Li, H.; Wang, J. Real-time global flood estimation using satellite-based precipitation and a coupled land surface and routing model. *Water Resour. Res.* **2014**, *50*, 2693–2717. [CrossRef]

31. Vergara, H.; Hong, Y.; Gourley, J.J.; Anagnostou, E.N.; Maggioni, V.; Stampoulis, D.; Kirstetter, Pi. Effects of resolution of satellite-based rainfall estimates on hydrologic modeling skill at different scales. *J. Hydrometeorol.* **2014**, *15*, 593–613. [CrossRef]

32. Mei, Y.; Nikolopoulos, E.I.; Anagnostou, E.N.; Borga, M. Evaluating satellite precipitation error propagation in runoff simulations of mountainous basins. *J. Hydrometeorol.* **2016**, *17*, 1407–1423. [CrossRef]

33. Yong, B.; Ren, L.; Hong, Y.; Wang, J.; Gourley, J.J.; Jiang, S.; Chen, X.; Wang, W. Hydrologic evaluation of Multisatellite Precipitation Analysis standard precipitation products in basins beyond its inclined latitude band: A case study in Laohahe basin, China. *Water Resour. Res.* **2010**, *46*. [CrossRef]

34. Downer, C.W.; Ogden, F.L. GSSHA: Model to simulate diverse stream flow producing processes. *J. Hydrol. Eng.* **2004**, *9*, 161–174. [CrossRef]

35. Afshari, S.; Omranian, E.; Feng, D. *Relative Sensitivity of Flood Inundation Extent by Different Physical and Semi-Empirical Models*; CUAHSI Technical Report No. 13; The Consortium of Universities for the Advancement of Hydrologic Science, Inc.: Washington, DC, USA, 2016; pp. 19–24.

36. Sharif, H.O.; Sparks, L.; Hassan, A.A.; Zeitler, J.; Xie, H. Application of a distributed hydrologic model to the November 17, 2004, flood of Bull Creek watershed, Austin, Texas. *J. Hydrol. Eng.* **2010**, *15*, 651–657. [CrossRef]

37. Caran, S.C.; Baker, V.R. Flooding along the Balcones escarpment, central Texas. In *The Balcones Escarpment-Geology, Hydrology, Ecology and Social Development in Central TEXAS*; Geological Society of America: Boulder, CO, USA, 1986; pp. 1–14.

38. Baker, V.R. *Flood Hazards along the Balcones Escarpment inCentral Texas Alternative Approaches to Their Recognition, Mapping, and Management*; Bureau of Economic Geology Circular; Bureau of Economic Geology, University of Texas at Austin: Austin, TX, USA, 1975.

39. Smith, J.A.; Baeck, M.L.; Morrison, J.E.; Sturdevant-Rees, P. Catastrophic rainfall and flooding in Texas. *J. Hydrometeorol.* **2000**, *1*, 5–25. [CrossRef]

40. Furl, C.; Sharif, H.O.; el Hassan, A.; Mazari, N.; Burtch, D.; Mullendore, G.L. Hydrometeorological Analysis of Tropical Storm Hermine and Central Texas Flash Flooding, September 2010. *J. Hydrometeorol.* **2015**, *16*, 2311–2327. [CrossRef]

41. Patton, P.C.; Baker, V.R. Morphometry and floods in small drainage basins subject to diverse hydrogeomorphic controls. *Water Resour. Res.* **1976**, *12*, 941–952. [CrossRef]

42. Furl, C.; Sharif, H.; Zeitler, J.W.; el Hassan, A.; Joseph, J. Hydrometeorology of the catastrophic Blanco river flood in South Texas, May 2015. *J. Hydrol. Reg. Stud.* **2018**, *15*, 90–104. [CrossRef]

43. Lin, Y.; Mitchell, K.E. 1.2 the NCEP stage II/IV hourly precipitation analyses: Development and applications. In Proceedings of the 19th Conference Hydrology, American Meteorological Society, San Diego, CA, USA, 9–13 January 2005.

44. Fulton, R.A.; Breidenbach, J.P.; Seo, Do.; Miller, D.A.; O'Bannon, T. The WSR-88D rainfall algorithm. *Weather Forecast.* **1998**, *13*, 377–395. [CrossRef]

45. Nelson, B.R.; Prat, O.P.; Seo, D.-J.; Habib, E. Assessment and implications of NCEP stage IV quantitative precipitation estimates for product intercomparisons. *Weather Forecast.* **2016**, *31*, 371–394. [CrossRef]

46. Huffman, G.J.; Bolvin, D.T.; Nelkin, E.J. Integrated Multi-satellitE Retrievals for GPM (IMERG) technical documentation. *NASA/GSFC Code* **2015**, *612*, 47.

47. Yong, B.; Liu, D.; Gourley, J.J.; Tian, Y.; Huffman, G.J.; Ren, L.; Hong, Y. Global view of real-time TRMM multisatellite precipitation analysis: Implications for its successor global precipitation measurement mission. *Bull. Am. Meteorol. Soc.* **2015**, *96*, 283–296. [CrossRef]

48. Hsu, K.; Gao, X.; Sorooshian, S.; Gupta, H.V. Precipitation estimation from remotely sensed information using artificial neural networks. *J. Appl. Meteorol.* **1997**, *36*, 1176–1190. [CrossRef]

49. Sorooshian, S.; Hsu, Ku.; Gao, X.; Gupta, H.V.; Imam, B.; Braithwaite, D. Evaluation of PERSIANN system satellite–based estimates of tropical rainfall. *Bull. Am. Meteorol. Soc.* **2000**, *81*, 2035–2046. [CrossRef]

50. Joyce, R.J.; Janowiak, J.E.; Arkin, P.A.; Xie, P. CMORPH: A method that produces global precipitation estimates from passive microwave and infrared data at high spatial and temporal resolution. *J. Hydrometeorol.* **2004**, *5*, 487–503. [CrossRef]

51. Huffman, G.J.; Bolvin, D.T.; Nelkin, E.J.; Wolff, D.B.; Adler, R.F.; Gu, G.; Hong, Y.; Bowman, K.P.; Stocker, E.F. The TRMM multisatellite precipitation analysis (TMPA): Quasi-global, multiyear, combined-sensor precipitation estimates at fine scales. *J. Hydrometeorol.* **2007**, *8*, 38–55. [CrossRef]

52. Downer, C.W.; Ogden, F.L. *Gridded Surface Subsurface Hydrologic Analysis (GSSHA) User's Manual*; Version 1.43 for Watershed Modeling System 6.1.; Engineer Research and Development Center Coastal and Hydraulics Lab.: Vicksburg, MS, USA, 2006.

53. Ogden, F.L.; Saghafian, B. Green and Ampt infiltration with redistribution. *J. Irrig. Drain. Eng.* **1997**, *123*, 386–393. [CrossRef]

54. Rawls, W.J.; Brakensiek, D.L.; Miller, N. Green-Ampt infiltration parameters from soils data. *J. Hydraul. Eng.* **1983**, *109*, 62–70. [CrossRef]

55. Zambrano-Bigiarini, M. Package 'hydroGOF': Goodness-of-Fit Functions for Comparison of Simulated and Observed Hydrological Time Series. R package Version 0.3-8. Available online: https://cran.r-project.org/web/packages/hydroGOF/hydroGOF.pdf (accessed on 8 February 2018).

56. Su, F.; Hong, Y.; Lettenmaier, D.P. Evaluation of TRMM Multisatellite Precipitation Analysis (TMPA) and its utility in hydrologic prediction in the La Plata Basin. *J. Hydrometeorol.* **2008**, *9*, 622–640. [CrossRef]

geosciences

MDPI

Article

Modelling the Present and Future Water Level and Discharge of the Tidal Betna River

M. M. Majedul Islam [1,*], Nynke Hofstra [1] and Ekaterina Sokolova [2]

[1]   Environmental Systems Analysis Group, Wageningen University and Research,
     6708 PB Wageningen, The Netherlands; nynke.hofstra@wur.nl
[2]   Department of Architecture and Civil Engineering, Chalmers University of Technology,
     412 58 Gothenburg, Sweden; ekaterina.sokolova@chalmers.se
*   Correspondence: majed25bd@gmail.com

Received: 14 June 2018; Accepted: 23 July 2018; Published: 24 July 2018

Abstract: Climate change, comprising of changes in precipitation patterns, higher temperatures and sea level rises, increases the likelihood of future flooding in the Betna River basin, Bangladesh. Hydrodynamic modelling was performed to simulate the present and future water level and discharge for different scenarios using bias-corrected, downscaled data from two general circulation models. The modelling results indicated that, compared to the baseline year (2014–2015), the water level is expected to increase by 11–16% by the 2040s and 14–23% by the 2090s, and the monsoon daily maximum discharge is expected to increase by up to 13% by the 2040s and 21% by the 2090s. Sea level rise is mostly responsible for the increase in water level. The duration of water level exceedance of the established danger threshold and extreme discharge events can increase by up to half a month by the 2040s and above one month by the 2090s. The combined influence of the increased water level and discharge has the potential to cause major floods in the Betna River basin. The results of our study increase the knowledge base on climate change influence on water level and discharge at a local scale. This is valuable for water managers in flood-risk mitigation and water management.

Keywords: flood; precipitation; water level; discharge; General Circulation Models (GCM); MIKE 21 FM model

1. Introduction

Floods often cause devastating effects on human life and properties worldwide. Climate change increases floods because of change in precipitation patterns and sea level rise (SLR) [1]. Bangladesh is extremely vulnerable to the impacts of climate change because of its low and flat terrain, high population density, high poverty levels, and the reliance of many livelihoods on climate-sensitive sectors [2]. The Betna River basin in the southwest of Bangladesh has been experiencing both fluvial flooding due to extreme precipitation during the monsoon and storm surge flooding due to cyclones originating from the Bay of Bengal. Floods hit this area almost every year of the last decade, causing loss of life and economic damage [3].

Increased precipitation and flooding cause increased runoff that brings pollution from the land into the river. Outbreaks of waterborne diseases, such as diarrhoea, are very common after flooding events [3] and cause serious public health risks in this area [4]. More frequent and intense flooding is likely to occur in this area in the future due to climate change and SLR [5]. Therefore, the impact of flooding could well become more severe, particularly due to the low lying areas, high population density, inadequate flood protection infrastructure, low level of social development, and high dependence on agriculture [4].

The fifth assessment report (AR5) of the Intergovernmental Panel on Climate Change [6] concluded that the projected adverse impacts of climate change on deltas would be mainly due to floods associated with extreme precipitation, increases in temperature and SLR. The future precipitation in Bangladesh is expected to increase between 5 and 20%, and average temperature is expected to increase between 2 and 4 °C by 2100 [7]. Trend analysis of SLR along the southwest coast of Bangladesh shows an annual increase of 5.5 mm based on tidal gauge record during 1977–1998 [8]. Although the impact of extreme precipitation and SLR on flood risks is substantial, current water management practices are not robust enough to cope with climate change consequences. The information about present climate variability and future climate scenarios needs to be better incorporated into planning and management of water bodies [8].

Process-based modelling is often performed to understand hydrodynamics of a basin and to assess the effectiveness of future flood protection infrastructure [9,10]. Several modelling and climate change studies were performed for South-Asian river basins [11–14]. These studies were mainly based on two climate change variables (temperature and precipitation) and reported that increases in future floods are very likely in this region, since increased temperatures and monsoon precipitation will likely impact river flows. However, in-depth studies on assessing combined impact of climate change and SLR on river hydrodynamics are lacking. Moreover, most of the climate change studies have been conducted in large or regional river basins like the Ganges, Brahmaputra and Meghna (GBM) River systems of India and Bangladesh [9,11–13]. No such studies have been performed for a relatively small basin; thus, the influence of climate change on water levels and discharge at the local scale remains unclear. Our study on the Betna River basin reduces this knowledge gap. The Betna River basin was selected for this study since it floods almost every year due to combined effects of extreme precipitation, storm surges and SLR. Moreover, the diversified water uses in the area (e.g., domestic, irrigation, shellfish growing, and bathing) require adequate water management.

The aim of the study was to assess the present and future water level and discharge in the Betna River. This was achieved by applying a process-based hydrodynamic model (MIKE 21 FM) to simulate water level and discharge under different future climate conditions. The MIKE 21 FM model for the Betna River was set up, calibrated and validated using the observed water level and discharge data. The model was then used to project the future (2040s and 2090s) water level and discharge. The output of this study is likely helpful in addressing frequent and intense flooding induced by climate change in the study area. The findings and model can be transformed to other basins of the world with similar characteristics.

2. Materials and Methods

2.1. Study Area

The study area covers an area of 10,706 hectares in the Betna River basin, located in the Satkhira district of southwest Bangladesh. The river has a total length of about 192 km with an average width of 125 m. The study focused on approximately 30 km of the downstream part of the river (Figure 1); the upper part of the river was not included in the model, since it becomes almost dry during dry months. The river is hydrologically connected with the Bhairab River near the Jessore district in the north and the Kholpetua River near Assasuni of the Satkhira district in the south. The Betna River has tidal influence, which is the predominant factor for its sustainability, because during the dry season the fresh water inflow from upstream areas becomes very limited. The tide generates from the Bay of Bengal and propagates to the north until the upstream boundary of the study area. Like most coastal areas in Bangladesh, this study area is governed by the semidiurnal tide. The usual range of fluctuation of the water level is 0.7 m during neap tide and 3.0 m during spring tide [15].

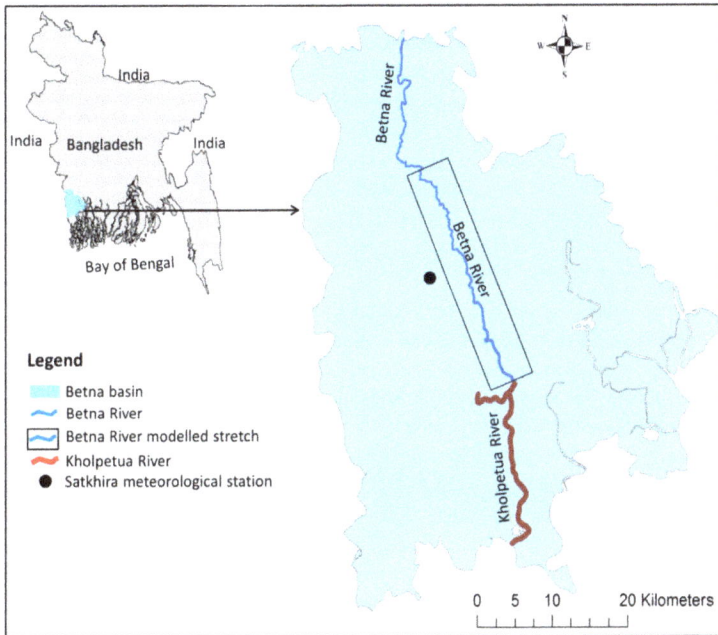

Figure 1. Study area, the Betna River basin in the southwest of Bangladesh.

The study area has a rainy season (monsoon) during June–September, followed by a cool and dry period during October–February and a hot season with frequent cyclones (pre-monsoon) during March–May. Mean annual rainfall in the area is about 1800 mm, of which approximately 70% occurs during the monsoon season [15]. This area is affected by inland flooding due to heavy incessant rainfall during the monsoon in August–September and by storm surge flooding during cyclone season in April–May [3,16]. Relative humidity of the area varies from about 70% in March to 90% in July. Mean annual air temperature is 26 °C with peaks of around 35 °C in May–June. Temperature in winter may fall to 10 °C in January [15]. Wind in the region shows two dominant patterns, i.e., south-westerly monsoon wind during June–September and north-easterly wind during November–February. Other months show no distinct wind direction pattern.

The study area is mainly a flat terrain with some low-lying depressions and many tidal channels and creeks that crisscross the area. The soils are mostly clay and loam. Land use is dominated by paddy cultivation and shrimp culture. About 8% of the total area is homesteads and settlements, about 10% is water bodies, 61% is agriculture and the remainder is wetlands used for aquaculture or integrated paddy shrimp culture [3,17].

2.2. Hydrodynamic Model Set Up

The two dimensional (2D) hydrodynamic model MIKE 21 FM [18] was applied to simulate future water level and discharge in the Betna River. This model was selected because it can simulate the hydrodynamic situation in a tidal river and generate outputs of high temporal and spatial resolution. The 2D model was used because the river is not very deep (maximum depth 9 m), and vertical mixing happens fast. The model simulates unsteady 2D flows in one vertically homogenous (depth averaged) layer and assumes that large flow gradients are absent in the vertical direction of the water column [1].

The hydrodynamic (HD) module of MIKE 21 FM simulates variations of water level and flows in response to several forcing functions on a rectangular or triangular grid of the study area when provided with the bathymetry, bed resistance coefficients, forcings and boundary conditions [19]. The model is based on a 2D numerical solution of Reynolds averaged Navier–Stokes equations. In the model, the Boussinesq simplifying approximation is used, and hydrostatic pressure is assumed [18]. The 2D model consists of vertically integrated momentum equations, continuity equation, advection-diffusion equations for temperature and salinity, and equation of state. The water density depends on temperature and salinity only [18]. The input and validation data required for the modelling process are river bathymetry, hydrodynamic and meteorological data. The detailed description of the model and the governing equations are presented in Danish Hydraulic Institute (DHI) [18] and Uddin et al. [19].

2.3. Bathymetric Survey and Mesh Generation

Modelling Bangladeshi river basins, including the Betna River, is difficult due to the scarcity of data. However, a comprehensive data collection survey was carried out in the Betna River system by the Institute of Water Modelling (IWM), Bangladesh, in 2012 to collect primary data, such as river bathymetry, water level and discharge. The distance between the surveyed river cross-sections varied from 400 to 500 m. The depth data were referred to meter Public Works Datum (mPWD) of Bangladesh (which is 0.46 m below mean sea level) using water level observed at the gauges within the survey area [15].

Mesh generation was done using the MIKE Zero mesh generator (Figure 2). A flexible mesh size with triangular elements was used, and the triangulation was performed with delaunay triangulation [18]. The mesh size was decreased and resolution increased where the river is narrow. The mesh consists of 4089 nodes and 6628 elements. The smallest element area is 42.5 m^2 and the largest area is approximately 498 m^2. In the modelling domain, intertidal zones were flooded and dried during every tidal phase to mimic natural conditions. The river is connected with some small drains, which have no water flow during dry weather and were thus considered as the land boundary in the model. However, during wet weather, stormwater runoff through the drains were included as a source in the model: four main drains were considered in the model. The runoff volume was estimated applying the runoff curve number method developed by the US Department of Agriculture [20].

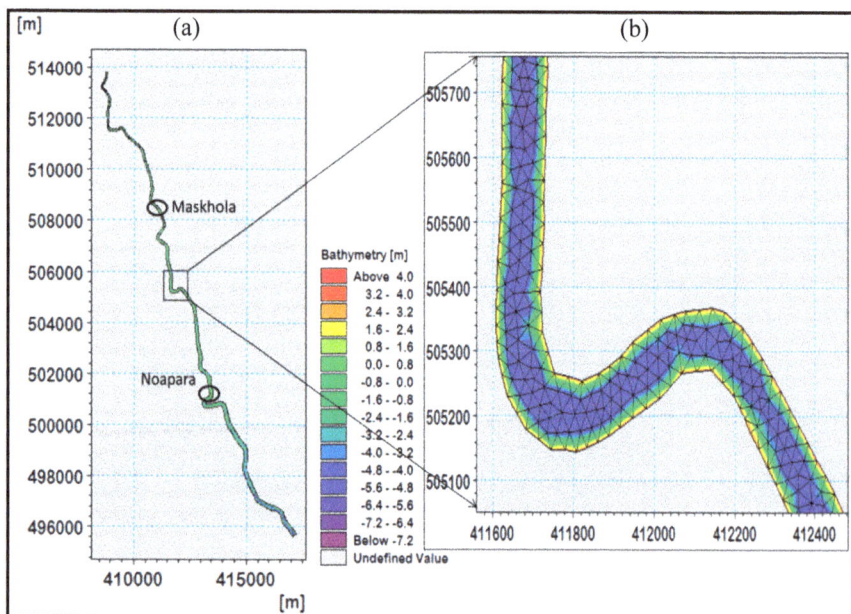

Figure 2. Bathymetry (**a**) and mesh (**b**) for the hydrodynamic model of the Betna River.

2.4. Meteorological and Hydrodynamic Data

Precipitation, wind speed and direction, air temperature (maximum and minimum) and relative humidity data were collected from the Satkhira meterological station (Figure 1, Table 1). To calibrate the hydrodynamic model, water level and river discharge data were collected from IWM (Table 1). Water level data were collected with 0.5 h interval at three locations (Maskhola, Noapara and the downstream boundary, (Figure 2) along the Betna River from 1 August to 10 October 2012. The observed minimum and maximum water levels at Maskhola and Noapara in the Betna River were 1.55 and 3.48 mPWD, and −2.10 and 3.50 mPWD, respectively. Discharge measurements were carried out near Noapara (Figure 2) and the upper boundary by IWM in September 2012 for 13 h with 0.5 h interval both in spring and neap tide. The observed maximum discharge at Noapara during spring tide was 277 and 392 m^3/s at the time of ebbing and flooding, respectively. For the baseline simulation, water level and discharge data were gathered from the Bangladesh Water Development Board (BWDB) for the period 2014–2015 (Table 1).

Table 1. Input data used for the hydrodynamic model.

Data Type	Resolution	Period	Location	Source
River bathymetry	Cross-section 400–500 m	2012	Modelled stretch	IWM [a]
Water level for calibration period	0.5 h	2012	Maskhola, Noapara and lower boundary	IWM [a]
Water level for validation period	3 h	2014–2015	Near Noapara and lower boundary	BWDB [b]
Discharge for calibration period	0.5 h	2012	Near Noapara and upper boundary	IWM [a]
Discharge for validation period	1 week	2014–2015	Near upper boundary	BWDB [b]
Precipitation	1 day	2012–2015	BMD Satkhira	BMD [c]
Air temperature	1 day	2012–2015	BMD Satkhira	BMD [c]
Wind speed and direction	3 h	2012–2015	BMD Satkhira	BMD [c]
Relative humidity	1 day	2012–2015	BMD Satkhira	BMD [c]

[a] Institute of Water Modelling, Bangladesh; [b] Bangladesh Water Development Board; [c] Bangladesh Meteorological Department.

The boundary conditions were described using time-series for discharge and water level at the upstream and downstream boundaries, respectively. The initial conditions were specified using the measured data for water level, and initial water velocity was set to zero. The flooding depth 0.05 m, drying depth 0.005 m and wetting depth 0.1 m were set in the model. In the model, a constant horizontal eddy viscosity (0.28 m^2/s), a constant clearness coefficient (70%), and default parameterisation (Table 2) for heat exchange were used. The model by default calculates evapotranspiration using air and water temperature data.

Table 2. Parameter values used in the hydrodynamic modeling.

Parameter Type	Formulation	Calibrated Value	Note
Horizontal eddy viscosity	Smagorinsky formulation Constant value Range	0.28 m^2/s 1.8×10^{-06}–1.0×10^{8} m^2/s	
Vertical eddy viscosity	k-epsilon formulation Range	1.0×10^{-07}–2.0×10^{-04} m^2/s	Selected after multiple simulations where different values were tested.
Bed roughness	Roughness height Constant value	60 $m^{1/3}/s$	
Wind friction	Constant value	0.001255	
Heat exchange	Constant in Dalton's law Wind coefficient in Dalton's law Sun constant, "a" in Ångström's law Sun constant, "b" in Ångström's law	0.5 0.9 0.176 0.37	
Light intensity	Light extinction coefficient	0.5	

2.5. Calibration, Validation and Sensitivity Analysis

The model was calibrated by adjusting the parameter values (Table 2) to obtain output results that match measured data. After calibration the modeled and measured values of water level and discharge were compared at Maskhola and Noapara. The calibration period from 26 August to 15 September 2012 was selected because the bathimetric survey and measurements of hydrodynamics were conducted in that period. The calibration period covers the important monsoon period, when floods occur, and hydrodynamic variability is higher. A sensitivity analysis was also performed to estimate the rate of change in model output with respect to change in model inputs. The sensitivity analysis was performed by changing the following model inputs and parameters: water level and discharge at the model boundaries, water velocity, river bed roughness, wind speed, eddy viscosity, clearness coefficient, and heat exchange rate. The calibrated model was then applied to simulate water level and discharge for a typical year October 2014 to September 2015 to represent the baseline conditions.

The model output for the baseline year was validated using the measured water level. The model performance was assessed using two statistical parameters: coefficient of determination (R^2) and the Nash-Sutcliffe efficiency (NSE). Generally, model performances are called satisfactory if $R^2 > 0.60$ and NSE > 0.50; the closer the model efficiency is to 1, the more accurate the model is [21]. After validation, the model was applied to predict future water level and discharge under climate change situation.

2.6. Future Scenario Development

There are many General Circulation Models (GCMs) that describe past, present and future changes in climate on a global scale. With the help of GCMs, variations in surface air temperature, precipitation and sea level were computed; and the output from individual GCMs and averages of future climate conditions can be applied in climate change impact analysis [22]. In the current study, outputs from GCMs used in the fifth phase of the Climate Model Intercomparison Project (CMIP5) were used. CMIP5 was utilised by the IPCC in AR5 [23]. In IPCC AR5, four Representative Concentration Pathways (RCPs) were used for climate change projections [6]. In the current study, a relatively low emission pathway (RCP 4.5) and a high emission pathway (RCP 8.5) were used. Two GCMs, MPI-ESM-LR (Max Planck Institute for Meteorology) and IPSL-CM5A-LR (Institute Pierre-Simon Laplace), were used. These models were selected, since they have been used widely in this region and best represent precipitation and surface air temperature for this study area [8].

At first, the CMIP5 daily climatic data for both GCMs and RCPswere downloaded from the Earth System Grid Federation Portal (http://cmip-pcmdi.llnl.gov/cmip5/; https://esgf-data.dkrz.de/search/cmip5/). Then, the daily GCM data were downscaled and bias corrected using the "Delta method" with "quantile-quantile" correction, as described by Liu, Hofstra [24]. This is a relative change method, which is widely used in bias correction and downscaling GCM data with different grid size. The GCM's gridded data have less variability and less local characteristics than the point input required for the model. Taking into account local variability is important for climate change impact studies, specifically in the case of a small river basin. This requires a downscaling method that generates point data from gridded data. The delta change method can efficiently produce point data [24]. The method includes quantile-quantile correction to ensure that extreme events become more extreme in the future climate data. In this study, daily observed air temperature and precipitation data from 1986 to 2005 were used as a basis for this downscaling. Two future time periods, 2031–2050 (2040s) and 2081–2100 (2090s), were considered for the two RCPs.

To project future precipitation, a 20-year average total monthly precipitation was computed for the observed and future downscaled data. Then, the percentage change between future and observed data was calculated for each month. The daily observed precipitation data for the baseline year 2014–2015 were modified by this percentage change and subsequently used as model input for the simulation of future hydrodynamic conditions. Fresh water inflow from upstream Gorai River basin areas were on average projected to increase during the monsoon period (driven primarily by increased basin precipitation) by 9% [25] and decrease during dry periods by 6% [4] by the 2050s. Similar change was assumed for the 2090s. These changes in discharge values along with the precipitation percentage changes were used to modify the observed current discharge data. The resulting discharge data were used to define the upstream boundary conditions in simulations for future prediction. The future stormwater runoff was estimated by the curve number method [20] using future precipitation data. For air temperature, daily average values were used.

SLR combined with the RCP scenarios was used to investigate the impact of climate change on the hydrodynamic characteristics of the Betna River. For RCP 4.5, the estimated global mean SLR is 0.24 and 0.4 m for the time horizons 2040s and 2090s respectively, relative to the average sea level for 1986–2005. For RCP 8.5, the estimated global mean SLR is 0.3 and 0.63 m for the time horizons 2040s and 2090s respectively [26]. Recent projections from the IPCC AR5 suggest that sea level in the northern Bay of Bengal (close to the study area) may rise between 0.1 and 0.3 m by 2050 and between 0.3 and 0.6 m by 2100, relative to the baseline period 1991–2010, without including local effects, such as land subsidence [27]. Based on these studies, the annual SLR for the study area was estimated. This estimated SLR and the annual land subsidence of 2.5 mm [27,28] together lead to the estimated relative mean SLR of 0.26 and 0.42 m for RCP 4.5, and of 0.44 and 0.76 m for RCP 8.5 for the 2040s and 2090s, respectively. These SLR values were added to the observed water level data to define the downstream boundary conditions in simulations for future prediction. Finally, the model was applied to simulate future (for the 2040s and 2090s) water level and discharge, and the results were compared with the modelling results for the baseline year (2014–2015) (Table 3).

Table 3. Projected change (%) in seasonal mean water level and discharge at Noapara in the Betna River: Comparison of the two Representative Concentration Pathway (RCP) scenarios for the near (2040s) and far (2090s) future with the baseline year. The monsoon season occurs during June–September and the dry season during November–February.

GCMs	Near Future (2040s)								Far Future (2090s)							
	RCP 4.5				RCP 8.5				RCP 4.5				RCP 8.5			
	Monsoon		Dry		Monsoon		Dry		Monsoon		Dry		Monsoon		Dry	
	Flood Tide	Ebb Tide	Flood Tide	Ebb Tide	Flood Tide	Ebb Tide	Flood Tide	Ebb Tide	Flood Tide	Ebb Tide	Flood Tide	Ebb Tide	Flood Tide	Ebb Tide	Flood Tide	Ebb Tide
Water level (% change)																
IPSL-CM5A	11.1	11.3	10.6	10.7	15.5	15.1	14.2	13.9	15.6	15.4	14.5	14.4	21.4	21.7	19.7	19.9
MPI-ESM	11.4	11.7	11.1	11.3	15.8	15.4	14.5	14.1	16.1	15.9	14.8	14.9	22.1	22.7	20.3	20.5
Discharge (% change)																
IPSL-CM5A	7.6	7.2	1.2	1.3	11.6	11.4	0.8	0.6	14.7	14.9	0.4	0.5	20.2	20.5	−1.8	−1.6
MPI-ESM	8.2	8.1	0.8	0.9	12.4	12.7	0.7	0.7	16.1	16.2	−1.3	−1.2	21.3	21.1	−2.2	−2.3

3. Results

3.1. Model Calibration, Validation and Sensitivity Analysis

Through calibration, a very good agreement with the measured data for both water level and discharge was achieved (Figure 3), with R^2 0.92 and 0.83, and NSE 0.81 and 0.66, respectively. Bed roughness (i.e., Manning's number) is a calibration parameter which has the largest impact on water level; after calibration, a constant Manning's number of 60 m$^{1/3}$/s was used. The agreement between the modelled and measured water level for the baseline year (validation) was very good (R^2 = 0.89, NSE = 0.76); the comparison for the discharge was not possible due to the lack of measured data. The model calibration and validation showed that the timing of the peaks was captured well, but the model slightly overestimated water level at low tides and underestimated discharge at high tides (Figures 3 and 4).

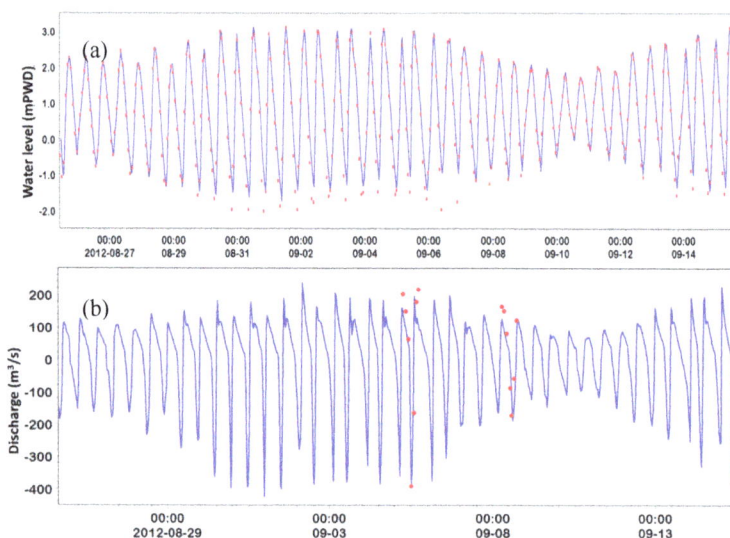

Figure 3. Model calibration: simulated and measured tidal water level (**a**) and discharge (**b**) at Noapara in the Betna River. The lines and dots represent the modelled values and measured values, respectively. The negative discharge values mean that during flood tide the flow is from the opposite direction.

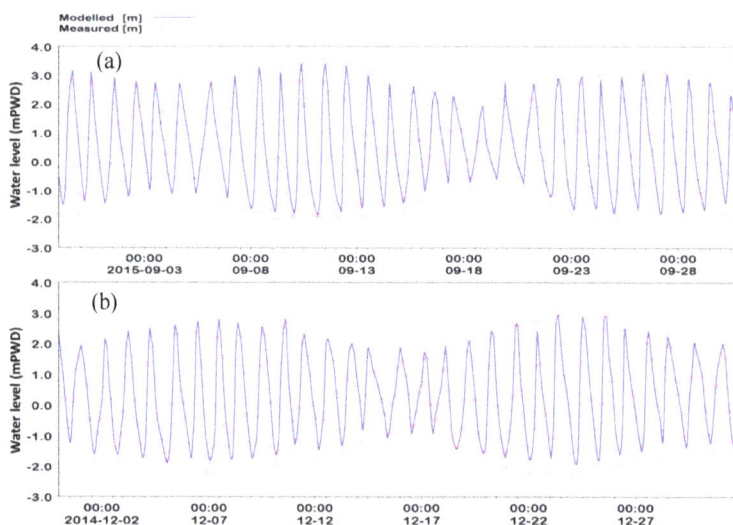

Figure 4. An example of comparison of modelled and measured tidal water level at Noapara in the Betna River. (**a**) represents a wet month (September 2015), and (**b**) represents a dry month (December 2014).

The sensitivity analysis showed that the model output was mostly influenced by the water level and discharge at the model boundaries. The increase in water level at the downstream boundary due to SLR greatly influenced the model output. The model output was also sensitive to wind, which influenced the water level. The average wind speed of 3 m/s had no strong influence on the water level, while the wind speed of 15 m/s (the strongest observed) caused decreased water level in the upstream parts of the river.

3.2. Future Projections

Future air temperature, precipitation and discharge projections show substantial variation between the two scenarios (RCP 4.5 and RCP 8.5) and two GCMs (Figure 5). Average air temperatures are expected to increase by 2 °C to 4 °C by the 2040s and 2090s respectively, compared to the observed baseline condition; the temperatures were consistently increased throughout the year, with greater increases by the 2090s compared to the 2040s. Precipitation is projected to increase during the monsoon season (June–September) by 3–28% and 5–32% by the 2040s and 2090s, respectively. Monthly average discharge shows increases in the near and far future compared to the observed discharge for both GCMs and both scenarios (Figure 5); the discharge is expected to increase in the monsoon periods and slightly decrease in dry periods (November–February).

The seasonal changes (percentage) in the mean water level and daily maximum discharge for the two GCMs and different scenarios are presented in Table 3. The mean change in water level at Noapara relative to the baseline year would be about 11% for the RCP 4.5 scenario and near future (2040s) and 23% for the RCP 8.5 scenario and far future (2090s), with increases being slightly larger in the monsoon than in the dry season. For the future discharge, the largest increase is expected in the monsoon period (June–September), with the data for the dry season showing no noticeable changes. The increase in the daily maximum discharge during the monsoon compared to the baseline year would be up to 8 and 16% in the near and far future respectively for RCP 4.5, and up to 21% by the end of century for RCP 8.5 (Table 3). The mean seasonal water level and discharge values for the two GCMs were not very different during the flood and ebb tide periods. The MPI-ESM GCM showed slightly higher water level and more extreme discharge in the Betna River compared to the IPSL-CM5A GCM.

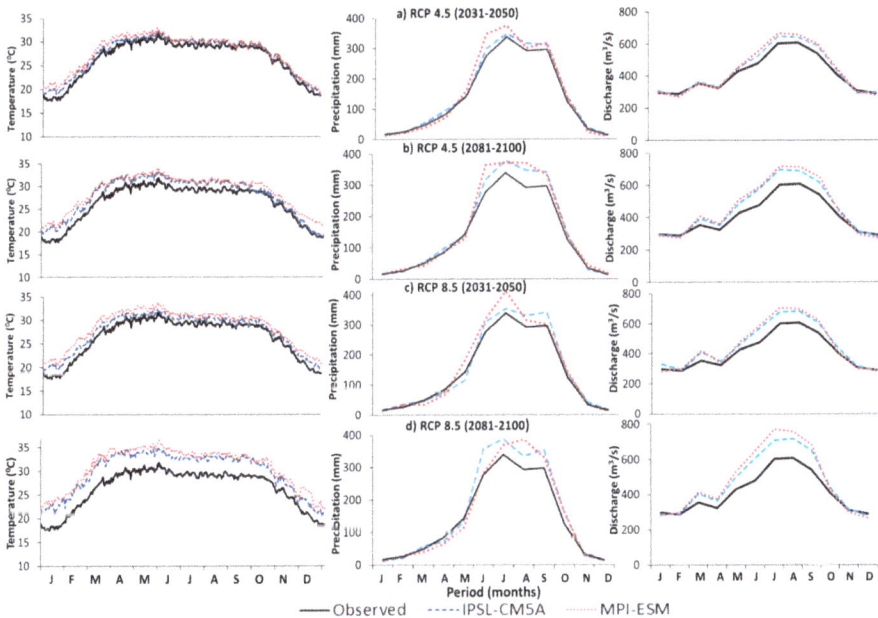

Figure 5. Daily average air temperature (left), monthly average precipitation (middle), and monthly average discharge (right) projections for the Betna River basin for two RCPs, RCP 4.5 (**a,b**) and RCP 8.5 (**c,d**) and two future periods, near future (2031–2050; **a,c**) and far future (2081–2100; **b,d**).

The impact of climate change and SLR was also assessed by comparing probability density functions (PDFs) developed for the baseline year and for the future conditions. Fitting the data to PDFs allowed identification of the trends and tendencies in changes of the river water level and discharge caused by climate change. Modelling results regarding hourly water level and daily maximum discharge for all studied scenarios were fitted to a non-parametric probability distribution (Figure 6). Compared to the baseline year, the frequencies of high water level and daily maximum discharge are expected to increase in the future. The comparison of the RCP 4.5 scenarios with the baseline year showed a little increase in water level and daily maximum discharge in the near future (2040s), while in the far future (2090s), comparatively higher water level and daily maximum discharge can be expected. The comparison of the RCP 8.5 scenarios with the baseline year showed that a comparatively higher (than the RCP 4.5) water level and discharge would be expected for both the near and far future. Larger change was projected for the results from the MPI-ESM model compared to the IPSL-CM5A model.

To understand the individual impacts of SLR and the increase in upstream discharge on the water level in the Betna River, the model was run with and without SLR and the increase in upstream discharge during the monsoon. With the increase in upstream discharge, but without SLR, the future water level was found to be almost the same as in the baseline period (water level increased overall by around 1%). On the other hand, with SLR but without the increase in upstream discharge, the future water level was found to increase by 10–15% in the 2040s and 14–21% in the 2090s. Thus, for this tidal river, SLR had a stronger impact on the water level compared to the impact of the increase in upstream discharge, because the direction of the flow in the river is dominated by the tide. However, when the increased upstream discharge during the monsoon season coincides with high tide, the impact on the water level is greater.

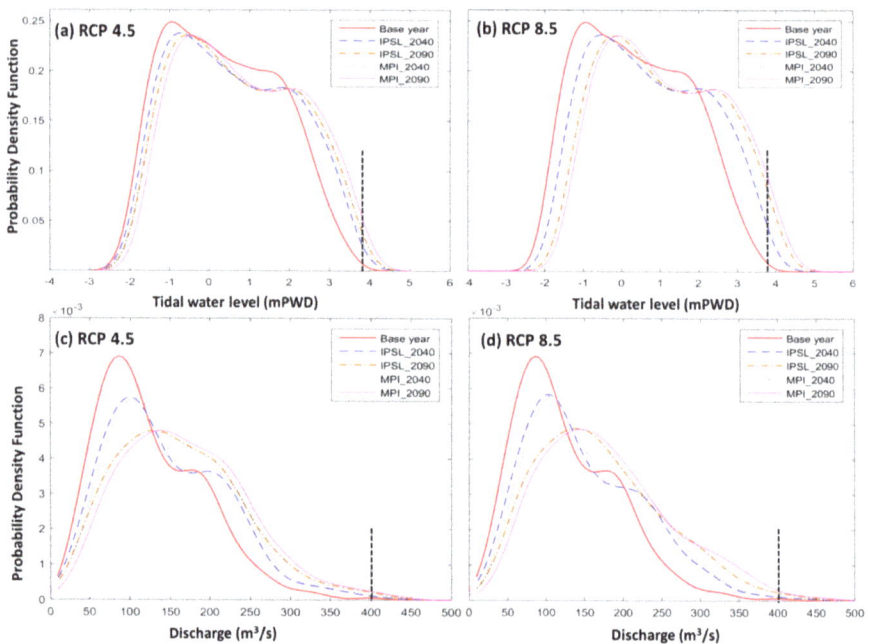

Figure 6. Probability density functions (PDFs) of hourly water levels (**a,b**) and daily maximum discharges (**c,d**) in the Betna River at Noapara for the baseline year (2014–2015) and in the near (2040s) and far (2090s) future under RCP 4.5 (**a,c**) and RCP 8.5 (**b,d**) for the two General Circulation Models (GCMs). The vertical dotted lines represent flood danger water levels (**a,b**) and extreme discharge levels (**c,d**).

The highest water level of the last 25 years in the Betna River at Benarpota (near Noapara) was 4.05 mPWD, and the flood danger level for 2014 was 3.84 mPWD [15]. The modelling results revealed that duration above the current flood danger level will increase up to 15 days by the 2040s and 34 days by the 2090s (Table 4). The duration of extreme discharge, >400 m^3/s (defined as an extreme based on modelled and measured data) will increase up to 14 days by the 2040s and 33 days by the 2090s.

Table 4. Change in duration above the current (2014–2015) flood danger level and extreme discharge at Noapara in the Betna River for two RCPs, two GCMs and two future periods.

GCMs	Scenarios	Duration above Flood Danger Level	Duration above Discharge of 400 m^3/s
Baseline (2014–2015)		13 days	11 days
IPSL-CM5A	RCP 4.5 (2040s)	25 days	21 days
	RCP 8.5 (2040s)	27 days	24 days
	RCP 4.5 (2090s)	41 days	35 days
	RCP 8.5 (2090s)	45 days	46 days
MPI-ESM	RCP 4.5 (2040s)	25 days	22 days
	RCP 8.5 (2040s)	28 days	25 days
	RCP 4.5 (2090s)	43 days	36 days
	RCP 8.5 (2090s)	47 days	46 days

4. Discussion

The results revealed an increase in air temperature, and increased monsoon precipitation and discharge, which are in good agreement with other studies of this region [9,12,13,29]. The increased monsoon precipitation and discharge indicate a wetter monsoon season in the future. However, during the dry season, precipitation and discharge are mostly expected to decrease in the future. This means that less water will be available during the dry season. The decreased precipitation and discharge during the dry season are unlikely to lead to droughts in the study area, because of the continuous tidal water inflow from the bay. This could exacerbate the existing salinity intrusion problem [28]. However, the decreased dry weather precipitation and associated salinization were not the focus of this study.

The future scenario analysis reveals that the mean change in water level and discharge relative to the baseline year would be up to 23% and 21% respectively, by the end of century. This increased discharge combined with the high water level would likely worsen the flooding situation in southwest Bangladesh and cause major flooding problems in the Betna River basin. Whitehead and Barbour [11] also indicated in their study that a 15% increase of discharge in the GBM River systems by the 2050s would have the potential to increase flood risk within Bangladesh. The increased river discharge in the monsoon season in the future is consistent with the future projected increased precipitation. The increase in monsoon discharge can also be attributed to the increased monsoon discharge in the upstream rivers of the Betna River. The increasing trend of monsoon precipitation and river discharge in South Asia is also evident from other studies [9,12,13,25,30,31].

The dry season data shows no noticeable change in the future discharge (Table 3) due to lack of connectivity with upstream rivers during the dry months. This is in agreement with Zaman, Molla [29] in their study in the GBM River systems of Bangladesh. Zaman and Molla [29] also found increased monsoon discharge and little change in the dry season discharge due to siltation at the upstream river mouth. The dry season discharge is expected to reduce by maximum 2% in the far future and this is unlikely to substantially change the present situation of the Betna River because of the tidal influence.

The results also revealed that SLR had a stronger impact on the water level compared to the impact of upstream discharge change. The upstream discharge is dependent on precipitation; during the monsoon, fresh water inflow from upstream areas increases. In a tidal river like the Betna River, the impact of discharge on the water level is limited, because the direction of the flow in the river is dominated by the tide. However, when the increased upstream discharge coincides with high tide, the impact on the water level is greater. The increased duration of the water danger level combined with increased frequency of extreme discharge (Table 4 and Figure 6) can severely affect the surrounding agricultural land by prolonged inundation during the monsoon and cause losses in terms of lives and livelihoods. When the water danger level coincides with the extreme discharge event, it would cause disastrous floods in the Betna River basin.

The MIKE 21 FM model was chosen to study the Betna River, because this model is suitable for tidal rivers and produces results at high temporal and spatial resolution. However, this choice comes at a cost of high computational requirements, which limited the opportunities to run the model for many years and many different GCM outputs. However, using several GCMs is recommended for further studies, since using a full ensemble of climate models would provide a better understanding of uncertainty.

The model was successfully calibrated for a monsoon season and validated for a full year, as has been done in other studies [4,25,26,28]. A possible reason for the slight overestimation of the water level at low tides can be the simplification of the river cross-section during mesh generation; thus better bathymetric data would be beneficial for further improvement of the model. This overestimation may also be the case for the results of future scenarios. However, as the impact is likely to be the same for the baseline period and future scenarios, and the future scenarios are compared with the baseline period, this slight overestimation should not affect the outcomes of this study. The slight underestimation of the discharge at high tides, which occurs in many hydrological models [32] and can possibly be explained by the lack of an extreme flood during the calibration period, could result in a stronger underestimation for a more extreme year. This may also mean that for the future, the peaks could be more underestimated than for the present.

To simulate the future scenarios, the discharge projections for the upstream Gorai River [25] were considered representative for the Betna River, due to the lack of local future discharge projections. Even though the modelling results were sensitive to the discharge at the upstream boundary, for the future conditions, the SLR proved much more important than the increase in the upstream boundary discharge. Therefore, the choice for the future discharge projections is relatively unimportant.

Identifying the trends and future scenarios for hydrodynamic characteristics triggered by climate change is required for the effective management of water bodies. This information on present and future hydrodynamic conditions can assist policy makers and water managers in planning flood risk mitigation and designing adequate flood protection structures. This information can be used to formulate a project for strengthening climate resilience. Although some other factors such as, changes in water use upstream, land-use change, population growth, socio-economic development and river dredging activities could significantly influence future impacts, focus should not be on climate change only; other changes that could have major impacts on the sustainability of water management infrastructure should not be overlooked [4]. Flood forecasting, i.e., early warning system should be developed to inform local farmers about their roles in an adverse weather condition. Construction of reservoirs near the river bank to store excess flood water for uses in the dry season can be another effective adaptation measure. The developed model, climate change scenario analysis approach and results of this study can potentially be useful for other river basins with similar geographic settings.

5. Conclusions

This study assessed the present situation and the future climate change impacts on the water level and discharge in the Betna River by applying a hydrodynamic model MIKE 21 FM. Based on the obtained modelling results, we conclude that:

- increased precipitation and SLR are expected in the Betna River basin in the near and far future under both RCP 4.5 and RCP 8.5;
- in RCP 8.5, water level and discharge in the Betna River are expected to increase up to 16 and 13% for the 2040s, and up to 23 and 21% for the 2090s, respectively;

- in RCP 4.5, although the expected increase in river discharge is relatively low (i.e., between 7 and 16%), the increased discharge combined with an increased water level is likely to cause major floods in the Betna River basin;
- the modelling results suggest that during the dry season, a small decrease in discharge (up to 2%) is expected for the 2090s;
- SLR explains a larger part of the future increase in water level than increasing upstream discharge; and
- in the future, the duration above the current flood danger level and of the extreme discharge events is expected to increase by half a month (per year) in the 2040s and by more than one month in the 2090s, causing prolonged inundation in the river basin, particularly during the monsoon.

Author Contributions: M.M.M.I.; Data curation, M.M.M.I. and E.S.; Formal analysis, M.M.M.I., N.H. and E.S.; Investigation, M.M.M.I.; Methodology, M.M.M.I., N.H. and E.S.; Project administration, M.M.M.I.; Resources, N.H.; Software, E.S.; Supervision, N.H.; Writing-original draft, M.M.M.I.

Funding: This research was funded by Ministry of Science and Technology, Government of Bangladesh, grant number 39.000.014.03.539 and the APC was funded by ESA group, Wageningen University and Research.

Acknowledgments: We thank the Ministry of Science and Technology of the Government of Bangladesh for providing a fellowship to pursue this study. We thank the Bangladesh Water Development Board, Meteorological Department and the Institute of Water Modelling for kindly providing river bathymetric and hydro-meteorological data and giving permission to use the data in the modelling work. We especially thank the DHI for providing the licence for the MIKE Powered by DHI software. Furthermore, we thank Rik Leemans (Wageningen University) for his help in critical review of the manuscript.

Conflicts of Interest: The authors declare no conflict of interest. The founding sponsors had no role in the design of the study; in the collection, analyses, or interpretation of data; in the writing of the manuscript, and in the decision to publish the results.

References

1. Webster, T.; McGuigan, K.; Collins, K.; MacDonald, C. Integrated river and coastal hydrodynamic flood risk mapping of the lahave river estuary and town of Bridgewater, Nova Scotia, Canada. *Water* **2014**, *6*, 517–546. [CrossRef]
2. Ahmed, A.U. Bangladesh Climate Change Impacts and Vulnerability: A Synthesis, Climate Change Cell Department of Environment, Govt. of Bangladesh. 2006. Available online: https://www.Preventionweb. Net/files/574_10370.Pdf (accessed on 25 March 2017).
3. CEGIS. *Environmental Impact Assessment of Re-Excavation of Betna River in Satkhira District for Removal of Drainage Congestion*; Bangladesh Water Development Board: Dhaka, Bangladesh, 2013; pp. 1–182.
4. Asian Development Bank (ADB). *Adapting to Climate Change: Strengthening the Climate Resilience of the Water Sector Infrastructure in Khulna, Bangladesh*; Asian Development Bank: Mandaluyong City, Philippines, 2011; pp. 1–40.
5. Karim, M.F.; Mimura, N. Impacts of climate change and sea-level rise on cyclonic storm surge floods in bangladesh. *Glob. Environ. Chang.* **2008**, *18*, 490–500. [CrossRef]
6. IPCC. Summery for policymakers. In *Climate Change 2014: Impacts, Adaptation, and Vulnerability. Part A: Global and Sectoral Aspects*; Field, C.B., Barros, V.R., Dokken, D.J., Mach, K.J., Mastrandrea, M.D., Bilir, T.E., Chatterjee, M., Ebi, K.L., Estrada, Y.O., Genova, R.C., et al., Eds.; Contribution of Working Group II to the Fifth Assessment Report of the Intergovernmental Panel on Climate Change; Cambridge University Press: Cambridge, UK; New York, NY, USA, 2014; pp. 1–32.
7. Mohammed, K.; Islam, A.S.; Islam, G.T.; Alfieri, L.; Bala, S.K.; Khan, M.J.U. Extreme flows and water availability of the brahmaputra river under 1.5 and 2 °C global warming scenarios. *Clim. Chang.* **2017**, *145*, 159–175. [CrossRef]
8. Bandudeltas. *Baseline Study on Climate Change in Bangladesh Delta Plan 2100 Project: Coast and Polder Issues, Planning Commision, Government of Bangladesh*; Bandudeltas: Dhaka, Bangladesh, 2015.

9.   Jin, L.; Whitehead, P.; Sarkar, S.; Sinha, R.; Futter, M.; Butterfield, D.; Caesar, J.; Crossman, J. Assessing the impacts of climate change and socio-economic changes on flow and phosphorus flux in the ganga river system. *Environ. Sci. Process. Impacts* **2015**, *17*, 1098–1110. [CrossRef] [PubMed]

10.   Kuchar, L.; Iwański, S. A modeling framework to assess the impact of climate change on river runoff. *Meteorol. Hydrol. Water Manag.* **2014**, *2*, 49–63. [CrossRef]

11.   Whitehead, P.; Barbour, E.; Futter, M.; Sarkar, S.; Rodda, H.; Caesar, J.; Butterfield, D.; Jin, L.; Sinha, R.; Nicholls, R. Impacts of climate change and socio-economic scenarios on flow and water quality of the ganges, brahmaputra and meghna (GBM) river systems: Low flow and flood statistics. *Environ. Sci. Process. Impacts* **2015**, *17*, 1057–1069. [CrossRef] [PubMed]

12.   Ghosh, S.; Dutta, S. Impact of climate change on flood characteristics in brahmaputra basin using a macro-scale distributed hydrological model. *J. Earth Syst. Sci.* **2012**, *121*, 637–657. [CrossRef]

13.   Apurv, T.; Mehrotra, R.; Sharma, A.; Goyal, M.K.; Dutta, S. Impact of climate change on floods in the brahmaputra basin using CMIP5 decadal predictions. *J. Hydrol.* **2015**, *527*, 281–291. [CrossRef]

14.   Alam, R.; Islam, M.S.; Hasib, M.R.; Khan, M.Z.H. Characteristics of hydrodynamic processes in the meghna estuary due to dynamic whirl action. *IOSRJEN* **2014**. [CrossRef]

15.   IWM. *Feasibility Study for Drainage Improvement of Polder 1, 2, 6–8 by Mathematical Modelling under the Satkhira District*; Institute of Water Modelling: Dhaka, Bangladesh, 2014; pp. 1–184.

16.   Hossain, A.N.H.A. *Integrated Flood Management: Case Study Bangladesh*; Bangladesh Water Development Board (BWDB): Dhaka, Bangladesh, 2003; pp. 1–14.

17.   Islam, M.M.M.; Hofstra, N.; Islam, M.A. The impact of environmental variables on faecal indicator bacteria in the betna river basin, Bangladesh. *Environ. Process.* **2017**, *4*, 319–332. [CrossRef]

18.   DHI. *Mike 21 and Mike 3 Flow Model FM*; MIKE by DHI: Horsholm, Denmark, 2011.

19.   Uddin, M.; Alam, J.B.; Khan, Z.H.; Hasan, G.J.; Rahman, T. Two dimensional hydrodynamic modelling of northern bay of bengal coastal waters. *Comput. Water Energy Environ. Eng.* **2014**, *3*, 140–151. [CrossRef]

20.   Cronshey, R.G.; Roberts, R.T.; Miller, M. Urban Hydrology for Small Watersheds. In *Procedding of the Specialty Conference, Hydrauolics and Hydrology in the Small Computer Age. Hydrology Division/ASCE, Lake Buena Vista, FL, USA, 12–17 August 1985*; Waldrop, W.R., Ed.; American Society of Civil Engineers: New York, NY, USA, 1985.

21.   Moriasi, D.N.; Arnold, J.G.; Van Liew, M.W.; Bingner, R.L.; Harmel, R.D.; Veith, T.L. Model evaluation guidelines for systematic quantification of accuracy in watershed simulations. *Trans. ASABE* **2007**, *50*, 885–900. [CrossRef]

22.   Christensen, N.S.; Lettenmaier, D.P. A multimodel ensemble approach to assessment of climate change impacts on the hydrology and water resources of the colorado river basin. *Hydrol. Earth Syst. Sci. Discuss.* **2007**, *11*, 1417–1434. [CrossRef]

23.   Taylor, K.E.; Stouffer, R.J.; Meehl, G.A. An overview of CMIP5 and the experiment design. *Bull. Am. Meteorol. Soc.* **2012**, *93*, 485–498. [CrossRef]

24.   Liu, C.; Hofstra, N.; Leemans, R. Preparing suitable climate scenario data to assess impacts on local food safety. *Food Res. Int.* **2015**, *68*, 31–40. [CrossRef]

25.   Billah, M.; Rahman, M.M.; Islam, A.K.M.S.; Islam, G.M.T.; Bala, S.K.; Paul, S.; Hasan, M.A. Impact of climate change on river flows in the Southwest region of Bangladesh. In Proceedings of the 5th International Conference on Water & Flood Management (ICWFM-2015), Dhaka, Bangladesh, 6–8 March 2015.

26.   Elshemy, M.; Khadr, M. Hydrodynamic impacts of egyptian coastal lakes due to climate change-example Manzala Lake. *Int. Water Technol. J.* **2015**, *5*, 235–246.

27.   Kay, S.; Caesar, J.; Wolf, J.; Bricheno, L.; Nicholls, R.; Islam, A.S.; Haque, A.; Pardaens, A.; Lowe, J. Modelling the increased frequency of extreme sea levels in the ganges–brahmaputra–meghna delta due to sea level rise and other effects of climate change. *Environ. Sci. Process. Impacts* **2015**, *17*, 1311–1322. [CrossRef] [PubMed]

28.   Dasgupta, S.; Kamal, F.A.; Khan, Z.H.; Choudhury, S.; Nishat, A. *River Salinity and Climate Change: Evidence from Coastal Bangladesh*; World Bank Policy Research Working Paper; World Bank Group: Washington, DC, USA, 2014.

29.   Zaman, A.; Molla, M.; Pervin, I.; Rahman, S.M.; Haider, A.; Ludwig, F.; Franssen, W. Impacts on river systems under 2 °C warming: Bangladesh case study. *Clim. Serv.* **2017**, *7*, 96–114. [CrossRef]

30.   Kirby, J.; Mainuddin, M.; Mpelasoka, F.; Ahmad, M.; Palash, W.; Quadir, M.; Shah-Newaz, S.; Hossain, M. The impact of climate change on regional water balances in Bangladesh. *Clim. Chang.* **2016**, *135*, 481–491. [CrossRef]

31. Mirza, M.M.Q.; Warrick, R.; Ericksen, N. The implications of climate change on floods of the ganges, brahmaputra and meghna rivers in Bangladesh. *Clim. Chang.* **2003**, *57*, 287–318. [CrossRef]
32. Kan, G.; He, X.; Ding, L.; Li, J.; Liang, K.; Hong, Y. Study on applicability of conceptual hydrological models for flood forecasting in humid, semi-humid semi-arid and arid basins in China. *Water* **2017**, *9*, 719. [CrossRef]

geosciences

MDPI

Article

An Attempt to Use Non-Linear Regression Modelling Technique in Long-Term Seasonal Rainfall Forecasting for Australian Capital Territory

Iqbal Hossain *, Rijwana Esha and Monzur Alam Imteaz

Faculty of Science, Engineering & Technology, Swinburne University of Technology, Melbourne, Hawthorn VIC 3122, Australia; resha@swin.edu.au (R.E.); mimteaz@swin.edu.au (M.A.I.)
* Correspondence: ihossain@swin.edu.au; Tel.: +61-3-9214-8120

Received: 9 June 2018; Accepted: 26 July 2018; Published: 28 July 2018

Abstract: The objective of this research is the assessment of the efficiency of a non-linear regression technique in predicting long-term seasonal rainfall. The non-linear models were developed using the lagged (past) values of the climate drivers, which have a significant correlation with rainfall. More specifically, the capabilities of SEIO (South-eastern Indian Ocean) and ENSO (El Nino Southern Oscillation) were assessed in reproducing the rainfall characteristics using the non-linear regression approach. The non-linear models developed were tested using the individual data sets, which were not used during the calibration of the models. The models were assessed using the commonly used statistical parameters, such as Pearson correlations (R), root mean square error (RMSE), mean absolute error (MAE) and index of agreement (d). Three rainfall stations located in the Australian Capital Territory (ACT) were selected as a case study. The analysis suggests that the predictors which has the highest correlation with the predictands do not necessarily produce the least errors in rainfall forecasting. The non-linear regression was able to predict seasonal rainfall with correlation coefficients varying from 0.71 to 0.91. The outcomes of the analysis will help the watershed management authorities to adopt efficient modelling technique by predicting long-term seasonal rainfall.

Keywords: non-linear model; seasonal rainfall; climate drivers; SEIO, ENSO; rainfall prediction

1. Introduction

Rainfall can be regarded as the most important climate element in the hydrological cycle that has considerable effects on the surrounding environment, including human lives. The spatial and temporal distribution of rainfall has significant impact on the water availability of earth surfaces, and hence on the agricultural activities. Since the agricultural activities and resulting crop production depends on the distribution of rainfall, prediction of monthly and seasonal rainfall is essentially important for the agricultural planning, flood mitigation strategies. However, accurate prediction of seasonal rainfall remains elusive to the scientists. Therefore, seasonal rainfall forecasting becomes plausible amongst the hydrologic researchers around the globe [1,2].

Seasonal forecasting can be classified into two broad categories: the statistical approach and the dynamic approach. In the statistical approach, the statistical relationships between the predictors and the predictands are investigated [3]. In the dynamic approach, seasonal meteorological estimates are used to build a hydrological model. However, there are methodological implications in using meteorological inputs in the current hydrological models [4]. The climate model produces the outputs based on coarse grid scales, which has the potential to capture forecasting uncertainties, and hence lead to bias. Furthermore, the data requirements of the dynamic models hinder the application of the modelling type. As a result, the statistical approach drew considerable attention to the practical users of the prediction models.

Long-term prediction of seasonal rainfall has the potential to help in the decision-making process for planning appropriate watershed management strategies [4]. Moreover, advanced prediction of rainfall can provide information to adopt the consequences of climate change [5]. As a result, the urge for the application of seasonal rainfall forecasting is increasing day by day. Therefore, seasonal forecasting is routinely performed by different research institutes, to have better understanding of climate change throughout the world. However, there exist limiting factors which act as the barriers for the wider application of the seasonal prediction models [6]. For example, the seasonal predictions are affected by the predictors, predictands, region and season [7]. Nevertheless, the chaotic dynamics of the atmosphere may lead to the erroneous prediction of seasonal rainfall [8]. The uncertainties in the model parameterization further hinder the prediction of seasonal rainfall.

Till today, precipitation is the most challenging climatic phenomena, which can be predicted with least accuracy [7]. On the other hand, most of the research studies on precipitation prediction have been conducted over a regional area of the world and in a particular season [9–12]. There exist only a few studies that concentrate on the precipitation analysis of the whole world [7,8,13]. Most of the studies conducted used number of various scores to evaluate predicted rainfall with the observed rainfall, such as, correlations, ranked probability score and Brier skill score. However, there is still doubt regarding the accuracy of the predictions of the seasonal models, which has immense implications in the decision-making process [14]. Nevertheless, there exists overconfidence and lack of reliability in the prediction of seasonal rainfall using the currently available models [15]. Therefore, it is necessary to assess the comprehensive performance of different models and their uncertainties in predicting seasonal rainfall.

It is well established that large-scale atmospheric circulation patterns significantly affect the annual precipitation around the globe including Australia. The atmospheric circulation configuration is dominated by the patterns of the sea surface temperature. Many researchers accept the capability of the El Nino Southern Oscillation (ENSO) in predicting time-series events. After analysing the role of ENSO on seasonal precipitation, Manzanas et al. [13] found that September to October is the most skillful season to predict rainfall around eastern Australia. Hossain et al. [9]; Hossain et al. [16] also identified the effects of ENSO and Indian Ocean Dipole (IOD) on West Australian rainfall. Therefore, the evaluation of the ENSO capability in time-series prediction is the fundamental requirement. Other climatic variables, such as sea surface temperature over the Atlantic and Indian Ocean have considerable impacts on the climate variability near the surrounding regions [17]. Recent studies also suggested that Indian Ocean Dipole (IOD) has the considerable effects on the climate variability in the continental regions including Australia [18,19]. Rasel et al. [20] revealed the effects of Southern Annular Mode (SAM) as a potential contributor of South Australian rainfall variability.

A number of studies have been examined to identify appropriate modelling technique for the prediction of seasonal rainfall. However, only a single climate driver is not capable of replicating the accurate precipitation characteristics. Multi-predictors models have the higher prediction skill than the single predictor models [21]. Nevertheless, there may exist dissimilar characteristics of seasonal rainfall patterns with the same rainfall totals [1]. On the other hand, there exist non-linear characteristics of the seasonal climate [8]. Therefore, a closer look at the appropriate mechanism of seasonal rainfall formation becomes essentially important.

This paper presents the efficiency of non-linear regression modelling technique in predicting long-term seasonal rainfall forecasting. The non-linear analysis was performed using the lagged (past) values of the climate indices as the potential predictors of long-term seasonal rainfall. Since there exist significant correlations between seasonal rainfall and two to three months average values of the climate indices, lagged values of the predictors were considered in this research. Furthermore, many researchers identified the ENSO and IOD as the most significant predictors of Australian seasonal rainfall. In this research, the efficiency of the climate indices in rainfall forecasting were assessed using the non-linear regression analysis technique. The non-linear analysis was performed considering South-eastern Indian Ocean (SEIO), Nino3.4 (sea surface temperature anomalies from 5° S to 5° N and

170° W to 120° W), southern oscillation index (SOI) and dipole model index (DMI) as the significant influential parameters of seasonal rainfall variation. The analysis was performed and applied to three rainfall stations in Australian Capital Territory (ACT). Seasonal rainfall forecasting can have practical implications to a wide range of users in diverse sectors, such as agriculture, energy, water supply and stormwater management [22]. The outcomes of the analysis may be the benchmark for future generations in predicting seasonal rainfall.

2. Study Area and Data Collection

The Australian Capital Territory (ACT) located in the south-east of the country is enclosed within the state of New South Wales (NSW). Unlike other Australian cities whose climates are moderated by the sea, the ACT experiences four distinct seasons. As a result, the inter-annual variation of precipitation in the ACT is higher. Annually, the ACT receives approximately 623 mm rainfall. The highest rainfall could be observed in spring and summer and the lowest in winter. This study concentrates on the application of non-linear regression modelling technique in the ACT for the prediction of long-term seasonal rainfall. The non-linear models were developed using the large-scale climate drivers. Therefore, the research requires both long-term rainfall data and climate indices data.

In Australia, the Bureau of Meteorology (BoM) collects and stores rainfall data from more than 2000 stations. For the achievement of the objectives of this paper, three rainfall stations located in the ACT were selected as a case study. Specific location of the rainfall stations is shown in Figure 1.

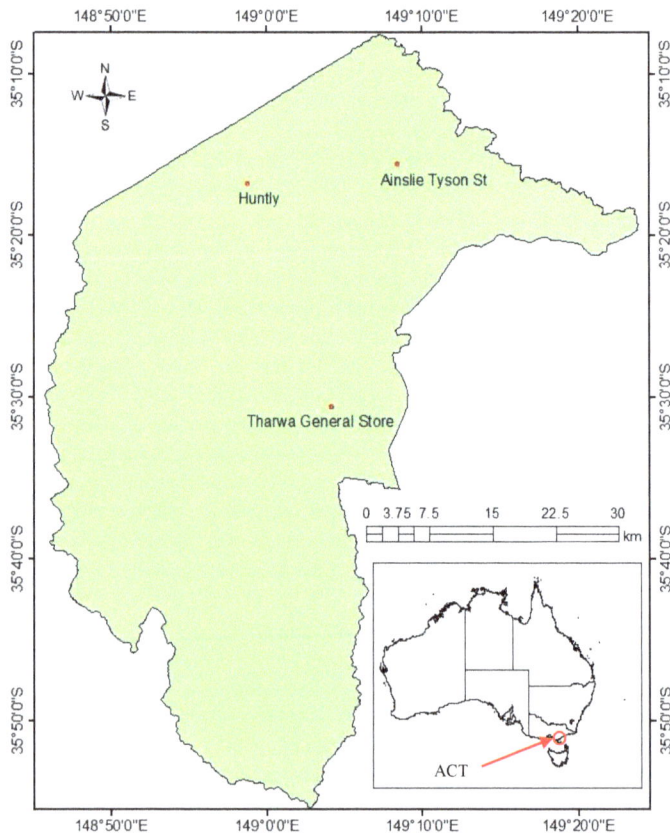

Figure 1. Selected rainfall stations in Australian Capital Territory.

Long-term monthly rainfall data from 1971 to 2017 were downloaded from Australian Bureau of Meteorology (http://www.bom.gov.au/climate/data/?ref=ftr). The rainfall stations were selected based on the availability of long-term data which have fewer missing values. Monthly variation of the rainfall for the selected rainfall stations throughout the study period is shown in Figure 2. The long-term variation of the same rainfall could be seen in Figure 3 (blue curve). Seasonal rainfall was estimated from the collected monthly rainfall data. In this paper, the average of the spring (September-October-November) rainfall data were used to perform non-linear regression analysis. The data collected were used not only to construct the non-linear regression models but also to validate the prediction capability of the developed models.

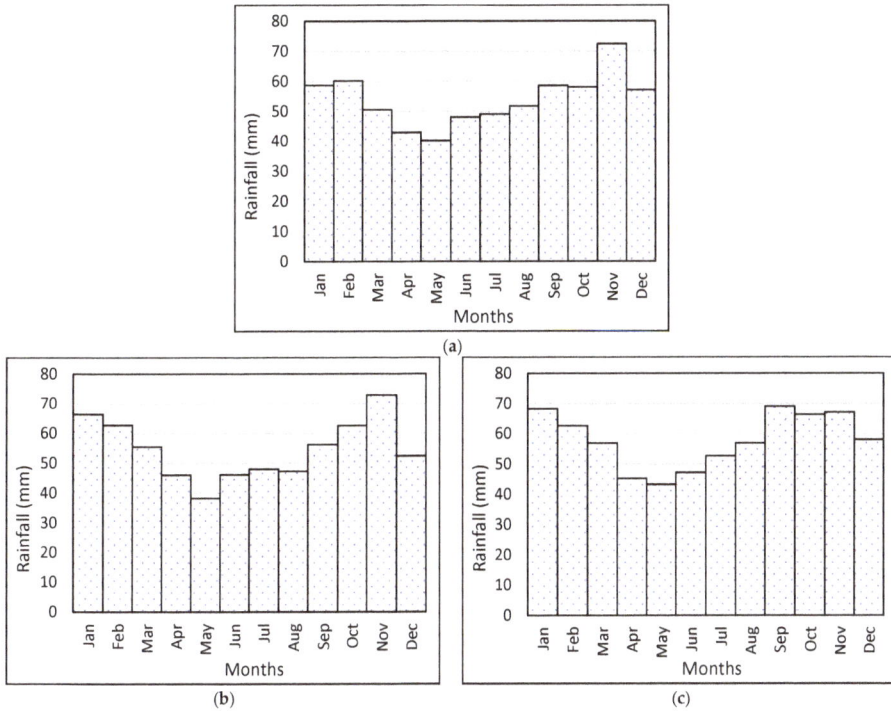

Figure 2. Monthly variations of rainfalls for the selected stations throughout the study period. (**a**) Ainslie Tyson St; (**b**) Tharwa General Store; (**c**) Huntly.

To replicate the appropriate characteristics of seasonal rainfall, it is required to identify which month's climate indices should be used in the analysis. Since the main focus of this paper is the efficiency of non-linear regression modelling technique in predicting long-term seasonal rainfall, the climate indices which have a significant correlation with rainfall were analysed and used for the construction of the non-linear models. Monthly values of the long-term climate indices data from 1971 to 2017 were collected from the climate explorer website (https://climexp.knmi.nl/start.cgi). Monthly values of the SEIO, Nino3.4, SOI and DMI were downloaded to achieve the objective of this research. In this paper, 90% of the data were used to construct the non-linear models and 10% of the data were used to assess the performance of the constructed models.

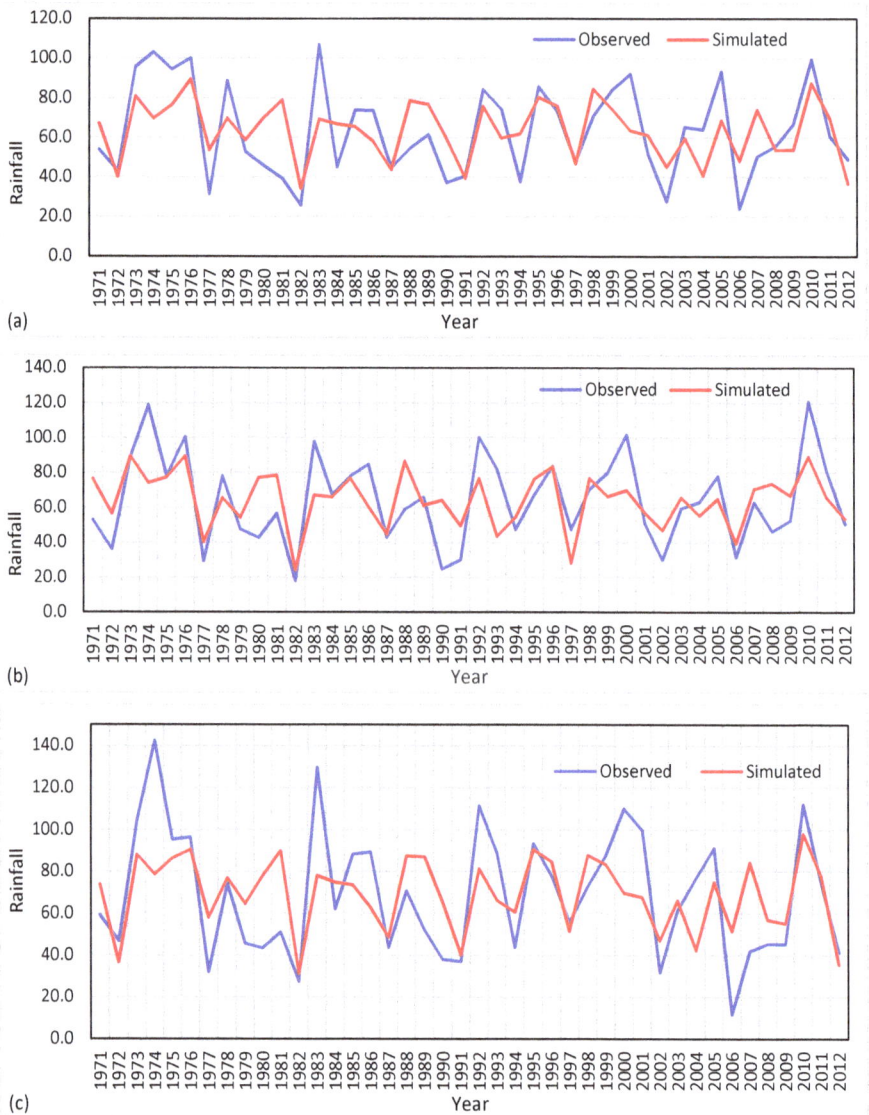

Figure 3. Comparison of the modelling output during the calibration period. (**a**) Ainslie Tyson St; (**b**) Tharwa General Store; (**c**) Huntly.

3. Methods

Traditional exploration of the relationship between two or more parameters are obtained by statistical regression analysis. In this study, non-linear regression analysis was performed to obtain steady relationship between long-term seasonal rainfall and large scale climate indices. In the non-linear regression technique, the arbitrary relationship between the predictands and predictors are obtained. One or more independent variables dictate the determination of non-linear relationship amongst the model parameters [23]. The general relationship of the non-linear regression can be explained according to Equation (1) [24]:

$$Y = b_1 X_i + b_2 X_j + b_3 X_i^2 + b_4 X_j^2 + \cdots + + b_n X_i X_j + e \tag{1}$$

where, Y is the dependent variable; $b_1, b_2, \ldots \ldots \ldots b_n$ are coefficients of the independent variables; n is the number of observations; X is the independent variables and e is model error. The fitted model has the potential to predict the value of Y for the additional observed values of X.

There may exist different non-linear functions which is suitable to replicate the appropriate rainfall pattern. To find out the suitable predicting model for streamflow, researchers have performed a series of simple regression analysis [25]. To recommend a suitable seasonal rainfall predicting model, six different functions: linear, quadratic, cubic, exponential, power and logarithmic were assessed in this study. The function which has the higher Pearson correlation with the rainfall was considered as the potential model for rainfall prediction. The long-term seasonal rainfall data were used for the estimation of the correlations. The predictions may be penalized due to the lack of understanding of the physical processes. The interactions amongst the sub-processes of the variables may further hider the predictive capability [26]. Therefore, individual correlations amongst the climate indices were performed to assess the significant correlations. The climate indices which have significant correlation amongst themselves were not used to develop the non-linear regression models.

Since the forecast verification assesses the capability of re-producing the observed data, the process is considered as an essential process of any model development [27]. In this research, the assessment of the forecast quality of the developed non-linear regression models were performed using the commonly used statistical errors, Pearson correlations, root mean square error (RMSE), mean absolute error (MAE) and index of agreement (d). The regression model which has the highest correlation between the seasonal rainfall and combined indices during the validation period (2013–2017) was considered as the recommended predictor model.

4. Results and Discussion

From the available linear and non-linear functions, the Pearson correlations between the climate indices and the ACT spring rainfall was determined from the fitted data sets. The function which has the higher correlation was considered as the suitable model for rainfall prediction. The correlations of the regression analysis is shown in Tables 1–3 for Ainslie Tyson St, Tharwa General Store and Huntly rainfall stations respectively. The month shown in the subscript is the value of the corresponding climate index for the specified month. The star (*) in the table refers that the correlation is significant at the 0.05 level.

Table 1. Pearson correlations of the regression analysis for Ainslie Tyson St station.

Index	Linear	Quadratic	Cubic	Exponential	Power	Logarithmic
DMI_{Jun}	0.083	0.097	0.154	0.082	0.182	0.187
$NINO3.4_{Jun}$	0.420 *	0.428 *	0.436 *	0.411 *	0.153	0.157
$NINO3.4_{Jul}$	0.462 *	0.462 *	0.495 *	0.456 *	0.056	0.061
$NINO3.4_{Aug}$	0.503 *	0.503 *	0.524 *	0.499 *	0.332	0.338
$SEIO_{Jun}$	0.127	0.265	0.267	0.135	0.331	0.312
$SEIO_{Jul}$	0.208	0.280	0.323 *	0.218	0.305	0.309
$SEIO_{Aug}$	0.236	0.242	0.243	0.232	0.099	0.098
SOI_{Jun}	0.380 *	0.385 *	0.389 *	0.374 *	−0.055	0.071
SOI_{Jul}	0.407 *	0.407 *	0.470 *	0.405 *	0.451	0.242
SOI_{Aug}	0.502 *	0.517 *	0.527 *	0.488 *	0.002	0.002

Table 2. Pearson correlations of the regression analysis for Tharwa General Store station.

Index	Linear	Quadratic	Cubic	Exponential	Power	Logarithmic
DMI_{Jun}	0.193	0.198	0.226	0.190	0.183	0.192
$NINO3.4_{Jun}$	0.457 *	0.480 *	0.507 *	0.440 *	0.232	0.240
$NINO3.4_{Jul}$	0.518 *	0.519 *	0.573 *	0.508 *	0.089	0.099
$NINO3.4_{Aug}$	0.542 *	0.542 *	0.589 *	0.533 *	0.298	0.303
$SEIO_{Jun}$	0.152	0.325 *	0.329 *	0.166	0.279	0.246
$SEIO_{Jul}$	0.195	0.253	0.301 *	0.204	0.226	0.230
$SEIO_{Aug}$	0.218	0.222	0.231	0.214	0.022	0.022
SOI_{Jun}	0.350 *	0.355 *	0.355 *	0.344 *	0.042	0.159
SOI_{Jul}	0.407 *	0.407 *	0.472 *	0.407 *	0.518 *	0.376
SOI_{Aug}	0.516 *	0.532 *	0.532 *	0.500 *	0.071	0.070

Table 3. Pearson correlations of the regression analysis for Huntly station.

Index	Linear	Quadratic	Cubic	Exponential	Power	Logarithmic
DMI_{Jun}	0.122	0.122	0.151	0.121	0.105	0.109
$NINO3.4_{Jun}$	0.383 *	0.387 *	0.391 *	0.376 *	0.288	0.283
$NINO3.4_{Jul}$	0.450 *	0.451 *	0.474 *	0.448 *	0.161	0.176
$NINO3.4_{Aug}$	0.476 *	0.476 *	0.510 *	0.476 *	0.375	0.380
$SEIO_{Jun}$	0.136	0.260	0.280	0.136	0.268	0.246
$SEIO_{Jul}$	0.216	0.235	0.280	0.221	0.231	0.236
$SEIO_{Aug}$	0.248	0.258	0.274	0.241	0.076	0.076
SOI_{Jun}	0.267	0.274	0.274	0.262	−0.097	0.019
SOI_{Jul}	0.322 *	0.322 *	0.399 *	0.321 *	0.495 *	0.320
SOI_{Aug}	0.467 *	0.477 *	0.478 *	0.453 *	0.011	0.010

For Ainslie Tyson St rainfall station, the cubic function has the maximum correlations between spring rainfall and all the climate indices except $SEIO_{Jun}$. This climate index has the maximum correlation with power function as shown in Table 1. Similarly for Huntly rainfall station, the cubic function has the maximum correlations between spring rainfall and the climate indices except SOI_{Jul}. As evidenced in Table 2, power function has the maximum correlation with rainfall and this predictor. However for Tharwa General Store rainfall station, cubic function has the maximum correlations between spring rainfall and all the climate indices as can be seen in Table 3. Generally, the cubic function is the best predictor and the logarithmic function is the least predictor for seasonal rainfall forecasting. A similar outcome was obtained by Esha and Imteaz [25] in predicting streamflow.

To develop a generalised non-linear model for the prediction of seasonal rainfall, the functions which have the maximum correlation were further analysed. Seventeen combined non-linear models were developed for each of the rainfall stations. The arrangement was selected in such a way that there is no significant correlation amongst the input combinations. The correlation coefficients for each of the developed models were also estimated. The combined indices that have been used to construct the non-linear regression models and their correlations are shown in Table 4 for all the selected three rainfall stations.

It is clear from Table 4 that only single model is not capable to predict seasonal rainfall with sufficient accuracy for all the rainfall stations. The table reveals that DMI-SOI based models are appropriate for predicting seasonal rainfall with maximum correlation 0.71. However, the appropriate combination is not in the same month for all the stations. For instance, DMI_{Jun} influence is dominant for all the three stations, whereas associated dominant indices are; SOI_{Jul} for Tharwa General Store station and SOI_{Aug} for Ainslie Tyson St station. For Huntley station, combined effect of $SEIO_{Jul}$ and $Nino3.4_{Aug}$ provided the highest correlation, with a Pearson correlation of 0.579. The outcomes support that the effects of climate indices vary spatially, and only single variable/index is not capable of predicting rainfall with sufficient accuracy. However, the combinations having higher correlations during the calibration were not considered as recommended models for rainfall prediction.

Table 4. Pearson correlations of the developed models for the selected rainfall stations.

Indices Combination	Correlations		
	Ainslie Tyson St	Tharwa General Store	Huntly
$SEIO_{Jun}$–$Nino3.4_{Jun}$	0.533 *	0.620 *	0.485 *
$SEIO_{Jun}$–$Nino3.4_{Jul}$	0.581 *	0.675 *	0.549 *
$SEIO_{Jun}$–$Nino3.4_{Aug}$	0.589 *	0.661 *	0.556 *
$SEIO_{Jul}$–$Nino3.4_{Jun}$	0.544 *	0.597 *	0.477 *
$SEIO_{Jul}$–$Nino3.4_{Jul}$	0.587 *	0.646 *	0.537 *
$SEIO_{Jul}$–$Nino3.4_{Aug}$	0.622 *	0.629 *	0.579 *
$SEIO_{Aug}$–$Nino3.4_{Jun}$	0.463 *	0.537 *	0.444 *
$SEIO_{Aug}$–$Nino3.4_{Jul}$	0.519 *	0.597 *	0.516 *
$SEIO_{Jun}$–SOI_{Jun}	0.472 *	0.489 *	0.400 *
$SEIO_{Jun}$–SOI_{Jul}	0.589 *	0.604 *	0.494 *
$SEIO_{Jun}$–SOI_{Aug}	0.602 *	0.616 *	0.530 *
$SEIO_{Jul}$–SOI_{Jun}	0.446 *	0.408 *	0.338 *
$SEIO_{Jul}$–SOI_{Aug}	0.619 *	0.625 *	0.547 *
$SEIO_{Aug}$–SOI_{Aug}	0.529 *	0.544 *	0.493 *
DMI_{Jun}–SOI_{Jun}	0.545 *	0.410 *	0.310 *
DMI_{Jun}–SOI_{Jul}	0.552 *	0.710 *	0.456 *
DMI_{Jun}–SOI_{Aug}	0.659 *	0.564 *	0.507 *

The combinations which produce maximum correlation during the validation period were considered to be the recommended models. For this case, $SEIO_{Jul}$–$Nino3.4_{Aug}$ is the best model for the Ainslie Tyson St and Huntly stations; whereas $SEIO_{Jul}$–SOI_{Aug} is the highest correlation producers for Tharwa General Store station. Therefore, three models that have the highest correlation between the spring rainfall and the climate variables during the validation period have been proposed. Derirved models are outlined in Equations (2)–(4) for Ainslie Tyson St, Tharwa General Store and Huntly rainfall stations respectively.

$$
\begin{aligned}
(\text{Rainfall})_{\text{Spring}} = {} & -74.0733 \times SEIO_{Jul}^3 + 69.3359 \times SEIO_{Jul}^2 + 11.5707 \times SEIO_{Jul} \\
& +6.21085 \times Nino3.4_{Aug}^3 - 1.57148 \times Nino3.4_{Aug}^2 - 26.6653 \\
& \times Nino3.4_{Aug} + 54.6311
\end{aligned} \tag{2}
$$

$$
\begin{aligned}
(\text{Rainfall})_{\text{Spring}} = {} & -91.0857 \times SEIO_{Jul}^3 + 71.6046 \times SEIO_{Jul}^2 + 12.3435 \times SEIO_{Jul} \\
& -1.5962 \times SOI_{Aug}^3 - 3.07473 \times SOI_{Aug}^2 + 20.7424 \times SOI_{Aug} \\
& +57.8264
\end{aligned} \tag{3}
$$

$$
\begin{aligned}
(\text{Rainfall})_{\text{Spring}} = {} & -91.5749 \times SEIO_{Jul}^3 + 69.4111 \times SEIO_{Jul}^2 + 20.8865 \times SEIO_{Jul} \\
& +8.33879 \times Nino3.4_{Aug}^3 - 1.96053 \times Nino3.4_{Aug}^2 + 33.6301 \\
& \times Nino3.4_{Aug} + 59.2245
\end{aligned} \tag{4}
$$

Since the cubic function has the potential to produce the higher correlation between seasonal rainfall and the considered climate indices, the equations have been developed for the cubic function. The combined capability of other functions will be assessed in future.

The plotted comparison of the analysis during the calibration period is shown in Figure 3. According to Figure 3, the non-linear regression models are not capable to replicate the actual seasonal rainfall with considerable accuracy. The statement is especially true for the extreme rainfall. When the rainfall is extremely high or extremely low, the approach is unable to capture the rainfall characteristics as evidence in Figure 3. More sophisticated analysis needs to be performed to replicate the extreme seasonal rainfalls. However, before concluding general remark, analysis on other area should be performed.

The plotted results of the prediction comparison during the validation period is shown in Figure 4. According to Figure 4, non-linear regression models should be used carefully to predict the seasonal rainfall with reasonable accuracy. To some extent, the approach is capable to predict the rainfall for some stations. For example, the approach is over predicting for Huntley rainfall station as evidence in Figure 4c. Therefore, other sophisticated modelling approaches should be explored for more accurate predictions of seasonal rainfall.

To evaluate the performance of the non-linear models developed, various statistical parameters were calculated. The outputs of the comparison are shown in Table 5. According to the table, the model with correlation more than 0.91 has higher RMSE and MAE than the model with correlation 0.71. In addition, models with correlation 0.86 is having more errors than the other two models. Similar outcomes were also observed for the index of agreement. Therefore, models which have higher correlation do not necessarily produce a lower error rate. The index of agreement close to one is considered to be the best predicting model. Therefore, the models could be used to predict seasonal rainfall with reasonable accuracy. However, the analysis should be performed with more rainfall stations in the same area and other states.

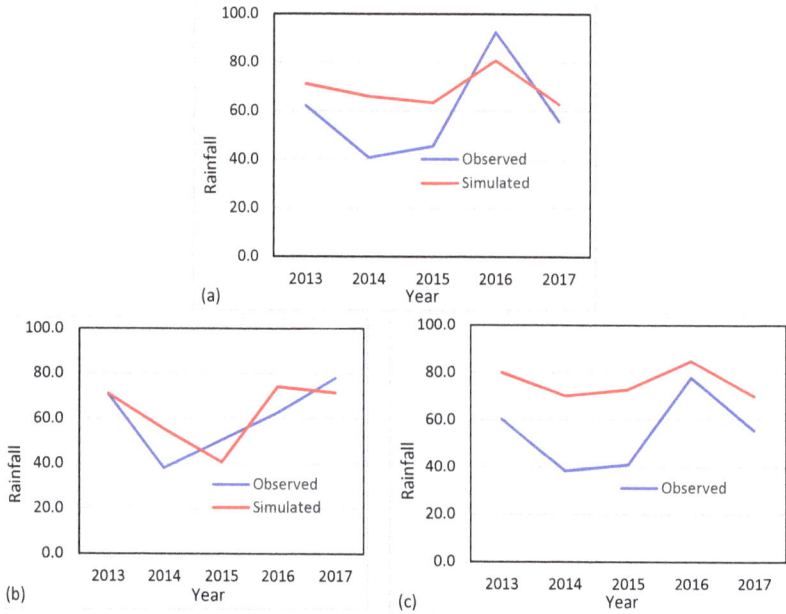

Figure 4. Comparison of the modelling output during the validation period. (**a**) Ainslie Tyson St; (**b**) Tharwa General Store; (**c**) Huntly.

Table 5. Estimated statistical parameters during validation period.

Parameters	Ainslie Tyson St	Tharwa General Store	Huntly
R	0.91	0.71	0.86
RMSE	15.62	10.68	23.08
MAE	14.2	9.0	20.9
d	0.71	0.82	0.56

5. Conclusions and Recommendations

Over the last two decades, prediction of seasonal time-series events were given considerable attention. As a result, many modelling approaches were developed and applied to predict seasonal rainfall. However, due to the spatial and temporal variation of rainfall, none of the available models are capable to predict seasonal rainfall with considerable accuracy.

In this research, the efficiency of non-linear regression models were assessed in predicting long-term seasonal rainfall. The non-linear models were constructed considering the lagged climate indices as the potential predictors of seasonal rainfall. Three rainfall stations located in the ACT were selected as a case study. The climate drivers SEIO, Nino3.4, SOI, and DMI were used and analysed in this study. The individual correlations between spring rainfall and the climate indices were determined for six functions (one linear and five non-linear). The functions which have the highest correlation between spring rainfall and the climate indices were further analysed to develop non-linear regression models. Seventeen combined non-linear regression models were developed and assessed to explore appropriate model(s) capable of predicting seasonal rainfall. The correlations between the outputs of the fitted models and the observed data were determined. The models which produce maximum correlation were considered as the potential model for seasonal rainfall forecasting. The accuracy of the predicted models' outputs was assessed by the widely used statistical parameters, R, RMSE, MAE and d. From the analyses of the current study, the following general conclusions could be made:

- Cubic function is capable of producing maximum correlation between seasonal rainfall and the climate indices.
- Logarithmic function produces the minimum correlations between seasonal rainfall and the climate indices.
- DMI-SOI based non-linear models are more suitable to predict seasonal rainfall, as they produce higher correlations.

However, before concluding a general remark, more rainfall stations should be analysed in this area and in other areas, which would be a part of a future study. Moreover, other non-linear modelling and genetic algorithm techniques will be explored, which are likely to be able to predict seasonal rainfalls with higher accuracy.

Author Contributions: Conceptualization, I.H.; Methodology, I.H.; Software, I.H.; Formal analysis, I.H.; Writing-Original Draft Preparation, I.H.; Writing-Review & Editing; I.H.; Visualization, I.H.; Methodology, R.E.; Software, R.E.; Formal analysis, R.E.; Conceptualization, M.A.I.; Supervision, M.A.I.; Writing-Original Draft Preparation, M.A.I.; Writing-Review & Editing; M.A.I..

Funding: This research received no external funding.

Conflicts of Interest: The authors declare no conflict of interest.

References

1. Tennant, W.J.; Hewitson, B.C. Intra-seasonal rainfall characteristics and their importance to the seasonal prediction problems. *Int. J. Climatol.* **2002**, *22*, 1033–1048. [CrossRef]
2. Frias, M.D.; Iturbide, M.; Manzanas, R.; Bedia, J.; Fernandez, J.; Herrera, S.; Cofino, A.S.; Gutierrez, J.M. An R package to visualize and communicate uncertainty in seasonal climate prediction. *Environ. Model. Softw.* **2018**, *99*, 101–110. [CrossRef]
3. Jenicek, M.; Seibert, J.; Zappa, M.; Staudinger, M.; Jonas, T. Importance of maximum snow accumulation for summer low flows in humid catchments. *Hydrol. Earth Syst. Sci.* **2016**, *20*, 859–874. [CrossRef]
4. Crochemore, L.; Ramos, M.H.; Pappenberger, F. Bias correcting precipitation forecasts to improve the skill of seasonal streamflow forecasts. *Hydrol. Earth Syst. Sci.* **2016**, *20*, 3601–3618. [CrossRef]
5. Winsemius, H.C.; Dutra, E.; Engelbrecht, F.A.; Archer Van Garderen, E.; Wetterhall, F.; Pappenberger, F.; Werner, M.G.F. The potential value of seasonal forecasts in a changing climate in southern Africa. *Hydrol. Earth Syst. Sci.* **2014**, *18*, 1525–1538. [CrossRef]

6. Goddard, L.; Aitchellouche, Y.; Baethgen, W.; Dettinger, M.; Graham, R.; Hayman, P.; Kadi, M.; Martinez, R.; Meinke, H. Providing seasonal-to-interannual climate information for risk management and decision-making. *Procedia Environ. Sci.* **2010**, *1*, 81–101. [CrossRef]
7. Barnston, A.G.; Li, S.; Mason, S.J.; Dewitt, D.G.; Goddard, L.; Gong, X. Verification of the First 11 Years of IRI's Seasonal Climate Forecasts. *J. Appl. Meteorol. Climatol.* **2010**, *49*, 493–520. [CrossRef]
8. Wang, B. Advance and prospectus of seasonal prediction: Assessment of the APCC/CliPAS 14-model ensemble retrospective seasonal prediction (1980–2004). *Clim. Dyn.* **2009**, *33*, 93–117. [CrossRef]
9. Hossain, I.; Rasel, H.M.; Imteaz, M.A.; Mekanik, F. Long-term seasonal rainfall forecasting: Efficiency of linear modelling technique. *Environ. Earth Sci.* **2018**, *77*, 28. [CrossRef]
10. Mekanik, F.; Imteaz, M.A.; Gato-Trinidad, S.; Elmahdi, A. Multiple linear regression and artificial neural network for long-term rainfall forecasting using large scale climate modes. *J. Hydrol.* **2013**, *503*, 11–21. [CrossRef]
11. Kim, H.M.; Webster, P.J.; Curry, J.A. Seasonal prediction skill of ECMWF System 4 and NCEP CFSv2 retrospective forecast for the Northern Hemisphere winter. *Clim. Dyn.* **2012**, *39*, 2957–2973. [CrossRef]
12. Lim, E.P.; Hendon, H.H.; Anderson, D.L.T.; Charles, A.; Alves, O. Dynamical, statistical-dynamical, and multimodel ensemble forecasts of Australian spring season rainfall. *Mon. Weather Rev.* **2011**, *139*, 958–975. [CrossRef]
13. Manzanas, R.; Frias, M.D.; Cofino, A.S.; Gutierrez, J.M. Validation of 40 year multimodel seasonal precipitation forecasts: The role of ENSO on the global skill. *J. Geophys. Res. Atmos.* **2014**, *119*, 1708–1719. [CrossRef]
14. Rayner, S.; Lach, D.; Ingram, H. Weather forecasts are for wimps: Why water resource managers do not use climate forecasts. *Clim. Chang.* **2005**, *69*, 197–227. [CrossRef]
15. Langford, S.; Hendon, H.H. Assessment of international seasonal rainfall forecasts for Australia and the benefit of multi-model ensembles for improving reliability. In *the Centre for Australian Weather and Climate Research Technical Report No. 039*; The Centre for Australian Weather and Climate Research: Victoria, Australian, 2011.
16. Hossain, I.; Rasel, H.M.; Imteaz, M.A.; Moniruzzaman, M. Statistical correlations between rainfall and climate indices in Western Australia. In Proceedings of the 21st International Congress on Modelling and Simulation, Gold Coast, Australia, 29 November–4 December 2015; pp. 1991–1997.
17. Goddard, L.; Mason, S.J.; Zebiak, S.E.; Ropelewski, C.F.; Basher, R.; Cane, M.A. Current approaches to seasonal to interannual climate predictions. *Int. J. Climatol.* **2001**, *21*, 1111–1152. [CrossRef]
18. Saji, N.H.; Yamagata, T. Interference of teleconnection patterns generated from the tropical Indian and Pacific Oceans. *Clim. Res.* **2003**, *25*, 151–169. [CrossRef]
19. Ashok, K.; Guan, Z.; Yamagata, T. Influence of the Indian Ocean Dipole on the Australian winter rainfall. *Geophys. Res. Lett.* **2003**, *30*, 1821. [CrossRef]
20. Rasel, H.M.; Imteaz, M.A.; Mekanik, F. Evaluating the effects of lagged ENSO and SAM as potential predictors for long-term rainfall forecasting. In *Proceedings of the International Conference on Water Resources and Environment (WRE 2015), Beijing, China, 25–28 July 2015*; Miklas, S., Ed.; Taylor & Francis Group: London, UK, 2015; pp. 125–129.
21. Liu, Y.; Fan, K. An application of hybrid downscaling model to forecast summer precipitation at stations in China. *Atmos. Res.* **2014**, *143*, 17–30. [CrossRef]
22. Manzanas, R.; Gutierrez, J.M.; Fernandez, J.; van Meijgaard, E.; Calmanti, S.; Magarino, M.E.; Cofino, A.S.; Herrera, S. Dynamical and statistical downscaling of seasonal temperature forecasts in Europe: Added value for user applications. *Clim. Serv.* **2018**, *9*, 44–56. [CrossRef]
23. Bilgili, M. Prediction of soil temperature using regression and artificial neural network models. *Meteorol. Atmos. Phys.* **2010**, *110*, 59–70. [CrossRef]
24. Adamowski, J.; Chan, H.F.; Prasher, S.O.; Ozga-Zielinski, B.; Sliusarieva, A. Comparison of multiple linear and nonlinear regression, autoregressive integrated moving average, artificial neural network, and wavelet artificial neural network methods for urban water demand forecasting in Montreal, Canada. *Water Resour. Res.* **2012**, *48*. [CrossRef]

25. Esha, R.I.; Imteaz, M.A. Seasonal streamflow prediction using large scale climate drivers for NSW region. In Proceedings of the 22nd International Congress on Modelling and Simulation, Hobart, Australia, 3–8 December 2017; pp. 1593–1599.
26. Pegion, K.; Kirtman, B.P. The impact of air-sea interactions on the simulation of tropical intraseasonal variability. *J. Clim.* **2008**, *21*, 6616–6635. [CrossRef]
27. Jolliffe, I.T.; Stephenson, D.B. *Forecast Verification: A Practitioner's Guide in Atmospheric Science*; John Wiley & Sons: Hoboken, NJ, USA, 2003.

geosciences

MDPI

Case Report

A Novel Method for Evaluation of Flood Risk Reduction Strategies: Explanation of ICPR FloRiAn GIS-Tool and Its First Application to the Rhine River Basin

Adrian Schmid-Breton [1,*], Gesa Kutschera [2], Ton Botterhuis [3] and
The ICPR Expert Group 'Flood Risk Analysis' (EG HIRI) [4,†]

[1]   International Commission for the Protection of the Rhine, 56068 Koblenz, Germany
[2]   Research Institute for Water and Waste Management at RWTH Aachen (FiW) e. V,
      Department of Innovation and Knowledge Transfer (Project coordination Africa), 52056 Aachen, Germany;
      kutschera@fiw.rwth-aachen.de
[3]   HKV Lijn in Water, Security and crisis management, 8203 Lelystad, The Netherlands; ton.botterhuis@hkv.nl
[4]   Expert Group, "Flood Risks" (EG HIRI) of the International Commission for the Protection of the
      Rhine (ICPR), Germany; sekretariat@iksr.de
*   Correspondence: adrian.schmid-breton@iksr.de; Tel.: +49-261-9425-222
†   The members of this expert group are listed by name at the end of the paper under "Acknowledgments".

Received: 17 August 2018; Accepted: 3 October 2018; Published: 6 October 2018

Abstract: To determine the effects of measures on flood risk, the International Commission for the Protection of the Rhine (ICPR), supported by the engineering consultant HKV has developed a method and a GIS-tool named "ICPR FloRiAn (Flood Risk Analysis)", which enables the broad-scale assessment of the effectiveness of flood risk management measures on the Rhine, but could be also applied to other rivers. The tool uses flood hazard maps and associated recurrence periods for an overall damage and risk assessment for four receptors: human health, environment, culture heritage, and economic activity. For each receptor, a method is designed to calculate the impact of flooding and the effect of measures. The tool consists of three interacting modules: damage assessment, risk assessment, and measures. Calculations using this tool show that the flood risk reduction target defined in the Action Plan on Floods of the ICPR in 1998 could be achieved with the measures already taken and those planned until 2030. Upon request, the ICPR will provide this tool and the method to other river basin organizations, national authorities, or scientific institutions. This article presents the method and GIS-tool developed by the ICPR as well as first calculation results.

Keywords: GIS; tool; flood risk analysis; transboundary flood risk assessment; flood risk management; effects of measures; effectiveness of measures; Rhine; ICPR; International Commission for the Protection of the Rhine; ICPR FloRiAn

1. Introduction

In the past, several important flood events occurred in the Rhine river basin (cf. Figure 1) and are the reason for why the nine countries of the basin are working together within the International Commission for the Protection of the Rhine (ICPR) [1] on the topic of transboundary flood risk management. The first results of this cooperation are the Action Plan on Floods (APF) [2,3] in 1998 and the first Flood Risk Management Plan (FRMP) for the international river basin district Rhine (IRBD) according to the "Floods Directive" of the European Union (Directive 2007/60/EC) in 2015 [4,5]. In the APF of 1998, one of the four objectives set out by the Rhine bordering states was to reduce the risk of flood damage by 10% by 2005, and by 25% by 2020, in comparison to the 1995 figures.

On the other hand, the most important objective of the Floods Directive (FD) in force since 2007 is the reduction of the adverse consequences of flooding upon human health, the environment, cultural heritage and economic activity. To help assess and monitor the effects and effectiveness of implemented flood risk management measures to verify and determine the risk and damage reduction resulting from the implementation of the APF and FRMP, the ICPR—supported by the engineering consultant HKV—developed a specific tool running in a geographic information system (GIS) named "ICPR FloRiAn (Flood Risk Analysis)" [6]. The tool is the result of a cooperation of several authorities of different nationality within the Rhine River Basin. The Technical report (ICPR report no. 237) [7] describes the method and calculations and the Synthesis report (ICPR report no. 236) [8] contains a summary of the method and describes the results of calculations undertaken using the tool.

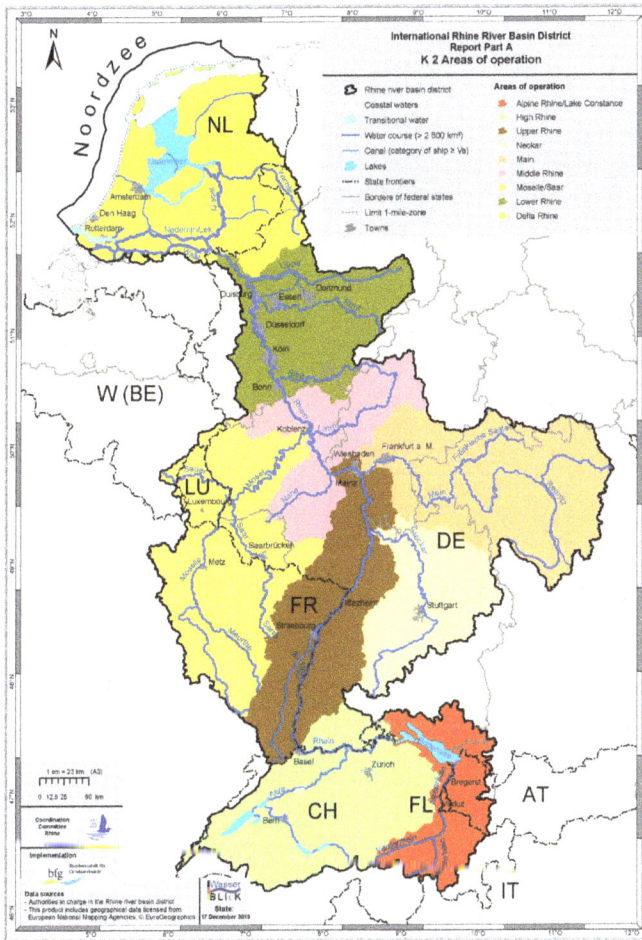

Figure 1. Rhine river basin [1].

Although other useful methods and tools exist [9–20], ICPR FloRiAn was specially tailored to the ICPR's wishes. As a result, the tool meets the needs and requirements of the Rhine bordering states, which had an impact on various parameters. For example, the tool is based on data available in the

Rhine bordering countries and, during the development of the tool, great importance was attached to create a link with the FD. This is also reflected in the type of measures and the data used.

As stated above, in addition to ICPR FloRiAn, there are several other methods, models and (GIS) tools that deal with the simulation of flood events and their consequences as well as the assessment of flood risks [9–15]. The specificity of ICPR FloRiAn is, however, to extend flood risk analysis to the effects or effectiveness of flood risk management measures on the development or reduction of damages or risks. The quantification of non-structural measures and their combination with each other is particularly innovative. For example, flood forecasting and sensitization measures have positive influence on one another and can also have an effect on the proportion of taking precaution for building protection. Another novelty is the consideration of other receptors as solely the economic activities: people, environment, and cultural heritage. Special reflections have been made to create appropriate methods for these receptors. Moreover, contrary to other methods or GIS-based applications, the aim of ICPR FloRiAn is not a cost-benefit analysis (only the economic damage is monetarized and costs of measures are not considered), but to identify a general damage or risk reduction (with or without the impacts of measures). Like many other models, the tool is able to carry out theoretical calculations (sensitivity analysis) at different (administrative) levels. Finally, ICPR FloRiAn differs significantly from instruments for crisis management [16–20], the focus being here on prevention measures.

Within the GIS toolbox ICPR FloRiAn (see Section 3 for an extract of the tool), flood hazard and risk maps (e.g., developed under the FD, see explanation in Section 2.1) [21] are input for the calculation. The tool consists of three interacting modules resulting in an overall damage or risk assessment for four receptors (or types of adverse consequences of floods) defined by the FD: human health, environment, culture heritage and economic activity. The ICPR has used this tool to assess the risk evolution along the Rhine from 1995 up to now (results are presented in Section 4) and has planned to use the tool to carry out regular reviews of the impacts of measures on flood risk reduction for the FRMP.

The tool ICPR FloRiAn, as well as the methods it is based on and a user guide, are available on simple demand at the ICPR (basic contract) and can be applied to other river basins by river organizations or national institutions [6], provided that basic GIS knowledge, GIS technical features (see Section 3), and the following required input data for the area under study are available in ESRI ArcGIS format: flood hazard (water depth grid), data related to receptors in flood prone areas (land uses, number of affected people, potentially polluting industries, nature protection areas, cultural heritage objects), damage functions, and various information on the implementation of measures. The instrument can also be used partially (for one or more modules or receptors), with less data or by using some ICPR data (such as damage functions) or even by using theoretical/dummy data. Although it was developed for a macroscopic level (the Rhine basin), tests were undertaken by extern users on a more local or regional level (City of Cologne, City of Rosenheim in Bavaria, German part of the Danube; not published) and gave interesting and logical results. Thus, the limit of applications to other areas is only given by available data, GIS system, and knowledge.

2. Description of the Method

2.1. Definitions and Basic Information

Mathematically, flood risk is defined as a product of probability of occurrence and the potential damage. The ICPR has developed specific methods, some of which are new, for determining the damage potential and the risk for the four receptors human health, environment, cultural heritage, and economic activities (cf. Sections 2.2–2.5) [7,8]. Furthermore, the effect of various measures can lead to changes in flood risk, which can be affected in two ways: by changing the flood probability and by influencing the potential damage (cf. Section 2.6). The modification of flood probability due to water level reduction measures such as retention measures and riverbed enlargement is described in the ICPR report no. 229 (cf. Section 2.6) [22]. The economic flood risk is calculated using the following formula [7,8,10]:

Flood risk [€/year] = Potential damage [€] × flood probability [1/year]

The damage potential of the receptors human health, environment, and cultural heritage is not calculated in monetary terms, so that the potential damage in € will be replaced by the number of persons or protected properties concerned (here as an example for the protection of human health [7,8,22]:

Flood risk [probability of being affected in inhabitants/year] = number of affected

inhabitants × (1 − safeguarding rate [inhabitants potentially evacuated or placed in

safety in % of total affected inhabitants]) × probability [1/year]. (see also Section 2.3)

The calculations of the flood risk are carried out using a GIS at the level of raster cells. During the evaluation, the results of individual raster cells are aggregated at the desired level in a table: e.g., stretches of the Rhine, municipality, district, region/federal state, or the whole Rhine catchment. The execution of calculations at different time horizons allows us to draw conclusions regarding the change of risk or the reduction of risk as a result of theoretically or actually implemented measures (cf. Section 4).

Figure 2 gives a general overview of the procedure for damage and risk calculation.

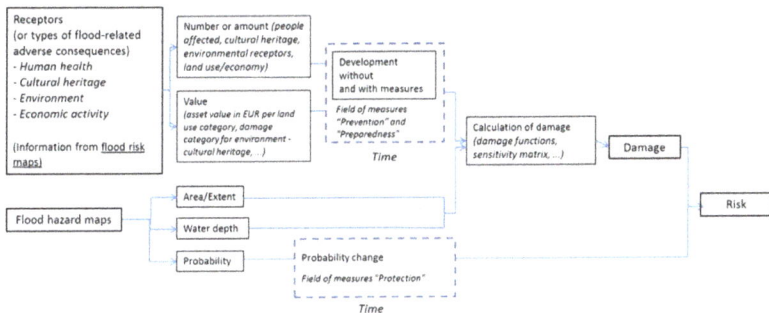

Figure 2. Overall risk analysis procedure [7,8].

Explanation of the calculation procedure and the data used presented in Figure 2 (see also sections hereafter):

- Flood risk maps: these maps are required by the FD and provide all the necessary information on receptors located in flood-prone areas (affected people, land use, etc.). A value or an amount (sum, number) is associated with these receptors. For economic damages, a damage function is associated with each type of land use (see Figure 3).
- Flood hazard maps: also required by the FD these maps provide all necessary information on the hazard (grid with inundation depth and flooding areas, flood probability in form of three scenarios: frequent, medium, and extreme floods). Hence, flood probability can be also entered separately in the tool.

The calculation of the damage (using special functions or sensitivity matrixes) and then the risks (combination with probability) (see Figure 2) can already be calculated with only input data from both types of maps mentioned above without including the influence of measures. To estimate the impacts of measures, one has to fill in the tool with information on their effects (entered and modifiable in the tool itself) and the level/number/percentage of measures implementation/realization (entered in specific shapefiles). An information on the realization would be; e.g., % of a municipality covered by risk-based urban plans or the number of sensitization campaigns in a certain period (see Table 1 in Section 2.6.1). The integration of the impacts of measures are explained in Section 2.6. For economic damages, for example, the damage function associated with each type of land use is modified where certain measures are being achieved (reduction of water levels or inundation depth, see Figure 3 and Section 2.6.1). The results of the calculations are given in the form of a GIS file and are given in euro/year or in number of people or objects affected/year.

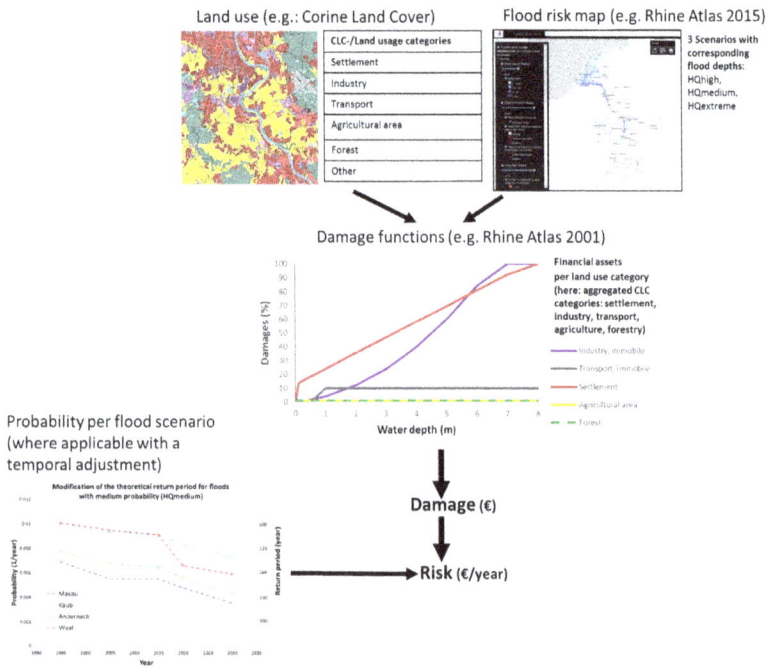

Figure 3. Approach for the analysis of the flood risk upon economic activity [7,8].

2.2. Receptor "Economic Activity"

The determination of the potential economic damage is based on the knowledge of the correlation between water depth and the resulting (relative) damage, the so-called damage functions. The direct economic damage potential is calculated in accordance with the methodology of the ICPR-Rhine Atlas 2001 [23]. Consequently, the damage potential is calculated on the basis of land use maps (Corine Land Cover 2006 in the case of the ICPR) included in flood risk maps (for the ICPR the ones of the Rhine Atlas 2015 [21]) for the three flood scenarios (frequent—HQhigh, medium—Hqmedium and extreme floods—Hqextreme) by means of damage functions and specific asset values for the 6 categories: settlement, industry, transport, agricultural areas, forest, and other (cf. Figure 3). Each cell in the map is reclassified to one of the 6 categories using the value of the land use as input. With this reclassification one of the 6 damage functions is coupled to a cell. Value of each cell in the flood map is reclassified with the damage function to a potential damage value (as a percentage of the asset value). The asset values (for property and immobile damage) are adjusted at a regional level on the base of economic growth or the consumer price index in order to adequately reflect the time horizon under consideration. As a result, the potential damage in €/m² per category and the integral/total damage are calculated.

2.3. Receptor "Human Health"

The damage to the receptor human health is defined as the number of people potentially affected in the flooded area (for the calculation formula, see Section 2.1).

The chosen method follows a two-staged approach (cf. Figure 4):

1. Determination of all people affected (per defined area, e.g., administrative district, municipality, . . .) regardless of water depth or other parameters for each flood scenario in total. In addition, the number of people affected can be established for the water level classes defined by the maps or the user.
2. Determination of the number of people who cannot get to safe places or be evacuated, using the approach of a state or area-specific minimal and maximal "safeguarding rate". This is the proportion of persons per region/area that could be evacuated or put in safety in advance of a potential flood and are therefore no longer in danger. The input of the safeguarding rate is given in % of the area under consideration (e.g., at municipality level) and is provided by the relevant countries for a reference time horizon (e.g., for 1995, minimal safeguarding rate of 20%; for 2015, maximal safeguarding rate of 80%). For the other time horizons (e.g., 2005), the safeguarding rate is calculated using a specific flow chart with an associated point system that considers the weighted effect of different prevention and preparedness measures according to their significance. This means that the safeguarding rate can be increased (e.g., in 2005 compared to 1995) by measures such as awareness rising, forecasting, warning and crisis management (cf. Table 1 in Section 2.6.1). The whole calculation procedure is precisely described in the ICPR report no. 237 [7].

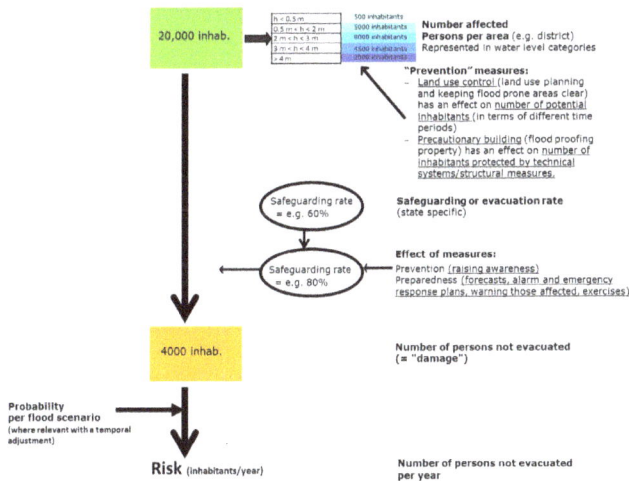

Figure 4. Approach for the analysis of damage to human health [7,8].

2.4. Receptor "Environment"

The method for assessing flood-related risks to the environment assumes that it is not the flood event itself, but rather the negative consequences triggered by the event that cause damage to surface water bodies that have a good or very good ecological status and to receptors/protected areas, in accordance with the FD. Negative consequences are understood to be the contamination of bodies of water via IPPC plants, SEVESO operation areas and waste water treatment plants due to flooding. The hazard or pollution potential resulting from the plant is defined in the tool on the basis of pollutant emission and transport models by means of an impact distance (distance between the source of danger and the receptor). Possible damages caused by the direct effect of flooding on the environment are not included in the study.

The environmental impact assessment is carried out in two stages (cf. Figure 5):

1. In the first stage, the contamination potential of the plant is combined with the water level category. The greatest contamination potential and the highest water depth present the highest threat. For each plant and each flood scenario, the respective threat is determined and assigned to a qualitative scale (1 to 5).
2. The second stage combines the ecological significance of a protected area with its threat.

This evaluation results in the three damage classes "low", "medium" and "high" and leads to an index per protected area. Within the framework of the calculations carried out, the damage indices per flood scenario and time horizon are summed (= aggregated damage index).

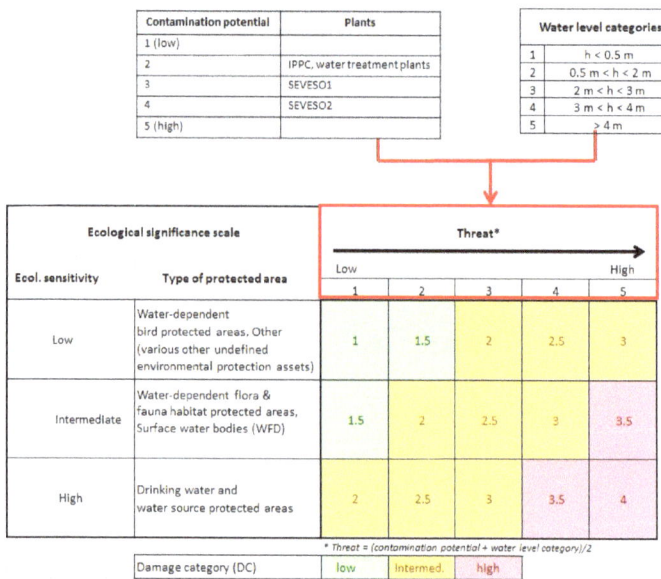

Figure 5. Approach for the analysis of damage to environment [7,8].

2.5. Receptor "Cultural Heritage"

Damage to cultural heritage can be approximated quantitatively by combining the significance of the cultural heritage (depending on the cultural heritage: UNESCO World Heritage Sites, protected urban areas, monuments) and water level.

By combining the defined value of a cultural asset with its water level, a specific matrix is created for assessing the damage to cultural assets. The matrix assessment results in a damage index for each object, to which one of the three damage categories is assigned, as in the case of the environmental damage. Cultural assets with low significance flooded by water levels of less than 2 m can expect a low level of damage, whereas water levels of 2 m or more lead to medium or high levels of damage.

2.6. Assessment of the Influence of Mitigation Measures—Elaboration of the Indicators

2.6.1. Change of Potential Damage

This section presents the measures in the areas of "prevention" and "preparedness" that impact the damage potential. Changes in the probability of flooding due to water level reduction measures, such as retention measures and widening of the riverbed (category of measure "protection"), were taken into account through the modification of the probabilities (cf. Section 2.6.2).

For the quantification of the impact of measures on the development of floods, risk indicators have been defined for the different receptors. The indicators should be representative, reproducible, and quantifiable for a group of measures (cf. list of measures and indicators in Table 1).

Table 1. Overview of the measures and indicators integrated into the tool and the calculations [7,8].

Type of Measure	Indicator	Unit and Scale of Indicator
Prevention		
Spatial planning, regional planning, and land use planning	Building regulations and codes/building development plans including requirements for flood protection (flood-adapted construction)	Expanse (m^2) of area (municipality or higher level) in which flood-adapted construction is regulated by building development plans [m^2] and percentage (%) of the municipality area for which development plans with these types of regulations exist.
Keeping flood prone areas open/clear (preventing the location of new or additional receptors) and adapted usage of areas	Modification of land use data (e.g., CLC data) within and outside of the flooding areas of the flood hazard map under analysis.	Modification of land use [m^2]
Flood-adapted design, construction, renovation	Measures implemented regarding flood-adapted development/building	Measures implemented/realized in the municipality (or higher level) in %
Precautionary building/flood-proofing property for households/municipalities	Protected areas due to precautionary building/flood-proofing property and/or mobile systems	Polygon with the area (in the municipality or higher level) protected by the flood-proofing of property or mobile systems [m^2]
Precautionary building/flood-proofing property in hazardous installations (IPPC plants, SEVESO operation areas and waste water treatment plants)	Protected installations due to technical protection, precautionary building/flood-proofing property and/or mobile systems	List of installations (IPPC, SEVESO, waste water treatment plant) that are protected/not protected
Flood-proof storage of water-polluting/hazardous substances for households/municipalities	Securing oil tanks and/or safe storage in upper floors	Number of households (as proportion of affected households in %), that have secured oil tanks or stored water polluting substances in upper storeys (per municipality or higher level)
Flood-proof storage of water-polluting/hazardous substances for hazardous installations (IPPC plants, SEVESO operation areas and waste water treatment plants)	Securing oil tanks and/or safe storage in upper floors	List of installations (IPPC, SEVESO, waste water treatment plant) in which secured oil tanks are safeguarded or pollutants are stored in upper storeys (unit: yes/no)
Provision of flood hazard and risk maps/establishing awareness in relation to precautionary behavior, education and preparation/preparedness for flood events	Frequency/update intervals with regard to information campaigns (incl. provision/presence of flood hazard and risk maps)	Update frequency of information campaigns (years) (in a municipality or higher level)
Protection		
Retention measures	Modification of probability (ICPR Report No. 229) [22]	Modification of probability and localization (stretch of river/gauge)
Dykes, dams, flood walls, mobile flood protection, ... / Maintenance/renewal of technical flood protection structures	For these measures, a probability is also indicated: Percentage evolution/change in flood probability between 1995 and present day due to improvements in protection. The information whether the area is protected/diked or non-protected/non-diked is relevant for the calculations.	Localization, renewals, modification of probability due to improvements in protection (%) (per measure or on a stretch of river)
Preparedness		
Flood information and forecast	Improvement in flood forecasting within a defined time-period	Forecast period in hours/days as well as further aspects (on a national level or for river stretches)
Alarm and emergency response planning (incl. recovery/aftercare)/warnings for those affected/exercises/training	Presence and update frequency of alarm and emergency response plans; number of warning systems (warning methods/ways and communication means), details of civil protection/crisis management exercises including frequency	Number of systems and update frequencies (on a municipality or higher level)
Safety/safeguarding/evacuation of (potentially) affected persons	Details of minimum and maximum safeguarding rate for those affected in a particular area	Minimum and maximum safeguarding in % on a national level or for river stretches (e.g., 70% can be evacuated, max. safeguarding rate = 70)

Based on a literature survey (see references in the Literature list of the report no. 237 [7]) and partly on expert knowledge, the maximum damage reduction, also referred to as the "effect" of a measure, was determined and defined per indicator. The degree of realization; i.e., which and how many measures have already been implemented/realized or will be implemented in the future (the information was provided by the delegates of the Rhine bordering states), has been included into the calculations.

Depending on the type of measure or indicator (influence on the damage potential or probability) and the considered receptor of the FD (human health, environment, cultural heritage, and economic activities), the impact of measures is calculated differently in the tool:

- Modification of the damage functions resulting from measures (receptors: economic activity and cultural heritage), as shown in Figure 6.
- Changes in the number of people due to evacuation combined with organizational measures (receptor: human health) (see Section 2.3 and Figure 4).
- Changes in the distance (buffer) of possible consequences arising from potentially hazardous facilities (receptor: environment).
- In the case of various measures, the effect is differentiated if the area is protected/embanked or unprotected/non-embanked. In general, it is assumed that, in unprotected areas which are more frequently flooded, potential victims have more flood experience and thus the reduction effect of potential damage is greater.
- In addition to the effect of individual measures, there are interdependencies/correlations between measures that are described in a dependency matrix for both embankment and non-embankment areas. Explanation: if several measures for one area that have an impact on the receptors economic activity and cultural heritage are combined, as a rule, the effect of the measure cannot be summed up in a simple manner, as there is the possibility that the effect would exceed 100%. Secondly, it is assumed that individual measures only have an effect when supplemented or used in combination with other measures (see examples and matrix in the report no. 237 [7]). The combination of measures that have an impact on human health has also been described in Section 2.3.

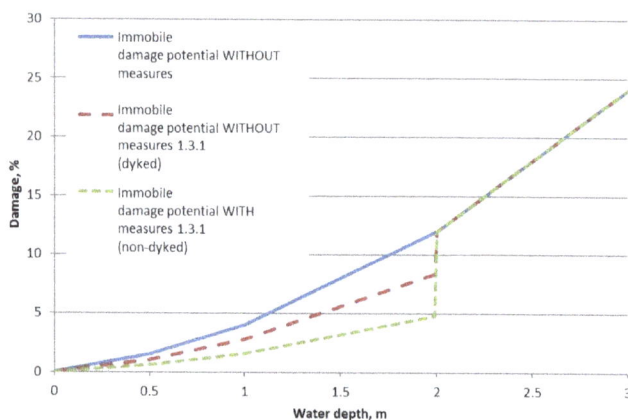

Figure 6. Modification of the damage function for immobile damage (industry) due to the measure "precautionary building" (= measure 1.3.1) for dyked and non-dyked areas [7].

2.6.2. Change of Flood Probabilities Due to Water Level Reducing Measures

Technical protection measures have an effect on the development of the flood risk, not only due to their influence on flood areas and depths, but also in the case of retention measures (e.g., retention basin, dyke relocation, measures from "Room for the River" in the Netherlands) and in the context of the ICPR, theoretically by changing the probability of flooding. For the calculations in the Rhine catchment area, retention measures already implemented and planned in the future were taken into account by changing the probabilities (cf. Section 4).

The effectiveness of implemented and planned flood-reducing/water level lowering measures on the Rhine was evaluated by an ICPR expert group which developed a specific method for estimating the change of flood probabilities [22,24]. The results of this method are changed return periods for floods with high, medium, and low probability for different time horizons or Rhine development states (1995, 2005, 2010, 2020 and 2030).

Figure 7 shows an example of the change in probability and return period for an extreme flood event at four selected gauging stations. The Waal (red dotted line) can be used to show the temporal change in the return period of an approximately 1000-year event in 1995 to a 2000-year event in 2030 for an extreme flood. This means that an extreme flood becomes less frequent due to the increase in return period.

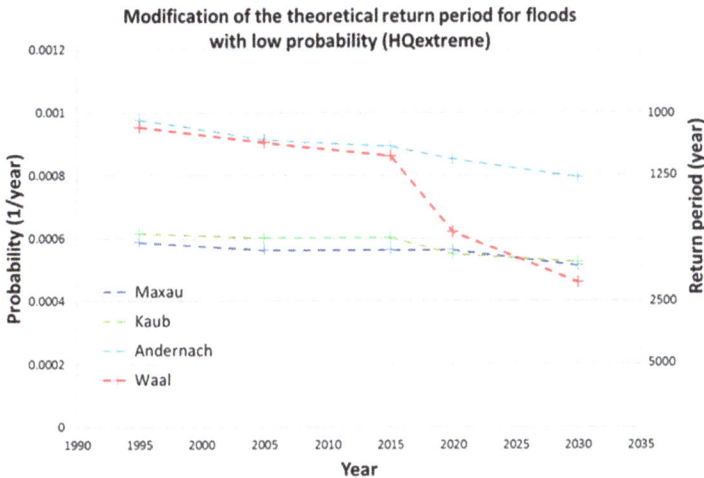

Figure 7. Change of probability (left hand *y*-axis) and return period (right y-axis) for the extreme flood event at four gauging stations [7,8,22].

3. Description of the Tool "ICPR FloRiAn"

ICPR FloRiAn was developed in English language and is running as a toolbox of custom tools (compiled C# code) under ESRI ArcGIS Desktop with the extension "ArcGIS Spatial Analyst". The method described in Section 2 is fully implemented in the tool and represents its backbone. The implementation in the GIS is carried out as a toolbox with four categories along the lines of the four risk receptors of the FD (cf. Figure 8). Each category consists of the following three interacting modules resulting in an overall damage or risk evolution/reduction assessment:

1. Module "Damage assessment": This module consists of one tool which calculates the damage using land use data, the extension of flood areas (maps), hydraulic data (water depth), asset values and damage functions. The output of this module is used in the next two modules.
2. Module "Measure summation" (this module is optional as the ICPR FloRiAn enables the calculation of flood damage or risk with or without measures): This module quantifies the impact of the different measures (which are introduced with a tool for each measure). Output is a damage reduction (on economic activity, human health, the environment, and cultural heritage) due to the implementation of measures. After the damage reduction for all measures is calculated,

a "summation" tool calculates the damage due to flooding after all measures are incorporated. This tool takes into account the interaction between different measures (the sum of the effect of two tools is not equal to the sum each individual effect) (see end of Section 2.6.1). The output can be used as an input for the next module.

3. Module "Risk assessment": This module calculates the risk by combining/multiplying the damage potential (output of "damage assessment" or "measure summation") with the flood probability.

4. The main outputs of the tool are maps with the damage values (actually grids with the damage values per pixel) and tables (*.dbf files) containing aggregated data for each administrative area as defined in the input (cf. Figure 8). Running the tool for different time horizons (with different input data as well as measures) and comparing the outputs results in the information of damage or risk changes over time.

In addition to the ICPR report no. 237 [7], a technical user guide/manual and a help function in the tool are available [25]. They contain detailed descriptions on the installation and running of the tool, required input data, individual toolboxes (calculation process) as well as the data structure. Originally the tool was developed for ArcGis Desktop 10.0 (ESRI, Redlands, United States), different users have operated the tool under versions 10.2, 10.3, 10.4 and 10.5.

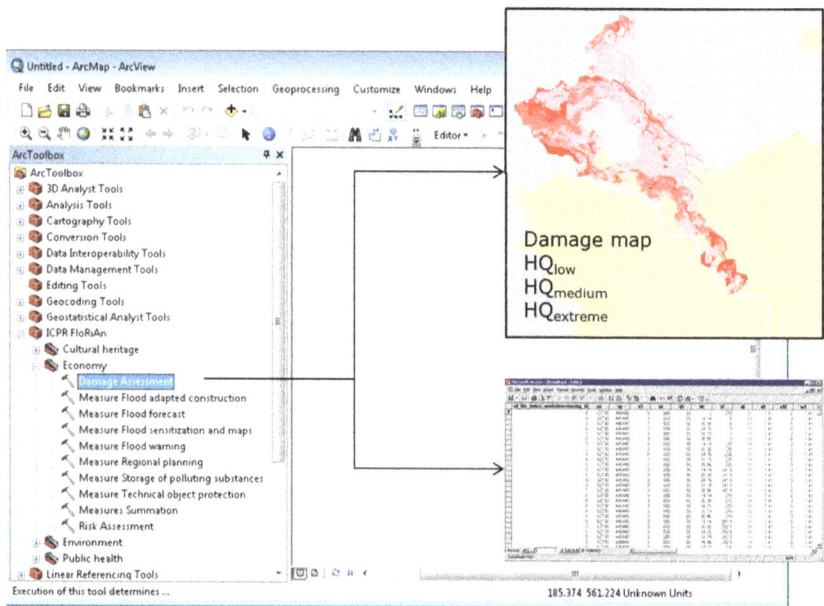

Figure 8. International Commission for the Protection of the Rhine (ICPR) FloRiAn as an ArcToolbox with the four receptors and different calculation modules as well as example of outputs (map and table).

4. Application of ICPR FloRiAn to the Rhine

This section presents the results of the ICPR FloRiAn calculations ran by the ICPR within the assessment of the damage and risk reduction objectives of the ICPR Action Plan on Floods for the four risk receptors human health, environment, cultural heritage and economic activity for the time horizons 1995, 2005, 2015, 2020 and 2030 (with implementation of the respective measures, cf. Table 1) [8]. Realized (until 2015) and planned measures (until 2030) along the Rhine were compiled from the Rhine States and included in the calculations (cf. Table 1). Detailed results and figures can be found in the

ICPR report no. 236 [8]. The assessment and calculations to demonstrate the evolution of the flood risk on the main stream of the Rhine and possible reduction during the period 1995–2030 have revealed the following:

1. When considering the risks to human health, it is apparent that measures such as safeguarding/evacuation of those potentially affected, raising awareness, flood forecasting and warning and alarm plans as well as the modification of the probability of flooding all help to mitigate the flood risk. Across the three flood scenarios, the measures can lead to an average reduction in the risk for human health of approximatively 70% to 80% (period 1995–2020) (cf. Figure 9) [8].

2. When assessing cultural heritage and the environment, based on the results of experimental methods (cf. Sections 2.4 and 2.5), the ICPR has found out that, due to the measures undertaken (for the environment: measures helping to mitigate damages of potentially polluting sites, and for cultural heritage: measures like the ones from economic activity, see below), over time, damage and risk to cultural heritage and the environment are reduced from 40% to 70% (period 1995–2020) across all damage categories and all flood scenarios (cf. Figures 10 and 11) [8].

3. In terms of economic activity it has been determined that the reduction by 25% before 2020 (target stated in the APF) compared to 1995 can be achieved. As in the case of the evaluation and calculation of the damage and risk for the other receptors, the ICPR has again performed a broad scale analysis. The latter showed that measures enabling water retention along the Rhine, such as the construction of flood retention areas, the relocation of dykes and measures that give more room to the river are most efficient with respect to changing the probability of flooding (cf. Section 2.6.2 and list of measures in Table 1). In addition, various other measures for prevention and preparedness, including flood forecasting, early warning systems and (pre-) crisis management have contributed to reduce the increase of damage in floodplain areas since 1995 (cf. Figure 12) [8].

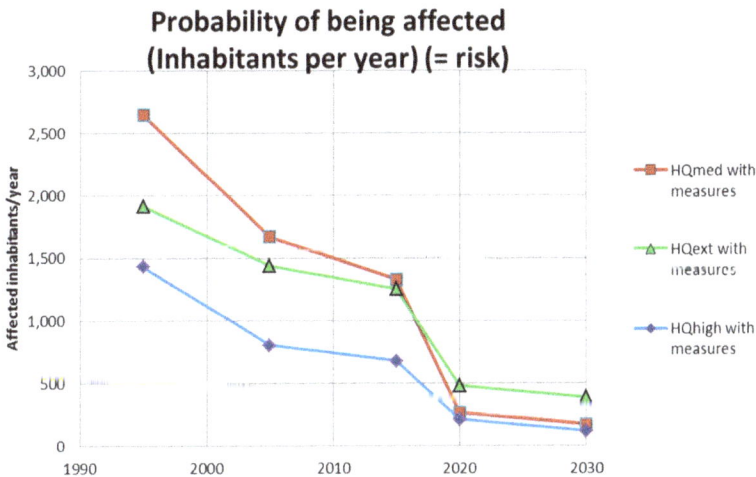

Figure 9. Probability of affected inhabitants with consideration of all measures (people affected/year) (= risk) [8].

Risk for cultural heritage (across all damage categories)

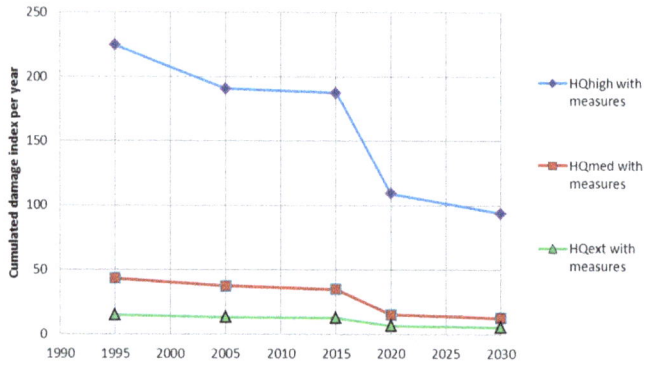

Figure 10. Risk evolution taking into account all measures (total damage index per year across all classes of damage) (*Y*-axis) [8].

Risk for environment (across all damage categories)

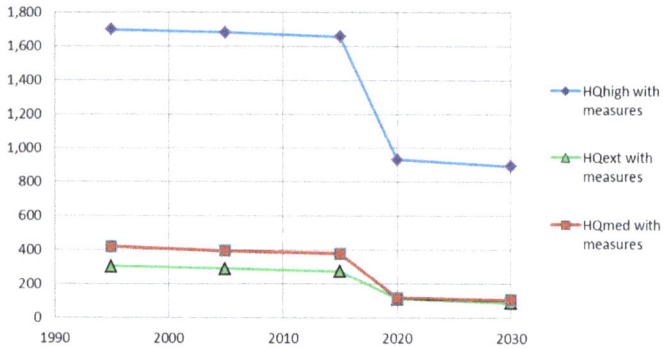

Figure 11. Risk evolution taking into account all measures (total damage index per year across all classes of damage) (*Y*-axis) [8].

Economic risk

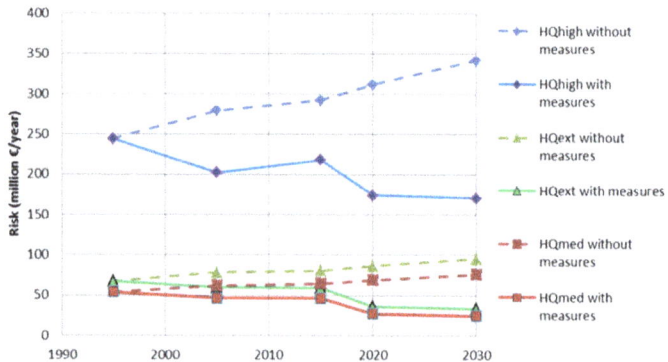

Figure 12. Development/Evolution of economic risk from 1995 to 2030 (€ million/a) [8].

5. Conclusions and Outlook

The ICPR has developed the GIS-toolbox ICPR FloRiAn that can be used to quantitatively assess and determine the impact of a set of realized or planned flood risk mitigation measures. Various assumptions were made and different methods were specified or newly developed, some of which are still strongly based on ICPR expert knowledge. In the future, the estimates and assumptions regarding the methods and measures underlying the tool should be optimized by gaining knowledge from the application of the tool by further users and improving input data. Nevertheless, the added value lies in the possibility of a macroscale (e.g., a river basin), temporally comparable and reproducible analysis. Calculations made by the ICPR identified—amongst other results—the reduction of flood risks by 25% between 1995 and 2020 for economic activities. On a broad scale, protection measures increasing water retention along the mainstream of the Rhine proved to be most efficient, but the computations also showed that over time, further non-structural prevention, and preparedness measures also contribute to reducing damage growth in the floodplain. In a nutshell, this means that the whole cycle of flood risk management with a range of preventive and protective actions should be addressed to reduce risks and damages.

The ICPR is planning to use the GIS tool "ICPR FloRiAn" in future to support the flood risk analysis and to determine the effectiveness of measures within the framework of the regular review of the FRMP of the Rhine river basin.

The developed tool can also be applied to other river basins. Upon request, the ICPR provides the tool ICPR FloRiAn and the methods it is based on to other river basin commissions, national authorities, or scientific institutions.

Author Contributions: All authors have worked on all parts of the manuscript.

Funding: This research was funded by the International Commission for the Protection of the Rhine (ICPR).

Acknowledgments: The contribution is based on the methodology and tool developed together with the HKV consortium and the ICPR expert group HIRI as well as on the calculations carried out. The results of the work are set out in the two technical ICPR reports no. 236 and no. 237 [7,8]. Both reports also indicate which institutions and persons have participated in and contributed to this work. Once again the authors would like to thank all the members of the HIRI expert group: Hendrik Buiteveld (Rijkswaterstaat, NL), Max Schropp (Rijkswaterstaat, NL), Frank Alberts (Rijkswaterstaat, NL), Jean-Pierre Wagner (Direction Régionale de l'Environnement de l'Aménagement et du Logement „Grand Est", FR), Régis Creusot (Direction Régionale de l'Environnement de l'Aménagement et du Logement „Grand Est", FR), Anne Toussirot (Direction Régionale de l'Environnement de l'Aménagement et du Logement „Grand Est", FR), Wolfgang Zwach (Regierungspräsidium Darmstadt, DE), Lennart Gosch (Ministerium für Umwelt, Klima und Energiewirtschaft Baden-Württemberg, DE), Jürgen Reich (Ministerium für Umwelt, Klima und Energiewirtschaft Baden-Württemberg, DE), Barbara Sailer (Ministerium für Umwelt, Klima und Energiewirtschaft Baden-Württemberg, DE), Holger Kugel (Struktur- und Genehmigungsdirektion Nord, Trier, DE), Reinhard Vogt (Stadtentwässerungsbetrieb Köln/Hochwassernotgemeinschaft Rhein, DE), Sabine Siegmund (Stadtentwässerungsbetrieb Köln/Hochwassernotgemeinschaft Rhein, DE), Urs Nigg (Bundesamt für Umwelt, CH), Markus Hostmann (Bundesamt für Umwelt, CH), Andreas Kaufmann (Bundesministerium für Land- und Forstwirtschaft, Umwelt und Wasserwirtschaft, AT), Clemens Neuhold (Bundesministerium für Land- und Forstwirtschaft, Umwelt und Wasserwirtschaft, AT), Gerard Huber (Abteilung Wasserwirtschaft Vorarlberg, AT), Dieter Vondrak (Abteilung Wasserwirtschaft Vorarlberg, AT), Emanuel Banzer (Amt für Bevölkerungsschutz—Landesverwaltung, LI), Catarina Proidl (Amt für Bau und Infrastruktur—Landesverwaltung, LI), Stephan Wohlwend (Amt für Bevölkerungsschutz—Landesverwaltung, LI).

Conflicts of Interest: The authors declare no conflict of interest.

References

1. International Commission for the Protection of the Rhine. Webpage of the International Commission for the Protection of the Rhine. Available online: https://www.iksr.org (accessed on 14 August 2018).
2. International Commission for the Protection of the Rhine. Action Plan on Floods (APF). 1998. Available online: https://www.iksr.org/en/international-cooperation/rhine-2020/action-plan-on-floods/ (accessed on 14 August 2018).

3. International Commission for the Protection of the Rhine. Brochure "The Rhine and its Catchment—A Survey" (Balance of the implementation of Rhine 2020 and the Action Plan on Floods 1995–2010). 2013. Available online: https://www.iksr.org/fileadmin/user_upload/DKDM/Dokumente/Broschueren/EN/bro_En_2013_The_Rhine_and_its_catchment.pdf (accessed on 4 October 2018).

4. International Commission for the Protection of the Rhine. Internationally Coordinated Flood Risk Management Plan for the International River Basin District of the Rhine ((Part A = Overriding Part). 2015. Available online: https://www.iksr.org/fileadmin/user_upload/Dokumente_de/Hochwasser/FRMP_2015__002_.pdf (accessed on 4 October 2018).

5. Directive 2007/60/EC of the European Parliament and of the Council of 23 October 2007 on the Assessment and Management of Flood Risks. 2007. Available online: https://eur-lex.europa.eu/LexUriServ/LexUriServ.do?uri=OJ:L:2007:288:0027:0034:en:PDF (accessed on 4 October 2018).

6. Webpage about "ICPR FloRiAn". Available online: https://www.iksr.org/en/topics/floods/flood-risk-tool-florian/ (accessed on 14 August 2018).

7. International Commission for the Protection of the Rhine. Report 237: Technical Report "Tool and Assessment Method to Determine Flood Risk Evolution/Reduction". 2016. Available online: https://www.iksr.org/en/documentsarchive/technical-reports/synoptical-table/ (accessed on 4 October 2018).

8. International Commission for the Protection of the Rhine. Report 236: Synthesis Report "Assessment of Flood Risk Reduction (APF) According to the Types of Measures and Risk Objects Covered by the FD". 2016. Available online: https://www.iksr.org/en/documentsarchive/technical-reports/synoptical-table/ (accessed on 4 October 2018).

9. Albano, R.; Mancusi, L.; Sole, A.; Adamowski, J. FloodRisk: A collaborative, free and open-source software for flood risk analysis. *Geomat. Nat. Hazards Risk* **2017**, *8*, 1812–1832. [CrossRef]

10. Albano, R.; Mancusi, L.; Abbate, A. Improving flood risk analysis for effectively supporting the implementation of flood risk management plans: The case study of "Serio" Valley. *Environ. Sci. Policy* **2017**, *5*, 158–172. [CrossRef]

11. Deckers, P.; Kellens, W.; Reyns, J.; Vanneuville, W.; De Maeyer, P. A GIS for flood risk management in Flanders. In *Geospatial Techniques in Urban Hazard and Disaster Analysis*; Springer: Dordrecht, The Netherlands, 2009; pp. 51–69.

12. Dottori, F.; Figueiredo, R.; Martina, M.L.V.; Molinari, D.; Scorzini, A.R. INSYDE: A synthetic, probabilistic flood damage model based on explicit cost analysis. *Nat. Hazards Earth Syst. Sci.* **2016**, *16*, 2577–2591. [CrossRef]

13. HAZUS Software (FEMA). Available online: https://www.fema.gov/hazus,https://www.fema.gov/hazus-software (accessed on 28 September 2018).

14. Moufar, M.M.M.; Edangodage, D.P.P. Floods and Countermeasures Impact Assessment for the Metro Colombo Canal System, Sri Lanka. *Hydrology* **2018**, *5*, 11. [CrossRef]

15. Samela, C.; Albano, R.; Sole, A.; Manfreda, S. A GIS tool for cost-effective delineation of flood-prone areas. *Comput. Environ. Urban Syst.* **2018**, *70*, 43–52. [CrossRef]

16. FLIWAS (Flutinformations- und Warnsystem). Available online: https://www.hochwasser.baden-wuerttemberg.de/flutinformations-und-warnsystem (accessed on 28 September 2018).

17. OSIRIS-Inondation. Available online: http://www.osiris-inondation.fr/index.php?init=1 (accessed on 28 September 2018).

18. SD-KAMA (Smart Data-Katastrophenmanagement). Available online: https://www.sd-kama.de/en/smart_data_disaster_management/ (accessed on 28 September 2018).

19. Peter, M. Dynamische Einsatzplanung–Big Data im Rettungsdienst. In *Herausforderung Notfallmedizin*; Springer: Berlin/Heidelberg, Germany, 2018; pp. 143–152.

20. Grossi, P.; Kunreuther, H.; Windeler, D. An introduction to catastrophe models and insurance. In *Catastrophe Modeling: A New Approach to Managing Risk*; Springer: Boston, MA, USA, 2005; pp. 23–42.

21. International Commission for the Protection of the Rhine. Rhine Atlas 2015 (Flood Hazard and Risk Maps of the International River Basin District 'Rhine'). Available online: https://www.iksr.org/en/documentsarchive/rhine-atlas/ and direct link http://geoportal.bafg.de/mapapps/resources/apps/ICPR_EN/index.html?lang=en (both accessed on 14 August 2018).

Geosciences **2018**, *8*, 371

22. International Commission for the Protection of the Rhine. Report 229: Assessment of the Modification of Probability Due to Flood Level Reduction Measures along the Rhine. 2015. Available online: https://www.iksr.org/en/documentsarchive/technical-reports/synoptical-table/ (accessed on 4 October 2018).

23. International Commission for the Protection of the Rhine. Rhine Atlas 2001 (Methodology) and Key Document for the Creation of the Atlas "Übersichtskarten der Überschwemmungsgefährdung und der Möglichen Schäden bei Extremhochwasser am Rhein—Vorgehensweise zur Ermittlung der Überschwemmungsgefährdeten Flächen Sowie Vorgehensweise zur Ermittlung der Vermögenswerte". Available online: https://www.iksr.org/fileadmin/user_upload/Dokumente_de/Rhein-Atlas/german/welcome_german.pdf (accessed on 14 August 2018).

24. International Commission for the Protection of the Rhine. Report 199: Evidence of the Effectiveness of Measures Aimed at Reducing Flood Levels of the Rhine. 2012. Available online: https://www.iksr.org/en/documentsarchive/technical-reports/synoptical-table/ (accessed on 4 October 2018).

25. International Commission for the Protection of the Rhine. User's Guide to ICPR FloRiAn. (not published). 2016.

geosciences

MDPI

Article

Analysis of Damage Caused by Hydrometeorological Disasters in Texas, 1960–2016

Srikanto H. Paul * and **Hatim O. Sharif**

Department of Civil and Environmental Engineering, University of Texas at San Antonio,
San Antonio, TX 78249, USA; hatim.sharif@utsa.edu
* Correspondence: srikantopaul@live.com

Received: 20 September 2018; Accepted: 18 October 2018; Published: 20 October 2018

Abstract: Property damages caused by hydrometeorological disasters in Texas during the period 1960–2016 totaled $54.2 billion with hurricanes, tropical storms, and hail accounting for 56%, followed by flooding and severe thunderstorms responsible for 24% of the total damages. The current study provides normalized trends to support the assertion that the increase in property damage is a combined contribution of stronger disasters as predicted by climate change models and increases in urban development in risk prone regions such as the Texas Gulf Coast. A comparison of the temporal distribution of damages normalized by population and GDP resulted in a less statistically significant increasing trend per capita. Seasonal distribution highlights spring as the costliest season (March, April and May) while the hurricane season (June through November) is well aligned with the months of highest property damage. Normalization of property damage by GDP during 2001–2016 showed Dallas as the only metropolitan statistical area (MSA) with a significant increasing trend of the 25 MSAs in Texas. Spatial analysis of property damage per capita highlighted the regions that are at greater risk during and after a major disaster given their limited economic resources compared to more urbanized regions. Variation in the causes of damage (wind or water) and types of damage that a "Hurricane" can produce was investigated using Hazus model simulation. A comparison of published damage estimates at time of occurrence with simulation outputs for Hurricanes Carla, 1961; Alicia, 1983; and Ike, 2008 based on 2010 building exposure highlighted the impact of economic growth, susceptibility of wood building types, and the predominant cause of damage. Carla and Ike simulation models captured less than 50% of their respective estimates reported by other sources suggesting a broad geographical zone of damage with flood damage making a significant contribution. Conversely, the model damage estimates for Alicia are 50% higher than total damage estimates that were reported at the time of occurrence suggesting a substantial increase in building exposure susceptible to wind damage in the modeled region from 1983 – 2010.

Keywords: natural hazards; hydrometeorological disasters; HAZUS; SHELDUS; Texas; property damage; economic loss

1. Introduction

Recent decades have witnessed a worldwide increasing trend in both the number of natural disasters and the resulting damages. In the two most recent years of the current study, the number of natural catastrophic events has increased from 730 events ($103 billion) in 2015 to 750 events ($175 billion) in 2016. Weather-related events such as severe storms and floods had the most significant increase in frequency. During the period 1980–2016, there were a total of 16,584 natural disaster events resulting in $4.3 trillion of damage worldwide, of which 80% were either hydrological or meteorological events and 20% climatological or geophysical events [1]. During the period 1980–2017, the U.S. experienced 233 weather and climate related disasters in which overall damages/costs reached or exceeded $1 billion (2018 CPI adjusted) for a total cost exceeding $1.5 trillion [2]. As of the date of this study, 2017 has set the record for the most expensive year for damage due to natural disasters in recorded history both globally and in the U.S.

Assessment of damages due to historic natural disasters can include direct and indirect replacement cost estimates and/or insurance payout information. The former use in this study is more applicable in longitudinal research since it is a function of available exposure value and includes all property whether insured. The Congressional Research Service reported that inflation-adjusted disaster appropriations have increased 46% from a median of $6.2 billion between 2000 and 2006 to $9.1 billion between 2007 and 2013. The hurricanes in 2017 were immense and had a much costlier impact as they collided with growing cities with higher exposure. As more people compete for real estate thereby pushing up the property values in disaster prone regions such as coastal Florida, Texas, and California, the level of property damage also increases [3]. Damage data are widely available from public sources such as Munich Re and SHELDUS but typically exclude long-term indirect costs such as healthcare discontinuity and investment opportunity cost. Therefore, there is significant uncertainty in the exact total costs of natural disasters especially when comparing damage over time and across areas of varying degrees of urban development.

There is general agreement among public and private organizations and governmental agencies including the Government Accountability Office (GAO) that the cost of natural disasters in the U.S. is increasing at a significant rate. However, there are different perspectives on whether the increase is due to more violent storms or if the increase is due to the increase in population and wealth of property that is susceptible to damage. The U.S., as well as many other countries around the world, has experienced a rise in the number of natural disaster events and losses in the last four decades primarily due to convective events which are disaster events developing out of thunderstorms, such as hail, heavy precipitation, tornadoes and strong straight-line winds. Gall et al. noted that direct losses from convective disaster events such as hurricanes, flooding, and severe storms are increasing and contribute about 75% of the total damage with hurricane and flood losses having tripled over the last 50 years [4]. A study by Sander et al. found that 80% of all losses in the U.S. from 1970 to 2009 were due to convective events that had normalized losses exceeding $250 million. The study also suggests that there is a correlation between the increase in losses and the changes in meteorological potential for severe thunderstorms driven by changes in the humidity of the troposphere [5].

The scientific community largely in agreement that the rise of humidity in the air over the last decades can be attributed to warming oceans and increased evaporation from their surfaces. The intuitive consequence is that the increase in disaster events due to climate change is responsible for the increased damage losses. However, although the relationship of climate change to disaster occurrence is accepted, the relationship of disaster occurrence and the increasing trend in property damage is still a debated topic. One perspective is that the reported increasing damage and losses from hurricanes are not necessarily evidence of any increase in hurricane or tropical storm activity but are due only to the changes in population and wealth of the impacted regions [6–8]. Klotzbach et al. reported that damage caused by tropical cyclones adjusted for inflation and normalized by regional wealth and population factors did not show an increasing trend from 1900 to 2016 in the U.S., suggesting that the increase in damages are more a function of the increased regional wealth and property exposure than the increase in number of cyclones [8].

Texas is second in population (2010 Census) only to California and has a large and diverse terrain that combines a gulf coastline that is extremely susceptible to tropical storms and hurricanes; flooding and flash flooding at the base of the Balcones Escarpment running through the mid-section of the state; heat and drought conditions in the south/southwest; and rural cold extremes in the northwest panhandle. Hydrometeorological events are the predominant disasters in Texas and have resulted in a high number of fatalities and losses to infrastructure [9,10]. The overall population growth coupled with the rapid urban and coastal development in recent decades have created an environment in which fatality rates are decreasing per capita due to population increases but property damage is increasing due to more people with more valuable property moving into more vulnerable (disaster prone) regions. This nexus of nature and society will continue to grow in Texas in the foreseeable future and warrants ongoing analysis to help policy and decision-makers identify and prioritize the social vulnerabilities that can be managed to reduce the risk to Texas life and property. This study is intended to provide a review of historic trends and types of damage and economic losses caused by hydrometeorological disasters impacting the coastal and inland property and infrastructure of Texas from 1960 to 2016. Spatial analysis of actual and normalized damage as well as a supplemental assessment of three major disasters causing extensive damage in Texas (Hurricanes Carla 1961, Hurricane Alicia 1983, and Hurricane Ike 2008) highlight the risk as a function of wind or flooding damage and the growth of exposure in hazard prone regions.

2. Study Area

Texas is the second largest state in the United States by population and area, with a population of 27,862,596 and a land area of 695,662 km². The southeast of Texas shares 591 km (367 miles) of coastline with the Gulf of Mexico and is susceptible to hurricanes and coastal flooding. A major topographical feature that affects weather disasters in Texas is the Balcones Escarpment that consists of a series of cliffs dropping from the Edwards Plateau to the Balcones Fault Line. This outer rim of the Hill Country is the formation point for many large thunderstorms, which frequently stall along the uplift and then hover over the region for prolonged rainfall. This flood prone region is known as "Flash Flood Alley" and includes counties having the fastest population growth rates in Texas [11].

Texas is the fastest growing state in the country by actual population and the fifth fastest by percentage. Between 1940 and 2010, Texas averaged 21.6% rate of growth per decade, compared to 13.3% for the U.S. Based on a conference presentation in 2013 by the Texas State Demographer's Office, the overall population is projected to increase to 55 million by 2050 assuming a continuation of the 2000–2010 migration pattern The split between the rural and the urban share of the population has experienced a complete reversal from 1910 to 2010 with nine out of ten Texans living in one of the state's 25 metropolitan areas and nearly two out of every three Texans living in Dallas-Fort Worth, Houston, Austin or San Antonio. Much of this growth is occurring within regions having high risk of hydrometeorological disasters such as hurricanes and flooding along the Texas Gulf Coast (e.g., Houston metropolitan area). Texas doubled the national job growth percentage in 2012 at 2.7% which translates to higher income and wealth exposure in preferred coastal regions of the state [12].

3. Data Sources

The primary source of property damage data is the Spatial Hazard Evaluation and Losses Database for the United States (SHELDUS) maintained by the Center for Emergency Management and Homeland Security at Arizona State University [13]. The database aggregates hazard losses across 18 different disaster categories. In Texas, the relevant disaster types are reduced to 15 with the omission of earthquakes, tsunamis, and volcanoes. SHELDUS losses are based on information from the National Centers for Environmental Information (NCEI, formerly the National Climatic Data Center, NCDC), the U.S. Geological Survey (USGS), and other credible sources. Building damage from the disaster case studies is generated from the HAZUS-MH hazard analysis model developed by the Federal Emergency Management Agency (FEMA) that contains detailed sociodemographic data for residences based on the 2010 census and Dun and Bradstreet data for commercial buildings [14]. The HAZUS hurricane model simulates the entire storm track with a series of engineering-based models and multiple nationwide inventory databases to develop damage and loss functions. The sub-models in the HAZUS hurricane model include a storm track model, a wind field model, a wind load model, a windborne debris model, a physical damage model, and building loss models [15,16]. The HAZUS hurricane model provides estimates of building damage and content losses and income-related business interruption due to the impact of wind damage to the infrastructure. The HAZUS model is a conservative estimate as noted by previous analytical comparisons [17–19].

4. Methodology

This study defines a hazard as a natural event that has the potential to cause harm and a disaster as the effect of the hazard on a community. Hydrometeorological disasters are defined as natural processes or phenomena of atmospheric, hydrological or oceanographic nature [20]. The analysis of annual distribution of damage for the entire state over the 57-year period (1960–2016) includes adjusted damage ($2016) and damage normalized by population and GDP. Before 1997, the basis of GDP was the Standard Industrial Classification (SIC). It transitioned from SIC to the North American Industry Classification System (NAICS) in 1997 resulting in two different GDP values for that year. The arithmetic average of the 1997 SIC GDP and 1997 NAICS GDP was used as the 1997 GDP in the analysis. The effect of population and GDP was also analyzed across metropolitan statistical areas (MSA) for the period (2001–2016) in which NAICS-based GDP data were available.

Normalization of the property damage provides an indication of the relationship between the damage, regional wealth, and population over time. The method used can be based on national adjustment factors as described by Pielke et al., in which the combined effects of inflation, wealth, and population were considered to adjust damage in the year occurrence to the perceived damage in the base year [7]. Wealth adjustment factor can be based on a number of metrics available through the U.S. Bureau of Economic Analysis (BEA) including Net Stock of Fixed Assets and Consumer Durable Goods or Gross Domestic Product (GDP). The current study used regional GDP made available by the USBEA by metropolitan statistical areas (MSA) as the wealth adjustment factor to normalize the damage trends. Normalization per capita was based on median population of either county or MSA with respect to the period analyzed.

Spearman's rho and kendall's tau were used for non-parametric correlation analysis to determine the statistical significance of property damage trends over time since both methods are conducive to environmental forensics but each has advantages and disadvantage. Spearman's rho is more sensitive to error and better for larger sample size and kendall's tau is less sensitive to error and more appropriate for smaller sample size [21]. Ranking of linear strength in the positive direction was selected based on general guidelines of correlation analysis: <0.3 (weak), 0.3–0.5 (moderate), and >0.5 (strong) with a similar opposite ranking for negative trends with a statistically significance based on a 5% significance level [22,23]. Quantitative boundaries for linear strength relationships were not adjusted for specific environmental forensics since the significance level represented by the *p*-value was the critical identifier of significance related for the analysis. The *p*-values associated with the correlation less than 0.05 were considered statistically significant with smaller *p*-values representing greater statistical significance.

5. Results

All property damage used in the historic trend analysis was inflation adjusted to $2016 unless otherwise specified as actual property damage in year of occurrence. Natural disasters caused more than $725 billion in property damage in the U.S. during 1960–2016 in which 8% ($54.2 billion) was due to Texas property damage. Convective storms such as thunderstorms, heavy precipitation, tornadoes and strong straight-line winds that cause extreme directional wind or object forces (hail), appear to result in a disproportionate amount of damage. Fifty-eight percent ($424 billion) of the total property damage in the U.S. was due to convective storm disasters. Texas property damage due to convective storm disasters accounted for 80% ($43 billion) of the total property damage. Hurricanes and tropical storms caused the greatest property damage in the U.S. and in Texas at 36% and 34% of their respective total property damage. Texas property damage was a significant percentage of the national property damage with hail accounting for 33%, followed by coastal property damage at 32% and drought damage at 28% (Table 1). In addition to the total extent of the property damage observed during the 57-year period, the current study analyzed temporal distribution of damage which indicated a distinct increasing annual trend as well as seasonal variation that may reflect some influence of climate change. As suggested by Sander et al., the changes in meteorological potential for severe thunderstorms driven by changes in the humidity of the troposphere (rise of humidity in the air over the last decades can be attributed to warming oceans and increased evaporation from their surfaces) may be contributing to the increase in hurricane winds and rainfall [5]. This increase in magnitude coupled with the increase in exposure will continue to result in escalating property damage particularly along hurricane prone regions.

Table 1. Total property damage caused by natural disasters in U.S. and Texas (1960–2016).

Disaster Type	TX ($2016)	US ($2016)
Hurricane/Trop Strm	18,325,926,908	258,877,397,419
Hail	12,301,331,729	37,295,600,013
Flooding	7,035,494,246	152,447,938,320
Severe /TStrm	6,298,129,432	35,434,257,137
Tornado	4,103,255,471	58,871,799,205
Wind	2,095,538,316	33,335,423,825
Drought	1,454,769,396	5,121,265,940
Coastal	940,709,789	2,955,956,179
Winter Weather	795,182,868	26,391,427,257
Wildfire	692,696,316	22,497,632,949
Lightning	159,383,681	3,101,409,332
Fog	6,885,490	58,142,521
Heat	1,794,114	504,346,195
Landslide	179,521	18,864,695,667
Avalanche	1,064	30,299,599
Grand Total	**54,211,278,340**	**655,787,591,561**

5.1. Annual Distribution of Damage

The average annual property damage between 1960 and 2016 is $951,075,059 ($2016) with a significant increasing trend (p-value < 0.0001 at the 5% significance level) (Figure 1a). Major disaster events such as hurricanes, tropical storms, widespread flooding in 1980, 1983 (Alicia), 1996, 2001 (Allison), 2005, and 2008 (Ike) resulted in the high variability of actual property damage over the 57-year period (coefficient of variation (COV) = 1.65) that is reduced to COV = 1.38 after adjusting the property damage to 2016 U.S. dollars.

Adjusted property damage per capita has a significant positive trend suggesting that the inflation rate outpaced the population growth. Actual property damage per GDP displays a non-significant positive trend display increasing trends over time highlighting substantial growth of the economy specifically building exposure susceptible to damage. The coefficient of variation of the damage per capita over the study period is 1.22 and has a statistically significant increasing trend (spearman's $\rho = 0.4194$, p-value = 0.0012) at the 5% significance level (Figure 1b). GDP appears to be a stronger contributor to the time series increase compared to population given the reduction in trend strength and significance (spearman's $\rho = 0.1906$, p-value = 0.1555) at the 5% level (Figure 1c). A secondary verification of linear strength and significance using Kendall's tau reduced the linear strength but did not change the significance in either normalization trends, respectively (*kendall* $\tau = 0.2897$, p-value = 0.0015 and *kendall* $\tau = 0.1278$, p-value = 0.1621). Furthermore, GDP normalization of damage results in the greatest reduction in variability (COV = 1.13).

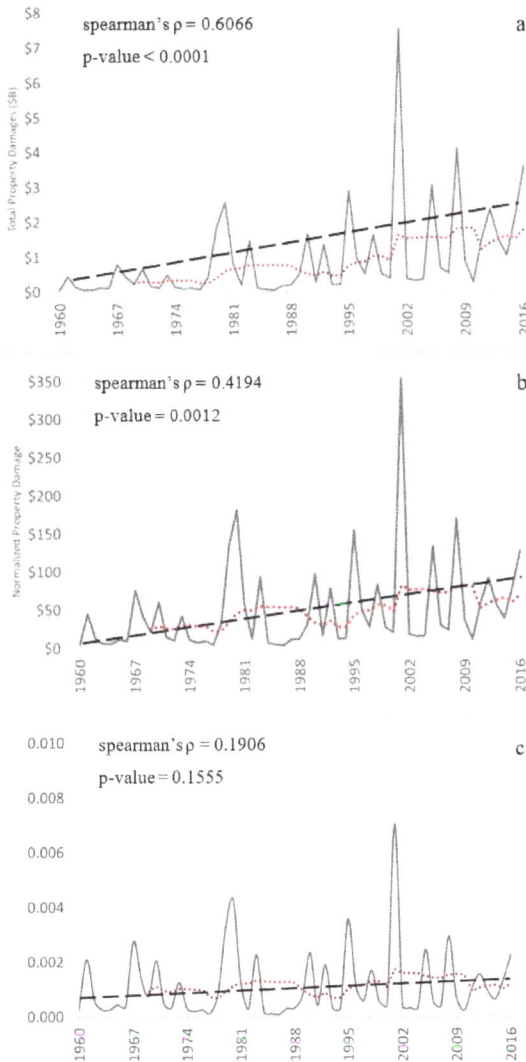

Figure 1. (a) Adj property damage caused by natural disasters in Texas (1960–2016); (b) Adj property damage per capita; and (c) Actual property damage by GDP. *Notation: 10-year moving average (red dots) and linear trend (black dash).*

5.2. Monthly Distribution of Damage

Monthly distribution of damage highlights a significant difference between damages incurring in Winter and those incurring in the other seasons. The winter months (December, January, and February) account for only 5% of the total damages. Spring is the costliest season (March, April, and May) accounting for 36% of the total damages followed by summer (34%) and fall (25%). The monthly analysis of property damage suggests that the property damage is in sync with the North American hurricane season. A tri-modal distribution with peaks in April, June, and October can be observed in Figure 2. Hail was the predominant cause of damage in April and hurricanes/tropical storms the dominant contributor to damage in June and September.

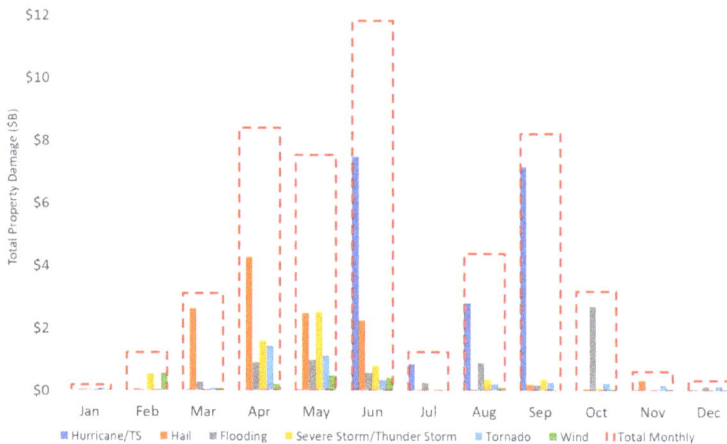

Figure 2. Top 6 disaster types monthly distribution of property damage (1960–2016).

Approximately 93% of the total property damage during 1960–2016 was caused by six of the 13 disaster types listed in Table 1. Monthly stratification of property damages caused by the hydrometeorological disasters shows that hurricane and tropical storms are the leading cause in four of the 12 months (June, July, August, and September). Hail damage is the leading cause of property damage in the months of March, April, May, and November. Flooding damage is highest in October. Hail, severe/thunderstorms, tornado, and wind overlap from February to June indicating the favorable seasonal conditions for storms that have intense wind and hail characteristics. The total damage for this period was $24.8 billion in property damage (46% of total). The bimodal annual characteristic of damage due to hurricanes and tropical storms is highlighted as well peaking in June and September when meteorological conditions are most favorable. The N.A. Atlantic hurricane season is 1 June–30 November and is closely aligned with the monthly damage totals suggesting a relationship between hurricanes and flooding. Flooding damage begins at the end of the June (peak hurricane month) and spikes in October toward the end of hurricane season which is likely a result of saturated conditions due to the cumulative amount of precipitation throughout the hurricane season.

5.3. Regional Normalized Trends and Spatial Distribution of Damage

Additional analysis of damage and risk to property as a result of natural disasters included groupings of counties known as metropolitan statistical areas (MSAs). The MSAs are currently the lowest level in which the U.S. BEA maintains GDP data. County level GDP is not yet available but is under development [24]. The MSAs represent economic subdivisions in which the wealth measured as GDP for each county within the MSA is combined and represented as the total GDP for the MSA. In Texas from 2001 to 2016, 18 of the 25 MSAs had greater average annual increase in GDP than the U.S. annual average (4.8%). The Texas MSAs with the highest relative GDP are centralized around the major cities of Houston, Dallas/Ft. Worth, San Antonio, Austin, and El Paso.

The limited time study of 16 years is not a large sample size, but can provide high-level perspectives on the effects of wealth on property damage across the geographically dispersed MSAs. Actual property damage normalized by the GDP for the MSAs follow increasing trends similar to the statewide annual trends (1960–2016) for 44% of the MSAs. Fourteen of the 25 Texas MSAs exhibited non-significant decreasing but trends at the 5% significance level and 11 MSAs exhibited increasing trend in property damage with only the Dallas MSA indicating a strong linear relationship and statistically significant trend (ρ = 0.8176, p-value = 0.0001). The Dallas and Houston MSAs had very similar total and annual average GDP yet had opposite trends suggesting that there is a significant difference in the periodicity of major damaging events over the 16-year period between the two regions. Dallas is an inland region that is prone to tornado and hail damage which caused almost all reported property damage in 2012 and 2016, while the property damage to the Houston regions is primarily due to hurricane winds and related flooding. The non-significant decreasing trend for the Houston coastal region reflects the extreme damage and irregular occurrence of hurricanes specifically in 2001 (TS Allison) and 2008 (Hurricane Ike) resulting in an inconsistent linear trend, while the tornado and hail damage events in the Dallas region are more frequent and continuous resulting in increasing in property damage due to increasing exposure.

Spatial comparison of property damage ($2016) for the period 1960–2016 shows regional differences in Texas with higher property damage concentrated in counties with high population and urban development. Not surprisingly, the regional density of property damage due to natural disasters is clustered around the four largest metropolitan areas (Houston, Dallas, Austin, and San Antonio) and is also in alignment with the raw fatality densities discussed in previous research [9,10]. This is expected since an area of greater property exposure is more susceptible to higher damage during a natural disaster event. Considering the property damage per capita and per GDP as the wealth indicator provides a closer look of the relationship of damage and exposure.

Distinct counties of high property damage without adjustment for population or wealth normalization over the 57-year period is depicted in Figure 3 (Left). The counties with the highest overall property damage were Harris, Dallas, Tarrant, Bexar, and Lubbock. Other than Lubbock, the counties with highest damage are counties with high population and wealth. Lubbock is unique in that, even though it is not a large urban center with high wealth, the county is affected by frequent damage due to multiple types of disaster, mostly hail, wind, severe storms, flooding, and tornados, that span over seasonal changes resulting in higher overall damages. Counties in coastal Texas and the "Flash Flood Alley" region show higher damage losses.

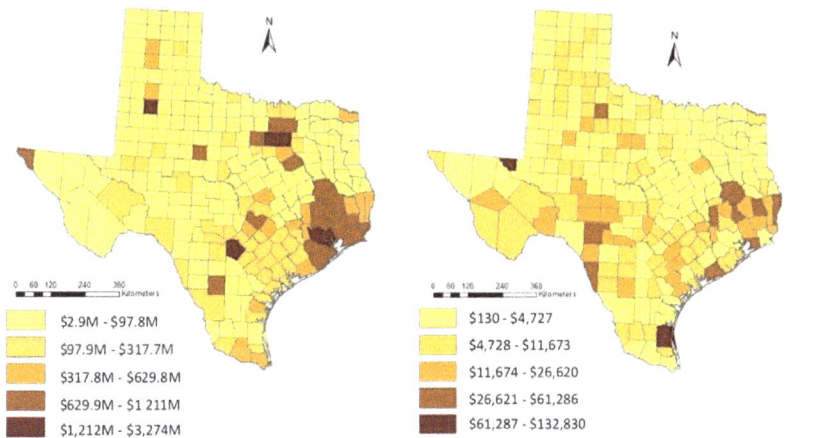

Figure 3. Spatial distribution of property damage ($2016) by Texas county (1960–2016): (**Left**) total property damage; and (**Right**) total property damage per capita.

Normalizing the property damage by median population (1988) reveals that the most affected counties with highly urbanized counties have below average normalized damage (Figure 3, Right). The most significant effect of this normalization is observed in the counties surrounding the Houston and Dallas metropolitan areas which have a much lower per capita damage due to the high population density in the counties compared to counties near the Gulf coast, along the southern edge of the Balcones Escarpment in north Texas, and other sparsely populated counties.

As noted previously, the effect of population and wealth on property damage cannot currently be evaluated at the county level since the lowest level of sub-state GDP data is at the MSA level. The total adjusted property damage ($2016) and total GDP for all MSAs during 2001–2016 was $30 billion and $19 trillion, respectively. The Houston and Dallas MSAs had the highest total GDP by a large margin (four times that of the Austin MSA or San Antonio MSA). The average *adjusted* property damage per capita highlights the regions of relatively high property damage with low population increase over the 16-year period. Normalized *actual* property damage by GDP highlights the regions (MSA) with relatively high property damage and lower GDP (Table 2).

Table 2. Normalized property damage and trend correlation for Texas MSAs (2001–2016). Ranked by decreasing linear trend strength.

MSA—Actual/GDP	16-year Average		16-year Annual Trend			
	Adj PD/Capita	Act PD/GDP	Spearman	*p*	Kendall	*p*
Dallas-Fort Worth-Arlington	$48.12	0.706	0.8176	0.0001	0.6500	0.0004
Lubbock	$393.40	9.672	0.4853	0.0567	0.3667	0.0476
Texarkana	$100.05	1.732	0.2796	0.2942	0.1757	0.3439
McAllen-Edinburg-Mission	$27.87	1.262	0.2264	0.4364	0.1868	0.3520
Tyler	$1.90	0.030	0.1319	0.6676	0.1026	0.6255
El Paso	$57.66	1.741	0.0682	0.8020	0.0342	0.8558
Austin-Round Rock	$43.82	0.727	0.0643	0.8199	0.0476	0.8046
Corpus Christi	$25.53	0.583	0.0440	0.8866	0.0769	0.7143
Killeen-Temple	$4.51	0.106	0.0382	0.8882	0.0000	1.0000
Sherman-Denison	$39.40	1.212	0.0382	0.8882	−0.0167	0.9283
Longview	$10.99	0.229	0.0059	0.9827	0.0084	0.9640
Laredo	$16.25	0.527	−0.0604	0.8445	0.0000	1.0000
San Angelo	$36.72	0.882	−0.1273	0.7092	−0.0909	0.6971
Waco	$5.13	0.122	−0.1471	0.5868	−0.1167	0.5285
Brownsville-Harlingen	$16.00	0.712	−0.1729	0.5377	−0.1409	0.4788
San Antonio-New Braunfels	$44.83	0.956	−0.2341	0.4010	−0.2297	0.2344
Amarillo	$221.04	4.698	−0.2357	0.3977	−0.1619	0.4002
Houston-The Woodlands-Sugar Land	$69.44	1.034	−0.2634	0.3242	−0.1925	0.2999
Odessa	$5.41	0.117	−0.2813	0.2912	−0.2017	0.2789
Victoria	$2.32	0.055	−0.3091	0.3848	−0.3333	0.1797
Beaumont-Port Arthur	$457.18	8.690	−0.3176	0.2306	−0.2667	0.1497
College Station-Bryan,	$6.04	0.168	−0.3370	0.2018	−0.2427	0.1912
Abilene	$194.88	4.499	−0.3626	0.2026	−0.2527	0.2080
Wichita Falls	$14.09	0.261	−0.4059	0.1188	−0.2833	0.1258
Midland	$91.18	0.693	−0.4794	0.0706	−0.3942	0.0420

* Green highlighted rows are decreasing trends.

More than 75% of the Texas MSAs exceeded the national average annual growth in GDP (4.8%) during 2001–2016. The highest annual growth in GDP was noted in the Odessa (10%), Midland (9.1%), Austin (8.9%), McAllen (8.3%), and Tyler (7.5%) MSAs. The top MSAs for total property damage in Texas were Houston ($5.6 billion), Dallas ($5.3 billion), Beaumont ($2.8 billion), Lubbock ($1.9 billion), and San Antonio ($1.6 billion). The regions with larger increases in average annual GDP suggest that these areas may have a higher risk of increased property damage in future disaster events. The GDP and property damage for each MSA clearly identifies the Houston and the Dallas MSAs as the wealthiest regions that also experienced the greatest property damage. Several of the MSAs in the western counties such as Amarillo, Lubbock, and El Paso have significantly lower GDP but relatively high property damage resulting in an appreciable normalized risk of economic loss (Figure 4, top).

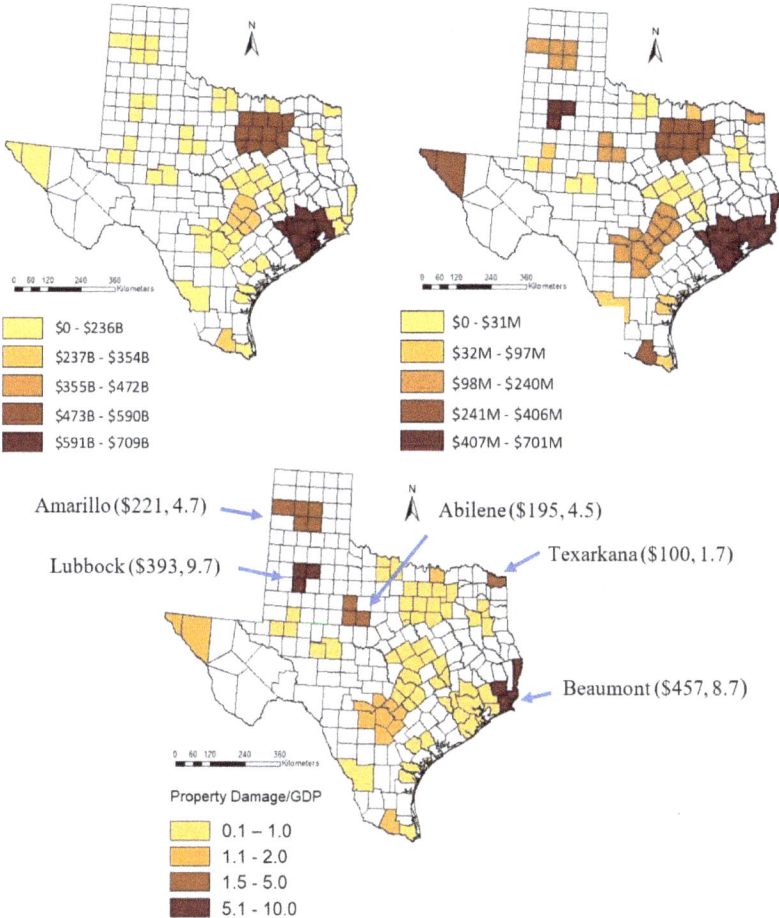

Figure 4. Spatial distribution of property damage for Texas MSAs (2001–2016): (**top left**) total GDP; (**top right**) total adjusted property damage; and (**bottom**) average actual property damage per capita and actual property damage by GDP ($\times 10^{-3}$).

The annual average adjusted property damage per capita and property damage normalized by GDP highlights regions of greater risk (Figure 4, bottom). The lower populations and relative wealth of these types of regions such as Lubbock (290,805/$171 billion), Beaumont (403,190/$324 billion), Abilene (165,252/$95 billion), and Amarillo (251,933/$172 billion) can create a greater critical need for assistance after a major disaster given their respective historic property damage ($1.9 billion, $2.8 billion, $524 million, and $860 million, respectively) and their limited resources compared to more urbanized regions.

5.4. Detailed Event Damage

The preceding review of historic damage estimates provides an assessment of damage trends, regional wealth diversity, and the relative impacts of different types of disasters on regions with different economic status. The damage estimates were culled from the SHLEDUS database that only includes direct property damage estimates adjusted to 2016 U.S. dollars. Natural disasters are difficult to categorize since the disaster types can overlap. In the current study, SHELDUS differentiates damage caused between hurricanes, severe thunderstorms, hail, flooding, and tornados among other categories. However, in reality, the damage due to a major natural disaster such as a hurricane will be a combination of the hurricane high winds, the extreme rainfall, possible hail, and the subsequent flooding. Additionally, hurricanes sometimes spawn tornados that cause extensive damage which may or may not be attributed to the original hurricane. It can be of value to model specific disaster types in a specific region of known economic wealth to obtain additional details of the damage across specific types of damage and types of property and disruption to the economy such as transport, communication, utilities, and displacement costs can provide perspective to benefit the development of risk mitigation strategies.

Extensive research and development are ongoing to develop or improve simulation and prediction modelling systems in the field of natural disasters. A paper published and presented at the 4th International Conference on Emerging Ubiquitous Systems and Pervasive Networks highlights the development emphasis with focus on an agent-based methodological framework that allows for customizable desired observables [25]. Researchers at the NOAA Geophysical Fluid Dynamics Laboratory are currently in the process of enhancing the predictability of major weather events by replacing existing models with new generation prediction model known as the FV3 [26]. Disaster modeling is also a critical component in emergency management that facilitates response planning and prioritization of resources based on predicted aftermath of a major disaster. FEMA has developed the HAZUS-MH Multi-hazard damage estimation tool which allows for modeling of known historic or probabilistic disaster events in any region of the U.S. to facilitate the estimation of property damage, business and infrastructure interruption in a user-defined region and 2010 building exposure and population [27].

The assessment of the historic hurricane damage during 1960–2016 highlighted the changes to specific types of infrastructure damage based on growth in population and development. The damage and loss information generated by HAZUS is a snapshot of the types of damage that can be caused by a hurricane to different types of building, facilities, and utility systems and can be used as a risk assessment surrogate for other regions. HAZUS-MH (v.4.0) was used to model the specific damages that resulted from three historical hurricanes but using 2010 population, building and infrastructure exposure based on the regional Level 1 inventory imbedded in Hazus model with the output adjusted to 2014 exposure valuation. Hurricane Carla made landfall in 1961, Alicia made landfall in 1983, and Hurricane Ike made landfall in 2008. The total economic losses output from HAZUS also includes cost of business interruption such as lost wages and, displacement costs. The modeled hurricanes are representative of major storm events that caused at least 80% of all property damage in Texas in the year of occurrence. The counties included in the simulation were those in the immediate vicinity of landfall for the hurricanes since they have the greatest potential for damage due to high winds which is the main contributor to building damage. These counties are also representative of the types of building and infrastructure given the population and extent of economic development of the region. Results of the current study indicate that hurricanes caused the greatest percentage of total economic losses in Texas with each event representing the majority of damage in the given year. Three hurricanes occurring 22 years and 25 years apart were simulated to provide insight into the relative risk and economic loss across building types as well as show the variation in damage estimates as a function of wind damage, flood damage, regional development, and data source variation. The storm tracks and residential exposure for the three hurricanes generated by the Hazus hurricane model are depicted in Figure 5.

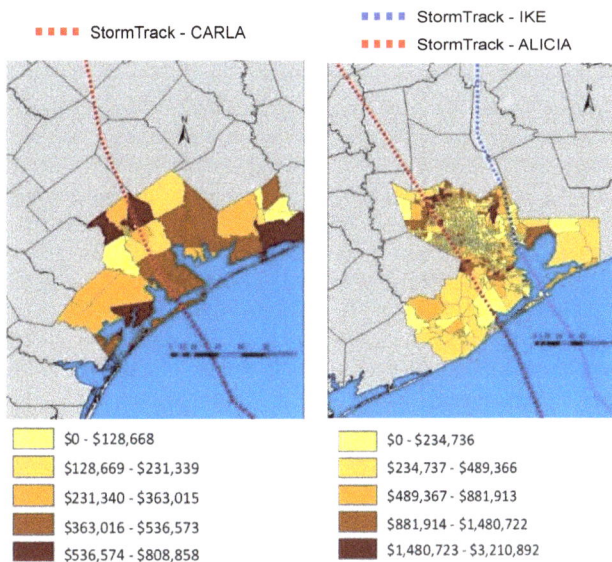

Figure 5. Storm tracks and residential (single-family) building exposure (2014): (**Left**) Hurricane Carla; and (**Right**) Hurricanes Alicia and Ike.

5.4.1. Hurricane Carla (2014 U.S. dollars)

Hurricane Carla made landfall on 11 September 1961 near Port O'Connor as a Category-5 hurricane but weakened quickly as it moved further inland in a northerly direction. Carla brought only small amounts of precipitation (3.15 in (80 mm)), but very strong wind gusts as high as 170 mph (280 km/h) in Port Lavaca. Since Hurricane Carla made landfall in proximity to Port O Conner (Calhoun county), the region selected for the model consisted of six counties in the immediate impact zone (Aransas, Calhoun, Jackson, Matagorda, Refugio, and Victoria) which is an area of 4468 square miles (11,572 km^2), and with 2010 census has 86,470 buildings, 71,000 households and a population of 189,492. Ninety-three percent of buildings and 80% ($17.1 billion) of the total building value are designated as residential. Commercial buildings comprise the majority (10.6%) of the non-residential buildings in the region. Essential facilities include 10 hospitals with a total bed capacity of 989 beds, 100 schools, 35 fire stations, 22 police stations, and no emergency operation facilities.

The model simulation of Hurricane Carla making landfall in 2010 at the same proximate location resulted in a total economic loss (property damage and business interruption) of $796 million representing 3.9 % of the total replacement value ($20.6 billion) of the region's buildings. The total losses consist of 85% direct property damage ($679 million) with the remaining 15% losses due to business interruption losses which include the inability to operate a business and temporary living expenses for those people displaced from their homes. The largest loss (both for property damage and business interruption loss) was sustained by the residential occupancies which made up over 86% of the total losses across all building types. The residential homes are primarily wood framed structures at an average cost of $213,207 per home. Seven percent of all wood structures would incur at least moderate damage.

5.4.2. Hurricane Alicia (2014 U.S. dollars)

Hurricane Alicia made landfall in August 1983 at Galveston Island as a Category-3 hurricane, with sustained winds of 100 mph (161 km/h) and gusts up to 127 mph (204 km/h). Alicia was not a particularly strong hurricane, but the area of maximum winds crossed a large metropolitan area

(Galveston-Houston) with a majority of the structural damage caused by high winds. The region selected for Model simulation included Brazoria, Chambers, Galveston, and Harris counties, an area of 4237 square miles (10,974 km^2), 1,435,808 buildings, 1,662,000 households and a population of 4,732,030 people. Ninety-one percent of buildings and 80% ($417.7 billion) of the total building value is designated as residential. Commercial buildings comprise the majority of non-residential buildings in the region. Essential facilities in HAZUS simulation include 76 hospitals with a total bed capacity of 15,898 beds, 1546 schools, 95 fire stations, 135 police stations and 11 emergency operation facilities.

The model simulation of Hurricane Alicia making landfall in 2010 at the same proximate location resulted in a total economic loss of $11.4 billion representing 2.2% of the total replacement value ($522 billion) of the area's buildings. The total losses consist of 89% direct property damage ($10.2 billion) with the remaining 11% losses due to business interruption losses. The largest loss (both for property damage and business interruption loss) was sustained by the residential occupancies which made up over 89% of the total losses across all building types. The residential homes are primarily wood framed structures at an average cost of $318,547 per home. Three percent of all wood structures would incur, at least, moderate damage.

5.4.3. Hurricane Ike (2014 U.S. dollars)

Hurricane Ike made landfall near the city of Galveston on 13 September 2008 as a strong Category-2 storm with a central pressure of 951.6 millibars and maximum sustained winds of 110 mph (177 km/h). Hurricane Ike differs from the two previous case studies in that more than 50% of the total damage was due to storm surge in coastal Galveston, Harris, and Chambers counties, whereas wind was the primary cause of damage in Hurricanes Carla and Alicia. Since Hurricane Ike made landfall in very close proximity to where Alicia made landfall, the same four county (Brazoria, Chambers, Galveston, Harris) area was selected for HAZUS simulation with geographical specifications and property exposure count and value.

The model simulation of Hurricane Ike making landfall in 2010 at the same proximate location resulted in a total economic loss is $7.2 billion representing 1.4% of the total replacement value of the region's buildings. The total losses consist of 91% direct property damage ($6.5 billion) with the remaining 9% losses due to business interruption losses. The largest loss (both for property damage and business interruption loss) was sustained by the residential occupancies which made up over 90% of the total losses across all building types. As was the case with Hurricane Alicia, the residential homes in the selected region are primarily wood framed structures at an average cost of $318,547 per home. Slightly less than 2% of all wood structures would incur, at least, moderate damage.

5.4.4. Variability and Limitations in Damage Estimates

The model damage estimates for the three hurricanes were adjusted from 2014 to 2016 US dollars to allow for comparison with each other and to other sources. The variability between the damage estimates provides some insight into the primary cause of damage (wind or flood), the level of property exposure in the modeled region, and the limitations of the modeling tool and database information used in the study (Table 3).

Table 3. Comparison of damage estimates for Hurricane Carla, Hurricane Alicia, and Hurricane Ike (*all dollar values adjusted to 2016 U.S. dollars*).

Disaster Event Name	Hurricane Carla	Hurricane Alicia	Hurricane Ike
Month/Days	September 8–14	August 15–21	September 11–12
Saffir-Simpson Classification	Cat-5	Cat-3	Cat-2
Year	1961	1983	2008
Landfall	Port O'Connor, Tx	Galveston, Tx	
Primary Cause of Damage	Wind/Storm Surge	Wind/Tornadoes	Flooding
Texas Almanac (total damage)	$2.4 billion	$7.2 billion	$15.6 billion
NWS/NOAA (total damage)	NA	$7.3 billion	$33.4 billion
SHELDUS (property damage)	$204.4 million	$1.23 billion	$3.8 billion
SHELDUS (crop damage)	$204.4 million	$1.23 billion	NA
HAZUS model counties	Aransas, Calhoun, Jackson, Matagorda, Refugio, Victoria	Brazoria, Chambers, Galveston, Harris	
Counties population (2010)	189,492	4,732,030	
Hazus model (2010 building exposure)	$20.9 billion (80% wood residential)	$529 billion (80% wood residential)	
Hazus model (property damage)	$689 million	$10.3 billion	$6.63 billion
Hazus model (business interruption)	$119 million	$1.23 billion	$697 million

First, even though all three of the historic disasters modeled were hurricanes, the type and extent of damage varies widely depending on the specific characteristics of the hurricane such as wind velocity, storm surge, and rainfall induced flooding. The HAZUS Hurricane model used in this study only calculates damages caused by the high winds recorded in the historic disaster parameter database. Hazus also has developed a Flood Model that could be used in conjunction with the Hurricane Model to aggregate damage estimates from both disaster types for the same region. This was not done in the current study since comprehensive damage prediction is not a focal point of the paper.

Secondly, the damage estimates are based on internal algorithms of structural design directly dependent on the types and number of structures are in the hurricane storm track. This structural inventory will vary depending on the selected region (one or more counties) and the level of urban development in these counties which increases with time in regions with population and GDP growth such as Texas. Since the model only analyzes wind damage across the user-defined subset of counties created for the model, the true actual damage across all regions affected by the storm track will always be greater.

Thirdly, the level of urban development is a dynamic variable that is critical in generating robust economic loss as a function of property damage and business interruption. Damage due to disasters is usually reported in current year dollars and then updated as new information is received and adjusted to the later year inflation factor (CPI). The version of the Hazus modeling tool used in the current study is developed with a structure inventory database that is based on 2010 census for residential buildings and 2010 Dun and Bradstreet for commercial buildings considered a Level 1 inventory. The model has the capability to be updated with more precise and current building data that would enhance the accuracy of the damage, repair and replacement cost estimates. Comparing the modeled damage estimates for Hurricanes Carla, Alicia, and Ike to other sources exemplifies the potential variation that can be attributed to one or more of these discontinuities.

For comparison to the Carla simulation, a *SHELDUS* data query for all property and crop damage caused by hurricane/tropical storm, coastal, flooding, hail, and severe thunderstorms for all Texas counties in the month of September 1961 reports $205 million property damage and an equal loss for crop damage. The *Texas Almanac* archives document total property and crop damage for Carla at $2.4 billion which is similar to the *NWS* estimate of $2.6 billion in which 2/3 was property and 1/3 crop damage. Storm surge played a major role in the damage for this hurricane which is not capture in the Hazus model. The NWS reports 1915 homes and 983 businesses, farm buildings, and other buildings were completely destroyed. Major damage occurred to 7398 homes and 2601 businesses, farm buildings, and other buildings. Minor damage was reported to 43,325 homes and 13,506 businesses, farm buildings and other buildings [28]. The Hazus model for Carla resulted in $688 million in property damage and $119 million in business interruption.

Similarly, for comparison to Alicia, a *SHELDUS* query for all property and crop damage caused by hurricane/tropical storm, coastal, flooding, hail, severe thunderstorms for all Texas counties in the month of August 1983 reports $1.2 billion in property damage and an equal loss for crop damage. The *Texas Almanac* archives report total property and crop damage for Alicia at $7.2 billion which is similar to the *NOAA* estimate of $7.3 billion. A report by the NRC Committee on Natural Disasters published in 1983 reported $602 million in economic loss for Galveston, Harris, Brazoria, and Chambers with about 50% due to property damage and the other 50% damage due to roads, utilities, agriculture, marine, and vehicles. Alicia was a high wind damage hurricane that spawned at least 22 tornadoes resulting in an additional 18 fatalities and property damage [29]. The Hazus model for Alicia resulted in $10.3 billion in property damage and $1.23 billion in business interruption.

The difference in actual to model was most pronounced with Ike. A *SHELDUS* query for property and crop damage caused by hurricane/tropical storm, coastal, flooding, hail, and severe thunderstorms for all Texas counties in the month of September 2008 reports $3.76 billion in property damage. The *Texas Almanac* archives report $15.6 billion in the counties of Harris, Chambers, Galveston, Liberty, Polk, Matagorda, Brazoria, Fort Bend, San Jacinto, and Montgomery, with an estimated $8.9 billion of that due to storm surge in coastal Galveston, Harris, and Chambers counties. The NOAA reports a much higher value of $33.4 billion but include all counties in the storm track suggesting that much of the Ike damage was not in these Texas counties. The Hurricane Ike Impact Report published by the Office of Homeland Security Division of Emergency Management in 2008 breaks out the damage as follows: housing damage ($3.8 billion), infrastructure repairs ($2.7 billion), Public buildings ($1.9 billion), hospital damage ($791 million), and transportation ($147 million) [30]. The Hazus model for Ike resulted in $6.63 billion in property damage and $697 million in business interruption (Table 3).

6. Discussion and Conclusions

The spatiotemporal trends of annual damage caused by natural disasters in Texas highlight the disproportionate impact of hurricanes, tropical storms, and hail accounting for approximately 70% of total property damage during 1960–2016. The extreme susceptibility of coastal property damage due to hurricane is explicit in Texas (34% due to hurricanes) and supported by national damage statistics in which 40% of all U.S. property damage during this period is due to hurricane or tropical storms. Given the ongoing progress in scientific understanding of climate change and the record setting temperatures around the world in recent years being investigated and made public by several scientific NGOs as well as the International Panel on Climate Change (IPCC) [31], it is a natural tendency to gravitate to the assumption that the primary driver of our increasing disaster losses is due to the increase in frequency and intensity of hydrometeorological events. The change in disaster severity is generally increasing but the change in frequency remains an open debate. Conversely, the increase in property damage due the disasters is most definitely increasing and not debated. This suggests that the increase in damage is due to other factors outside the disaster itself such as the exposure, wealth and level of development of the regions with high risk of hurricane activity (e.g., coastal communities).

Annual distribution of property damage (2016 U.S. dollars) due to hydrometeorological disasters during the 57-year study period exhibited a statistically significant increasing trend with high annual variability that decreases when normalized per capita. Actual property damage (year of occurrence) normalized by annual GDP further decreases to a non-significant statistical positive trend suggesting a relationship of population and wealth to the level of damage in the impacted region. Additional inference of wealth effects on property damage is observed with trend analysis of the 25 Metropolitan Statistical Areas (MSA) in Texas during 2001–2016. Although a 16-year study period is limited in prediction value for trend analysis, it does provide some high-level results that highlight the differences in regional wealth in Texas. Actual property damage normalized by the GDP for each of the MSA regions property damage results in increasing trends similar to the statewide annual trends (1960–2016) for 11 of the MSAs with only the Dallas MSA indicating a strong linear and significant correlation. Fourteen of the Texas MSAs exhibited non-significant decreasing trends based on non-parametric correlation analysis at a 5% significance level suggesting that growth in GDP is outpacing the actual property damage in those regions.

Simulation of three historic hurricanes using the HAZUS hurricane modeling estimation tool generated conservative estimates of total economic loss as expected given the limitation noted in Section 5.4.4. However, the simulation can be useful as a source of information for the types of buildings most susceptible to wind damage as well as an indirect indication of the number of counties impacted by the entire storm track. The model simulations provide a microcosm of lower level building level damage within one specific disaster event that may be representative of disaster damage that may occur in future disasters and therefore be useful as a loss predictor. The HAZUS modeling tool is designed and managed by FEMA to serve as a predictor of damage and resource disruption to aid emergency management personnel and is not intended to provide a precise account of the financial impact of a disaster. Within that context, the model output was substantially different from the total damage estimations produced by other sources that can provide perspective on the nature and extent of the modeled hurricanes.

Hurricane Carla made landfall at Galveston Island in 1961 and headed northwest into Harris county. Carla caused damage to many counties along its northeasterly track from Texas into the Great Lakes region resulting in $2.4–2.6 billion in total property and crop damage along its path (Texas Almanac and NWS). The SHELDUS aggregate data estimate for the entire month of September 1961 was $410 million and the Hazus simulation model using 2010 building inventory for six counties around the Texas landfall showed $688 million in property damage and $119 million in business interruption. The fact that the model output is higher than the other sources suggests that Carla was a combination of wind and water damage along the landfall counties due to storm surge and the building exposure has likely increased significantly from 1961 to 2010. Hurricane Alicia and Hurricane Ike made landfall within 40 km of each other around Galveston Island, albeit 25 years apart. These two hurricanes were drastically different in wind and rainfall which affected the type and location of damage. Alicia spawned at least 22 tornados causing excessive wind damage and Ike had heavy rainfall causing much more flood damage than wind damage. NOAA reported $7.3 billion in total damage for Alicia and $33.4 billion total damage for Ike which includes property and crops of all affected areas. SHELDUS data for the entire month of August 1983 indicated $2.4 billion (property and crop) and $3.8 billion (property) damage for September 2008. The Hazus simulation model (2010 inventory) was $11.5 billion (property damage and business interruption) for Alicia and $7.3 billion for Ike. In summary, both the Carla and Ike simulation models captured less than 50% of their respective estimates reported by other sources suggesting a broad geographical zone of damage with flood damage making a significant contribution. Conversely, the Alicia model damage estimates are 50% higher than total damage estimates reported at the time of occurrence by NOAA suggesting a substantial increase in building exposure in the modeled region that was damaged by high winds captured by the model.

In conclusion, it is apparent that damage estimates for historic disaster events can vary widely based on data source and methodology. Particularly in hurricane events where there is overlap in damage causality of wind, storm surge, and flooding, the estimates are very dependent on the categorization of the disaster type and the regions of impact. Deriving meaningful conclusions from historic data or simulation model data based on historic events should be tempered with the known limitations of the source data. For example, the damage reported by SHELDUS is spread out over the affected counties equally even if the event may have impacted one county disproportionately. In addition, prior to 1995, the database only reported an event that caused at least one death or over $50,000 in damage neglecting all smaller events [25]. Further research into historic property damage assessment may consider securing access to a level of SHELDUS data that is aggregated by disaster event and presidential disaster declaration to better relate the damage to the disaster. The hurricane modeling used by the HAZUS-MH hurricane model only includes damage and losses caused directly and immediately by the hurricane wind. Moreover, the default building, infrastructure, and terrain data in HAZUS-MH is based on the best information available at the time of the application release which, at best, includes 2010 U.S. census for residential buildings and Dun and Bradstreet estimates for commercial buildings. The accuracy of HAZUS simulation output can be greatly increased by modeling the historic disaster of interest for wind, hail, and flooding in high risk areas and using advanced regional inventory, utility, transportation, and economic data. The noted constraints that limits the quantitative accuracy of the Hazus model output can be mitigated by: (1) Using multiple disaster models to account for damage due to wind and flooding when considering a hurricane event. (2) Upgrading the Level 1 building inventory and infrastructure database with user defined building files obtained from the property tax assessment agency and current hazard maps that provide accurate flood depth grids and boundaries and hurricane wind fields. (3) Building the model region to include as many counties as practical in the storm track.

Author Contributions: H.O.S. provided research supervision and professional guidance in all aspects of writing this paper. S.H.P. performed the data mining, quantitative analysis, qualitative interpretation and discussion, and is responsible for all content in this paper.

Funding: This research received no external funding.

Acknowledgments: The authors thank David and Bethene Player for review of the manuscript and providing a sounding board for interdisciplinary perspective.

Conflicts of Interest: The authors declare no conflicts of interest.

References

1. Munich, R. NatCatSERVICE, Natural Catastrophe Know-how for Risk Management and Research. Natural Catastrophe Online Tool. Available online: http://natcatservice.munichre.com/ (accessed on 18 September 2018).
2. National Oceanic and Atmospheric Administration (NOAA). National Centers for Environmental Information. Storm Events Database 2017. Available online: https://www.ncdc.noaa.gov/billions/overview (accessed on 18 September 2018).
3. Irfan, U.; Resnick, B. Megadisasters Devastated America in 2017. And They're Only Going to Get Worse. Storms, Fires, Floods, and Heat Caused at Least $306 Billion in Destruction Last Year. 2018. Available online: https://www.vox.com/energy-and-environment/2017/12/28/16795490/natural-disasters-2017-hurricanes-wildfires-heat-climate-change-cost-deaths (accessed on 18 September 2018).
4. Gall, M.; Borden, K.A.; Emrich, C.T.; Cutter, S.L. The Unsustainable Trend of Natural Hazard Losses in the United States. *Sustainability* **2011**, *3*, 2157–2181. [CrossRef]
5. Sander, J.; Eichner, J.F.; Faust, E.; Steuer, M. Rising Variability in Thunderstorm-Related U.S. Losses as a Reflection of Changes in Large-Scale Thunderstorm Forcing. *Am. Meteorol. Soc.* **2013**. [CrossRef]
6. Pielke, R.A.; Landsea, C.W. Normalized Hurricane Damage in the United States: 1925–95. *Am. Meteorol. Soc.* **1998**, *13*, 621–631. [CrossRef]
7. Pielke, R.A.; Gratz, J.; Landsea, C.W.; Collins, D.; Saunders, M.A.; Musulin, R. Normalized Hurricane Damage in the United States: 1900–2005. *Nat. Hazards Rev. ASCE* **2008**, *9*. [CrossRef]
8. Klotzbach, P.J.; Bowen, S.G.; Pielke, R.A.; Bell, M. Continental U.S Hurricane Landfall Frequency and Associated Damage, Observations and Future Risks. *Am. Meteorol. Soc.* **2018**. [CrossRef]
9. Paul, S.; Sharif, H.; Crawford, A. Fatalities Caused by Hydrometeorological Disasters in Texas. *Geosciences* **2018**, *8*, 186. [CrossRef]
10. Sharif, H.; Jackson, T.; Hossain, M.; Zane, D. Analysis of Flood Fatalities in Texas. *Nat. Disasters Rev.* **2014**, *16*. [CrossRef]
11. Gaines, J.; Hunt, H. Tierra Grande. Housing Markets. Publication 2028. 2013. Available online: https://assets.recenter.tamu.edu/documents/articles/2028.pdf (accessed on 18 September 2018).

12. Potter, L. The Changing Population of Texas and West Texas. In Proceedings of the Cross Roads Healthcare Transformation in West Texas Conference, Lubbock, TX, USA, 5 June 2013; Office of State Demographer: Austin, TX, USA, 2013.

13. CEMHS. *Spatial Hazard Events and Losses Database for the United States, Version 16.1*; Center for Emergency Management and Homeland Security, Arizona State University: Phoenix, AZ, USA, 2018.

14. Schneider, P.; Schauer, B. Hazus-Its development and its future. *Nat. Hazards Rev.* **2006**, *7*, 40–44. [CrossRef]

15. Vickery, J.; Lin, J.; Skerlj, P.; Twisdale, L.; Huang, K. Hazus-MH hurricane model methodology I: Hurricane hazard, terrain, and wind load modeling." and "Hazus-MH hurricane model methodology II: Damage and loss estimation. *Nat. Hazards Rev.* **2006**, *7*, 82–93. [CrossRef]

16. Hahn, D.; Emmanuelle, V.; Ross, C. Multi-Hazard Mapping of the U.S. *J. Risk Uncertain. Eng. Syst. Part A Civ. Eng.* **2017**, *3*. [CrossRef]

17. Gordon, P.; Richardson, H.W.; Davis, B. Transport-related impacts of the Northridge earthquake. *J. Transp. Stat.* **1998**, *1*, 21–36.

18. Burrus, R.; Dumas, C.F.; Farrell, C.H.; Hall, W.W., Jr. Impact of low-intensity hurricanes on regional economic activity. *Nat. Hazards Rev.* **2002**, *3*, 118–125. [CrossRef]

19. Pan, Q. Case Study: Economic Losses from a Hypothetical Hurricane Event in the Houston-Galveston Area. *ASCE Nat. Hazards Rev.* **2011**, *12*, 146–155. [CrossRef]

20. Hydrometeorological Hazards. Disaster Risk Reduction. Natural Sciences. United Nations Education, Scientific, and Cultural Organization (UNESCO). Available online: http://www.unesco.org/new/en/natural-sciences/special-themes/disaster-risk-reduction/natural-hazards/hydro-meteorological-hazards/ (accessed on 18 September 2018).

21. Gautheir, T. Detecting Trends Using Spearman's Rank Correlation Coefficient. *Environ. Forensics* **2001**, *2*, 359–362. [CrossRef]

22. Wilson, L.T. Statistical Correlation. 2009. Available online: https://explorable.com/statistical-correlation (accessed on 8 October 2018).

23. Statistics Solutions, Complete Dissertation. Available online: https://www.statisticssolutions.com/kendalls-tau-and-spearmans-rank-correlation-coefficient/ (accessed on 8 October 2018).

24. Guci, L.; Mead, C.; Panek, S. A Research Agenda for Measuring GDP at the County Level, Bureau of Economic Analysis. 2016. Available online: https://www.bea.gov/system/files/papers/WP2016-4.pdf (accessed on 18 September 2018).

25. Mustapha, K.; Mcheick, H.; Mellouli, S. Modeling and Simulation Agent-Based of Natural Disaster Complex Systems. In Proceedings of the 4th International Conference on Emerging Ubiquitous Systems and Pervasive Networks (EUSPN-2013), Niagara Falls, ON, Canada, 21–24 October 2013.

26. FV3: Finite-Volume Cubed Sphere Dynamical Core, Geophysical Fluid Dynamics Laboratory (GFDL), Princeton University Forrestal Campus. Available online: https://www.gfdl.noaa.gov/fv3/ (accessed on 18 September 2018).

27. Federal Emergency Management Agency (FEMA). *U.S. Hazard Multi-hazard Estimation Tool, Hurricane Model*; HAZUS-MH 4.0; Federal Emergency Management Agency (FEMA): Washington, DC, USA, 2018.

28. National Weather Service (NWS). National Oceanic and Atmospheric Administration, Carla 50th Anniversary, Storm Summary. Available online: https://www.weather.gov/crp/hurricanecarla (accessed on 8 October 2018).

29. Savage, R.P. National Research Council (NRC), Committee on Natural Disasters, Commission on Engineering and Technical Systems. In *Hurricane Alicia: Galveston and Houston, TX, USA, 17–18 August 1983*; National Academy Press: Washington, DC, USA, 1984.

30. Colley, J.; DeBlasio, S.M. *Hurricane Ike Impact Report-December 2008*; Office of Homeland Security, Division of Emergency Management: Austin, TX, USA, 2008.

31. International Panel on Climate Change (IPPC). Summary for Policymakers of IPCC Special Report on Global Warming of 1.5°C Approved by Government. 2018. Available online: https://www.ipcc.ch/news_and_events/pr_181008_P48_spm.shtml (accessed on 8 October 2018).

MDPI

St. Alban-Anlage 66

4052 Basel

Switzerland

Tel. +41 61 683 77 34

Fax +41 61 302 89 18

www.mdpi.com

Geosciences Editorial Office

E-mail: geosciences@mdpi.com

www.mdpi.com/journal/geosciences

www.ingramcontent.com/pod-product-compliance
Lightning Source LLC
Chambersburg PA
CBHW051844210326

41597CB00033B/5764